WHAT DISTURBS OUR BLOOD

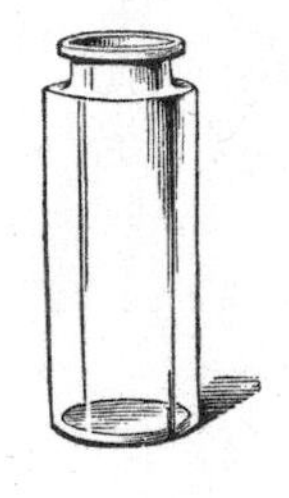

A SON'S QUEST TO REDEEM THE PAST

WHAT DISTURBS OUR BLOOD

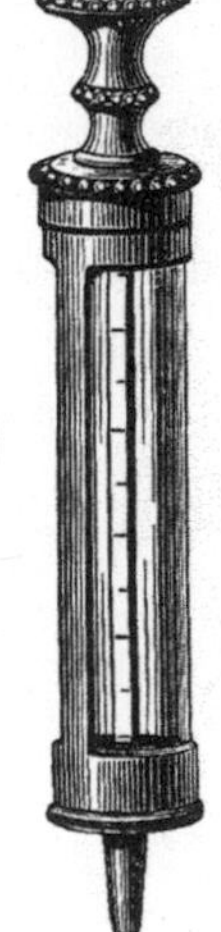

JAMES FITZGERALD

RANDOM HOUSE CANADA

 Published in 2010 by Random House Canada, a division of Random House of Canada Limited, Toronto. Distributed in Canada by Random House of Canada Limited.

www.randomhouse.ca

Pages 483-85 are a continuation of the copyright page.

LIBRARY AND ARCHIVES CANADA CATALOGUING IN PUBLICATION

FitzGerald, James, 1950–
What disturbs our blood : a son's quest to redeem the past / James FitzGerald.

Includes index.
ISBN 978–0-679–31315–1

1. FitzGerald, John Gerald, 1882–1940—Mental health. 2. FitzGerald, James, 1950——Family. 3. Connaught Laboratories—Officials and employees—Biography. 4. Pathologists—Canada—Biography. 5. Grandfathers—Canada—Biography. I. Title.

RB17.F58F58 2010 616.07092 C2009–905252–0

Design by Kelly Hill

Printed in the United States of America

2 4 6 8 9 7 5 3

This book is dedicated to my family,
with love and respect.

. . . what disturbs our blood
Is but its longing for the tomb.

WILLIAM BUTLER YEATS,
The Wheel

The man you know, assured and kind,
Wearing fame like an old tweed suit—
You would not think he has an incurable
Sickness upon his mind.

Finely that tongue, for the listening people,
Articulates love, enlivens clay;
While under his valued skin there crawls
An outlaw and a cripple.

Unenviable the renown he bears
When all's awry within? But a soul
Divinely sick may be immunized
From the scourge of common cares.

A woman weeps, a friend's betrayed,
Civilization plays with fire—
His grief or guilt is easily purged
In a rush of words to the head.

The newly dead, and their waxwork faces
With the look of things that could never have lived,
He'll use to prime his cold, strange heart
And prompt the immortal phrases.

Before you condemn this eminent freak
As an outrage upon mankind,
Reflect: something there is in him
That must for ever seek
To share the condition it glorifies,
To shed the skin that keeps it apart,
To bury its grace in a human bed—
And it walks on knives, on knives.

CECIL DAY LEWIS, *Almost Human*

CONTENTS

PROLOGUE 1

PART I

1. The Ghosts of Balmoral 5
2. Room at the Top 42
3. Daddy Ded 69
4. Dr. FitzGerald Must Learn to Cover Up His Feelings 89
5. In Dreams 111
6. Let Sleeping Dogs Lie 119
7. I'm My Own Grandpa 127
8. The Unpardonable Sin 141

PART II

9. Science Has No Fatherland 153
10. Crucible of Ambition 161
11. A Kind of Superman 169
12. Gospel Certainty of Relapse 177
13. I Am a Mass of Flesh, and You Are Another 194
14. Salvation Lies in the Most Concrete Records 206
15. The Travel Bug 217

PART III

16. The Miracle in a Stable 229
17. Born with the Dead 244
18. The Sugar Sickness 266
19. The Four Horsemen of the Apocalypse 289
20. *Mens Sana In Corpore Sano* 304
21. The High Priest of Quality Control 323
22. A Monument More Lasting Than Bronze 338

PART IV

23. Hell of a Good 353
24. The Living Lie 369
25. . . . and the Penalty Is Death 377
26. Damn Clever, These Spooks! 412
27. Occam's Razor 425
28. The Archaeology of Silence 444

ACKNOWLEDGEMENTS 469
SELECT BIBLIOGRAPHY AND SOURCES 472
PERMISSIONS 483
PHOTO CREDITS 485
INDEX 487

Prologue

> We have rated the powers of children too low and that there is no knowing what they cannot be given credit for.
>
> SIGMUND FREUD

On a bleak November day in 1953, my mother, a thwarted artist turned reluctant housewife, snapped the cold shutter of her camera and captured the image of my three-year-old self, wrapped in a hand-me-down corduroy coat with velvet collar. My charismatic, witty mother was a shape-shifter; I was an emotional weatherman, my head tilted like a radar dish, hyperattuned to her moods. Some believe all writers write for their mothers.

Decades passed before I began to plumb the meaning of the worried yet curious expression of the boy in the photograph, haunted by what he knows and is coming to know. Why did the nightmares of my childhood—slow suffocations, falling trees, flashing knives—routinely rumble through my unconscious like clues to an unsolved murder? What was I trying to figure out? What could I possibly make of the electrical storms that raged in my head? Even as a toddler, was I sensing that the gothic three-storey house where I slept was built by my long dead grandfather, a driven, eminent doctor of whom no one spoke? Was I already interpreting the silences that buffered a family history I was made to feel I must never question?

•

My father, too, was a fiercely dedicated doctor, and as I grew up, he acted as if his three children lived in another city. The only time he hugged me was after he suffered a breakdown in his fifties, at the peak of his professional success. Naturally I did not know how to respond to the sudden gesture. He made two suicide attempts and was fated to spend the last twenty years of his life alone in front of the television, neutralized by psychiatric drugs, quietly consumed by his regret for the past and his dread of the future.

My inability to feel like anything other than a spectator in my own life finally pushed me at the age of thirty-three into the hands of a psychotherapist. He knew that if I didn't pierce the mystery of my silently disintegrating father, I was doomed to re-enact the buried generational drama that was already undercutting my natural passions with an invisible hand.

And so I plunged into the hall of mirrors that is this book: men pushed by troubled and withholding fathers and differently troubled and withholding mothers into extraordinary accomplishments in the world that in the end they can no longer sustain. My recurring dreams, like cave paintings, guided me to the Pandora's box of repressed secrets, for of course there were more than one. I learned that the only way out of the haunted house of my childhood was to return to it, to struggle year by year to inhabit the alien skins of my father and grandfather.

In city archives, I was amazed to discover that my father's father was a remarkable man of science, an innovator and visionary who transformed the Canadian system of public health between the world wars, a heroic figure who eradicated the disease of diphtheria and made insulin available to the masses, a dynamo who travelled the world for the Rockefeller Foundation, bringing Canada's paragon of preventive medicine to the international community. His singular achievements saved countless lives and earned universal praise—until that summer day of horror and shame I was never meant to know about; the day that explained why the memory of such a man could be erased; the day that finally explained his son, my father, to me.

Comfort comes from validation: From the very beginning, the dark dreams and intuitions of the quizzical three-year-old boy in hand-me-down clothes, framed by his mother's distancing eye, embodied an uncanny form of knowing and, ultimately, healing. Truth will out, and there is no knowing what a child cannot be given credit for.

PART I

ONE

The Ghosts of Balmoral

> Those who know ghosts tell us that they long to be released from their ghost life and led to rest as ancestors. As ancestors, they live forth in the present generation, while as ghosts they are compelled to haunt the present generation with their shadow life.
>
> HANS LOEWALD,
> *On the Therapeutic Action of Psychoanalysis*

My story opens in the haunted house of my birth. Three storeys tall, nearly a century old, the place stands silent in my memory, as lean and austere as the midnight hands of a grandfather clock. Erected by my paternal grandparents at the outbreak of the Great War, the timbered beams, grey stuccoed walls, dormered windows, and chimneyed roof cast the sombre shadows of a past that holds me still. Night after night, my adult dreams still pull me down, back through its darkening staircases and corridors, nudging open the door of the nursery where I slept as a child.

Like my father before me, I was conceived in the second floor master bedroom. In my case, it was on a bleak January night in 1950, in the perfect middle of the century, in the perfect middle of the city of Toronto, the economic engine of English Canada. I arrived nine months later, a celiac baby afflicted with poor digestion; whatever I was asked to swallow, I spat out. Like all infants,

I was an open vessel, exquisitely attuned with innocent intuition. From the start, I knew in my bones that 186 Balmoral Avenue was inhospitable to children.

In her bed at the Toronto Western Hospital, my mother, Janet, exhausted by the ceaseless agitations of my twenty-one-month-old sister, Shelagh, sank into a postpartum depression, overwhelmed by the prospect of suckling a second bundle of raw infantile need. On the second day of my life, I was surrendered to Mothercraft, a child-care agency whose name emitted ominous overtones of sorcery. While my mother recovered, I spent the next two weeks of my existence in a grand, buff-brick mansion at 49 Clarendon Avenue, a two-minute stroll from the house on Balmoral, where I was rocked in the starchy arms of a succession of nurses. Occupying an impressive two-acre, stone-gated estate, Burnside was once owned by a granddaughter of Timothy Eaton, the bewhiskered Northern Irish patriarch of the department store empire, who virtually gave away her property to Mothercraft some years after she lost her sole offspring on the *Titanic*. Designed to help both healthy and convalescent mothers to raise their infants and preschoolers, the progressive agency promoted the immunological virtues of breastfeeding, confronting the stubborn view that it was a low-class occupation, performed by impoverished mothers who could not afford cow's milk. But by the time I was returned home, the milk of my ample-bosomed mother had dried up.

I was dipped in the lukewarm baptismal waters of Grace Church on-the-Hill, an austere High Anglican enclave of grey stone that gravely watched over neighbouring Bishop Strachan, the private girls' school my sister was destined to enter; henceforth, the rituals of my young life would continue to mesh with these institutional vestiges of the Family Compact, the nineteenth-century ruling class clique of the British colony of Upper Canada. When they first landed here in the 1790s, the founding generation of red-coated aristocrats claimed the high ground where the wind swept away the swarms of insects. They called it the Davenport escarpment, unaware that in centuries past, the shores of a vast glacial lake lapped up to its base. Displacing the Aboriginal tribes who loped through the wooded ravines where I would one day play as a boy, the blueblood fathers of a muddy frontier town imposed on unruly nature a rigid grid system of roads, and from the beginning saw themselves as innately superior members of the upper crust. Generations on, I knew without being told what I must believe. The "Upper Canadians" who occupied the neighbourhoods of Deer Park and Forest Hill crowning the Avenue Road hill were born and bred above it all, and the natives below were to be seen as beyond the pale, quite literally beneath us.

Our short, tree-lined block was composed of facing rows of gentrified, three-storey houses owned by hard-driving professionals—lawyers, doctors, judges, and stockbrokers—with sturdy WASP names like McMurrich, Macmillan, White, Horne, Wilson, Reid. The very name of the street, Balmoral, echoed Queen Victoria's craggy castle in the windswept highlands of Scotland; the fragment "moral" suggested a clear-cut, unequivocal sense of good and bad, right and wrong. Our own house had been built in 1914 by my grandfather, John Gerald "Gerry" FitzGerald, a fiercely ambitious, high-minded doctor of Irish Protestant blood who died in 1940, ten years before I was born. My father, Jack, followed his accomplished father into the ranks of the medical profession and, soon after his marriage, inherited the family home. Although Gerry had lived here for twenty-five years, with his wife, Edna, no portraits of him adorned the walls. As I grew up, my dead grandfather's name mysteriously failed to fall from my father's lips. Still, I sensed a compelling, invisible presence in the shadows of my earliest years.

186 Balmoral Avenue, built 1914

═══

Engulfed by a wave of postwar, martini-laden parties, my mother, Janet, learned she was pregnant two months before her wedding; to reveal the scandalous truth would have shocked her religious, Victorian mother into a state of catatonia. When Janet submitted to an illegal abortion in the back room of a dingy storefront on Queen Street West—a secret she revealed to me late in her life—the hemorrhaging that ensued nearly killed her. Yet alcohol continued to play a role in the conception of her legitimate children; she admitted that my younger brother, Michael, was, in that unhappy phrase, "an accident," conceived in a guest cottage in the wake of an inebriated summer party on the sultry beaches of Lake Simcoe. My father pushed for an abortion, but my mother, understandably, refused.

On a late March day in 1952, suffering the early throes of labour, Janet was driven pell mell down the Avenue Road hill to the Toronto Western Hospital. Behind the wheel sat her unflappable Scottish father, keeping his cool as she frantically urged him to run the red lights. Her husband was otherwise engaged, pressing his ear to the radio, listening to the annual Oxford-Cambridge boat race and getting blotto with his buddies. The mores of the time did not dictate that fathers (even fathers who were doctors) attend the reception of their offspring. Yet his selfishness planted in my mother's heart a silently festering virus of resentment.

With Michael's arrival close on the heels of my own, I was destined to inherit the perspective of a middle child, forever struggling for a sense of balance. Separated by only three and a half years, we three kids found ourselves compressed into the fuse box of a common environment, the quirks of our characters squirming for uniqueness. Shelagh tended to the volatile, Michael to the phlegmatic, James to something in-between. Today, whenever my siblings and I decompress our memories of Balmoral, we recall a tightly organized regimen of meals, naps, baths, and tears before bedtime. We share no recollections of carefree, spontaneous play together; a profound isolation divided us, even as a silent sympathy braided us together. Of the three of us, Balmoral made me the obsessive writer of furrowed brow, the psychic archaeologist trolling the cryptic rubble of my formative years.

═

One such dream, one of hundreds before and since, comes after writing the passage above, returning me to the nights of my childhood. I am standing in the front yard of Balmoral on a black, moonless night in winter, gazing at the giant, leafless maple tree. Suddenly it teeters and falls, crashing through the slate-shingled roof, bisecting the long, grey house as if it were a human body. The massive trunk—the family tree? the tree of knowledge?—lands squarely on the bed of a dreaming boy, killing him instantly. I wake with a violent start, sitting bolt upright, in the dawn light. Yet the pounding of my heart tells me I am still alive. . . .

═

My sister, Shelagh, was deemed "a handful." A whirlwind of midget willpower, she pulled medicine bottles out of the bathroom cabinet with wild abandon and gulped pills by the chubby fistful; at least once, our mother rushed her to the emergency ward to have her stomach pumped. She rolled milk bottles down the hardwood stairs and giggled as they smashed on the landing. When my mother, pregnant with me, poised herself to sit down at the Singer sewing machine, Shelagh pulled the chair out from under her and she collapsed to the floor. My sister proved an escape artist of Houdini-esque calibre: One winter day, my mother zipped her in a snowsuit and tightly harnessed her to the maple tree in the front yard. Later, glancing out the window, she saw her defiant daughter prancing through the snowbanks, as naked as the day she was born.

My terrible two-year-old sister was consumed by such restlessness that she roamed the rooms of the house at all hours of the night; my exasperated parents resorted to plying her with tranquillizing doses of Seconal and cocooning her in her bed by affixing the blankets to the mattress with thick safety pins. But my tenacious big-little sister continued to wriggle free, obsessively twirling tufts of hair with her fingers and pulling them out by the roots. Compounding her distress was my own recent arrival in the nursery, an unwelcome event that detonated a paleolithic sibling rivalry; one day my mother had to restrain Shelagh from crushing my baby-soft skull with an ashtray.

For all the Sturm und Drang, "She-She" showed early executive leadership abilities, waving her arm and famously exhorting her toddler brothers, "Follow me, men!" And follow her we did. Our house shared a driveway and a large two-car coach house with our neighbours, the Macmillans. On frosty winter mornings, the trio of FitzGerald children slipped out the back door, past the coach house where our father parked his sleek Studebaker and our mother her nimble Nash Rambler, for a spontaneous visit next door. I would consciously avoid looking upward at the loft where I knew a black wooden beam, originally designed for hoisting bales of hay for the carriage horses, stuck out like a gallows.

The Macmillan family felt like a Floridian sanctuary. "Jick" Macmillan was an easy-going industrial psychologist with Canada Packers; I imagine him busily salving the guilt complexes of butchers flecked with the blood of fatted calves. His wife, "Miggs," was a kindly, chain-smoking soul who seemed unperturbed by our routine invasion of her breakfast table. Compared to our strict parents, our neighbours seemed like carefree bohemians, letting their three kids and slobbering standard poodle have the run of the place. At their table, we jabbered freely, spinning the Lazy Susan with impunity. Our intermittent morning missions to the Macmillans felt redemptive, like warming chilled fingers by a pot-bellied stove; here, we felt welcome.

On the other side of our house stood a progressive nursery school, Windy Ridge, run by yet another psychologist, the controversial anti-Freudian Dr. Bill Blatz, a short, iron-willed son of German immigrants, who wore a toothbrush moustache disturbingly redolent of der Führer. Rumoured to be a proponent of wife swapping, Blatz—pronounced "blots"—loved to shock the conventional sensibilities of the Toronto middle class, declaring, for instance, that mother-love was highly overrated and that Santa Claus did not exist. I never set eyes on Dr. Blatz, but in later years my imagination conjured a heartless Snidely Whiplash figure waving pages of black ink blots and affixing electrodes to the temples of squirming toddlers. My older cousins Anna and Patrick had attended the school, but mysteriously I would not. Of course, I had no way of knowing that Windy Ridge, funded by the massive Rockefeller fortune, was the first establishment in the world expressly designed to study the "mental hygiene" of preschoolers, nor that it was patronized by Toronto's old money families, including the Gooderham distillery clan; eventually I would learn that both the Rockefeller and Gooderham family fortunes had played an integral part in my dead grandfather's story.

═══

My mother, Janet Grubbe, was born in December 1918 amidst the unchecked fury of the influenza pandemic that would kill untold millions of people worldwide. She was lucky to emerge unscathed; luckier still, over her lifetime, to seem immune to bouts of the flu or the common cold.

She grew up at 21 Delisle Avenue, a gloomy three-storey house only a few blocks east of Balmoral, the third and last child of Talbot Grubbe, a severe, poker-faced, Scots-blooded manager of a Royal Bank branch from whom I took my middle name. Her mother, Mabel Ewart Steele, a voracious reader and pious teetotaller, was raised in the cocoon of upper-middle-class privilege, a pioneer female graduate of University College, class of 1907, and a classically trained pianist. My mother liked to maintain she had enjoyed a carefree childhood, even as she let slip story after story that made me suspect otherwise.

A looming, square-jawed giant of six foot three and 220 pounds, my maternal grandfather was proud of his pioneer Toronto roots stretching back to the early nineteenth century, ancient by Upper Canadian standards. Talbot was born in his great-grandfather's

1834 stone farmhouse in Thistletown in northwest Toronto, presently owned by my brother; and although he entered the world in the same week in 1882 as my paternal grandfather, and lived only blocks apart, the two men never met. A major in the 48th Highlanders, Talbot stood grim and muscular in spats and sporran, his thick, tree-trunk legs shooting down from under a tartan kilt. He dedicated himself to local history, jigsaw puzzles, curling, and rowing; on the wall of his den hung framed black-and-white photos of him stroking hard for the Toronto Argonauts in the Henley Regatta of 1906.

And he was a bully. On Saturday mornings when I pushed the rotating blades of his mower over his front lawn, he would loom over me, demanding I work in a tight, clockwise square around the gnarled oak tree—deviate ever so slightly and there was hell to pay. One morning, when I showed up half an hour late, I trembled at the consequences of my tardiness. Sure enough, he glared down at me and spat out words I would never forget: "I'm ashamed your middle name is Talbot." So was I—I would have gladly swapped it for something Irish.

Most Sundays after church, we dutifully showed up at 21 Delisle. I remember rooting around in the backyard vegetable patch, pulling up fresh carrots by the stalks and carefully washing off the black dirt with the garden hose. The standard joke was to bite off a piece of the carrot, Bugs Bunny–like, and cry, "What's up, Doc?" Then we sat down for formal noonday dinners, my hated necktie feeling as tight as a noose. As my namesake hovered over the roast beef with the carving knife, I never saw him break a smile. My father was mysteriously exempt from this unbending ritual; after his wedding day, he never set foot in the house of his Scottish in-laws and at the time I didn't notice or care.

After graduating from St. Clement's, a private high school for girls, my mother asked her father if she could enrol in the Slade School of Fine Art in London. Artists, of course, stood a notch below parasitic ne'er-do-wells and he flatly refused. Instead, she followed her mother and took an arts degree at the University of

Toronto. As my mother confessed to me many years later, she was glad to escape 21 Delisle postuniversity. She hated the ritualized meals where she swung her head back and forth, watching her parents serve sarcastic volleys across the tablecloth. The constant tension spawned aches in her stomach. One day, when Talbot reduced Mabel to tears with yet another scorching barb, my twelve-year-old mother sided with her father and learned to despise such feminine displays of weakness; on the spot, she vowed never to cry again.

Suffering a nervous collapse, my middle-aged grandmother withdrew into the dark seclusion of her bedroom for months on end. Janet, in turn, escaped into theatrics, singing and dancing with Johnny Wayne and Frank Shuster—Canada's future, long-lived comedy team—in the University College Follies of 1936. My mother was hell-bent on puncturing the bilious gloom of her "square" Delisle home with brash outbursts of vivacity.

During the war, Janet defied her controlling parents and worked in the Rockefeller Center in Manhattan as a decoding clerk for the inscrutable Sir William Stephenson, "The Quiet Canadian," who ran the Allies' MI5 spy network for Winston Churchill. Because her mother and brother belonged to a pacifist organization, the Oxford Group, she was investigated by the FBI and RCMP before being cleared for duty. In 1942 she volunteered to work at an office in Guatemala City to help monitor suspected Nazi spies in Latin America. In a local bar, she met a pilot from Eau Claire, Wisconsin, and married him, swept up in the ongoing epidemic of impetuous, booze-fuelled war romances. Given that her anglophile father detested Americans on principle, it was not surprising that he failed to show up to give away the bride. When her husband's B-29 bomber disappeared in a long-distance air raid over Japan in 1944—no wreckage was ever found—my mother returned to Toronto, a widow of twenty-six. In November 1945, at the wedding rehearsal of mutual school friends, she met my father, a junior intern at the Toronto Western Hospital. Fixing his blazing gaze on her, Jack FitzGerald badgered her to break an upcoming weekend date and step out with him instead. She did—setting in mysterious motion the course of my own future existence.

The object of my father's desire stood over six feet tall in heels, buxom and statuesque as an Amazon, with fine skin and long, stemlike gams. Well-groomed, lively, and charismatic, she radiated an air of taste and sophistication that rarely failed to turn heads in a room. And she played hard to get. As part of his mating ritual, Jack jokingly filled out a psychiatric certificate committing her to the 999 Queen Street West mental hospital:

"I hereby certify that I have personally examined Janet Ewart Grubbe of Toronto and that the following signs and symptoms have been observed which indicate a psychopathic disability, namely, slurred speech, unsteady gait, inability to maintain a normal, healthy attitude towards life in general, and ME in particular!"

Possessed of unusual drive and intelligence, a love of travel and adventure, Janet could have run a corporation as well as any man, yet marriage compelled her to quit her job at American Airlines and submit to the unlikely and unsuitable role of housewife and mother. Like many bright women in those prefeminist days, she often played dumb, disguising the fact that she was as quick on the uptake as anyone. In a diary she bequeathed to me, she wrote she had thrown in with Jack FitzGerald because he was "sexy, witty and had good earning potential." Initially, she was charmed by the shabby gentility of 186 Balmoral Avenue and intrigued by the medical family who had built it. Only decades later, after my father's death, did she admit that early in the marriage, she realized she had made a mistake; living at Balmoral, she confessed to me, she had always sensed "something rotten under the floorboards." Yet there was no question of divorce; like her military father, she soldiered on, hoping for the best.

═

Early childhood memories situate my mother talking into the telephone to invisible strangers as I wished she'd bend down and speak to me instead. Repeated words like *Crippled Civilians* and *Women's Auxiliary* and *Art Gallery of Ontario* seemed to carry some import. A booming laugh would invariably lead to one of her favourite catchphrases: "Oh, honestly!" She devoured the novels of Graham Greene and thumbed the weekly issues of *The New Yorker*; she loved to dabble in photography, gardening, and interior decorating. Despite her father's tacit disapproval, she set up an easel and dabbed oils on a canvas, casually puffing Matinée cigarettes—then abandoned both habits.

As a toddler, I would plant myself on the floor a safe distance from my mother's feet, holding a picture book upside down, pretending to read, my back turned to her. Every so often I would swivel my head, catching the subtle expressions passing like clouds over her face, gauging how close I could get to her curving, silk-stockinged legs, even as I wriggled towards her, inch by inch, on my bum. I was forever sticking my baby finger into the wind of my mother's changing moods.

What was left out of this oft-repeated story was that by an astonishingly early age, I somehow divined that my mother, under her mask of self-possession, was overwhelmed by my sister's one-girl, placard-waving protest movement. Strategically, I gambled that the only way I could reel in a modicum of maternal physical affection was not to demand any. I would wait her out, hoping, one day, that she'd come across. I gambled and lost; my mother, relieved by my self-imposed hands-off policy, cast me as a "good boy" and the label stuck. I was perpetually and unnaturally quiet, never a bother, the one who never once asked to be lifted onto her lap. For my thoughtful acquiescence, she rewarded me with a passing smile.

Only much later would I come to interpret my mother's mercurial and contrary temperament as largely a consequence of the self-thwarting of her artistic impulses; or come to understand that she treated Shelagh more as a rivalrous sibling than a daughter.

But before pneumonia claimed her in her eighty-eighth year, my mother did manage to snatch some belated creative fulfillment, spending the last two years of her life writing and publishing a biography of her maternal great-great-grandfather, John Ewart, a pioneer Scottish builder and architect. Ewart was responsible for erecting an impressive array of gothic public edifices in early Toronto that included the city's first general hospital; a wing of the law society, Osgoode Hall; a wing of Upper Canada College (where his two sons excelled as head boys in the 1830s); a Catholic church that sheltered refugees of the Irish famine; and part of the notorious 999 Queen Street lunatic asylum. But as a child, of course, none of this meant a thing to me.

Occasionally, as a special treat, my mother orchestrated evening visits of we three kids to the living room. Clad in dressing gowns and slippers, squeaky clean from our nightly bath, we padded past the threshold of the French doors, the invisible barbed-wire fence of no kid's land. My parents reclined in easy chairs, stirring their cocktails with swizzle sticks, while jazz, cool and hot, jumped off the record player. In these isolated, electric moments, my cheeks flushed, my pulse quickened, my adrenalin flowed, engulfing my body in a rush of kiddie bliss. Yet all the while I knew the grandfather clock was ticking towards an arbitrary deadline—any second, we three mice might be shooed back to our black holes in the nursery.

From time to time, terse family anecdotes sifted down to our childish ears, and we picked up dropped names that meant nothing: An antique book cabinet with latticed glass windows once belonged to an Irish writer named Jonathan Swift; a winter landscape was painted by a Canadian doctor named Fred Banting; a sterling silver dinner knife, inscribed with an equine coat of arms, spoke of a medieval Irish knight to whom we seemed remotely tied. In the adjacent dining room, my eyes roamed over the mahogany sideboard, the ornate brass candlesticks, the hand-carved, drop-leaf rosewood dining table, the escritoire studded with tiers of tiny drawers—solemn artifacts that spoke of another age. I knew I may look, but I must not touch.

My curiosity was drawn most powerfully to a foot-tall plaster statue of a nine-year-old boy that stood fretfully on a windowsill, looking like Peter Pan, the boy who wouldn't grow up. He was wearing short pants, grey woollen knee socks, and the crested blazer of his prep boarding school, Upper Canada College, only three blocks distant. I vowed that one day, when no one was looking, I would break the rules and run my fingers over the contours of the talismanic object that, mysteriously, looked just like me. It was, I eventually learned, an unerring replica of my father, Jack, frozen in time—and boarding school—*circa* 1926.

Our brief, pre-bedtime forays to the living room—fifteen minutes? twenty minutes?—indelibly conditioned our young nervous systems. Emotionally egalitarian, my mother had we three creatures of desire scrambling for a single, scarce natural resource. Affection, like food, was given and withdrawn at will, a process of intermittent reinforcement—a kind of bizarre Skinner box experiment—that pressed a lid firmly on our ids. In adult life, no matter where I spent the night, I would forever feel like a perpetual guest, dogged by a subtle sense that my presence was being barely tolerated, like a servant or tenant or guest or pet. Our father, it seemed, wanted our mother all to himself.

Following our command performances in the living room—why did they call it a *living* room?—we headed back upstairs to be quarantined in the Siberian steppes of the nursery. In bed, I watched my little brother hold his hands over his ears as he mechanically rocked his head to and fro on his pillow, humming in the dark. Sometimes, when my parents threw cocktail parties—both had a happy talent for making friends—I'd slip out of bed, creep to the second floor landing, and sneak peeks through the bars of the wooden banister. With guilty pleasure, I inhaled the sensuous buzz below, the clinking glasses of martinis and grasshoppers, the diminuendos and crescendos of brassy big-band jazz, the sharp bursts of hysterical laughter cutting through the blue clouds of cigarette smoke.

No memories survive of my parents visiting the nursery to tuck us in or read us stories, although I suppose they must have. I do recall my perfumed, mink-stoled mother standing at the doorway before a night on the town, stretching an imaginary kiss from her lipsticked mouth, like a piece of taffy; using her fingers as imaginary scissors, she'd cut it off, roll it up into an invisible ball, then throw it at us. Then she'd turn off the lights and close the door.

Whenever one of us started crying, it triggered the other two. To divide and conquer our piercing wails, our mother sometimes shifted one of us from the nursery to the spare bedroom, from the spare room up the back stairs to the third floor. For years, she rotated us about the house like pieces of antique furniture, a strategy that I grew convinced was designed to quell the kid-induced anxiety mounting inside her like a riptide. We were, quite literally, being put in our place.

My dreams invariably pulled me down the deserted streets of our neighbourhood—Lynwood, Edmund, Farnham, Clarendon, Cottingham—where devouring nocturnal monsters lurked behind neatly clipped hedges. I dreamt of being chased, my legs turning to maple syrup; I dreamt of being impaled on a stake; I dreamt that my murdered corpse lay under the floorboards. I curled up into the fetal position and hugged myself tightly, as if to prevent my body from exploding into pieces, scattering among the painted wooden blocks that littered the nursery floor. Other nights, I feared I'd float out the gothic dormer window into the black, infinite, airless void of outer space, cut adrift like a birthday balloon.

Then, far too often, the night terrors came, far worse than bad dreams—dreadful, petrifying, ravenous as a mythical beast.

A thunderstorm crashes beyond the curtained windowsill, forked lightning stabbing at the curtains; I fear being consumed by the terrifying blackness of the night sky. What is this grave, unseen incubus pressing on my chest? My throat burns, as if coated with the child-killing germ of diphtheria. Dream and reality fuse. Flung into helpless confusion, I scream into the silence, but no one comes. I scream and scream; still no one comes. Is no one listening? Are my parents dead? Am I dead? No longer heard, I no longer speak. Driven inside myself, I lash myself with the accusation, over and over: *"Why is this happening? What have you done? What have you done? You must have done something terribly wrong."*

I survived such nights, but not unscathed. In time, I came to realize that in the house of my dead grandfather, my psychic apocalypse was nothing personal, merely the transmission of an ancient command: *Thou shalt not feel.* Most of all, I was not meant to feel the dark spirit of Dr. Gerry FitzGerald slipping under my skin and into my bloodstream, like a needle in a vein.

═

Where was my father in all of this? Straining to conjure my childhood impressions of him, I make out a classic concave Irish profile—a thin, balding man with prominent chin and ears, his nose crowned by a pair of horn-rimmed glasses. I recollect the dainty, near-feminine gestures; the clipped, eccentric turns of phrase; the repertoire of nervous twitches; the way he pursed his lips and pensively ran his fingers through his thinning strands of reddish-brown hair; his immersion in cryptic crossword puzzles; his tearing open of crinkling cellophane packs of Dad's Cookies; his habit of removing his glasses, breathing on the lenses, then rubbing the mist clean with his handkerchief; how he wore his silver-banded wristwatch with its face unconventionally turned inward, not outward; how his dressing-gowned body, just shy of six feet tall (the same height as our mother), flapped past me in open-heeled bedroom slippers, his domed head lost in thoughts I struggled to divine. These surface traces are what first appear; for, like a ghost, he shunned extended contact with flesh-and-blood offspring as if any second he might shatter like a frozen pane of glass. Perhaps to actually touch the bodies of his own children would push him perilously close to the lip of his own Balmoral boyhood.

I know the facts of my father's life far better than I ever knew him. John Desmond Leonard FitzGerald was born on May 29, 1917, the same day as John Fitzgerald Kennedy, the doomed American president of Irish blood. This was only one of many weird pieces of synchronicity that infused our family romance. My father intensely admired JFK and the media-generated myth of Camelot, yet never said why. I suppose there was something about their shared name and birthday that seemed to define him, and us; something about his identification with the wretched excesses of the Irish, tragically striving and failing, the flashes of wit and charisma, the liberal idealism, the isolating and sacrificial nature of manly success.

Why did my father become a doctor? In my childish, idealizing mind, I assumed he wanted to help people. Much later, as details of his early life emerged, I realized his motives sprang from a poisoned chalice.

In 1942, he graduated from the University of Toronto Medical School—no small feat for a chronically indifferent student. It was also the year Jack married—but not my mother. Brief and tempestuous, my father's first marriage burst with the stuff of Hollywood melodrama.

It all began with Peter Spohn, a close friend from Spokane, Washington, with whom Jack bonded during their boarding days at Upper Canada College. The two moved on to medical school together, both competing with the reputations of powerful doctor-fathers. Dogged by a depressive temperament, Peter Spohn would eventually suffer professional setbacks and drown himself in his early forties.

When Spohn's fiancée arrived in Toronto at Christmas 1941, she brought along a Spokane-born girlfriend, Caroline Leuthold, a slim, striking redhead of German blood who had attended the Baldwin School in Philadelphia, an exclusive all-girls boarding institution, and then the elite Smith College. The foursome double-dated and Caroline stayed at Balmoral as a house guest. Like her hostess, Mrs. FitzGerald, she was an heiress to a family fortune, and the two women got along famously. The velvet nights of jazz and drink made Jack voluble and witty in the presence of the glamorous yet naive American aristocrat born with a silver spoon in her mouth; the attraction proved instant and mutual.

Jack next saw Caroline in the spring of 1942, when she travelled to Toronto to attend his graduation from medical school. In May, he headed south to Baltimore to spend the next seven months interning at Johns Hopkins Hospital, tending to the afflictions of affluent southern gentry on a private ward during the week, and cutting a swathe through the jazz nightclubs of Manhattan on the weekends with the dazzling Caroline—now studying fashion design in New York—on his arm. Living in the same hospital where his dead father had once worked over thirty-five years earlier, Jack wrote to his mother of his loneliness among the Haitian, Chinese, Belgian, and Venezuelan resident interns, his infatuation with Caroline, his lapses into sudden fits of weeping. Swept up in the winds of a cocktail-laced wartime romance and a powerful sense of entitlement, he pressed his mother to let him marry Caroline as soon as possible. Further, he wanted Edna to support him financially until he got on his own feet.

Even as he wooed Caroline that spring and summer, my father became enamoured with one of his instructors, a married physician; she and her husband had both graduated from the Hopkins medical school two years earlier. Within two weeks of meeting "Dr. X," as she was later named in court documents, Jack slipped secretly into her apartment one afternoon when the husband was out of town. Here, under the breastlike dome of the most famous hospital in North America, here in the gothic cradle that gave modern scientific medicine its name, my father abandoned himself to one last fling.

The affair lasted six weeks. Then, in September, he married Caroline. Edna had urged her twenty-five-year-old son to postpone the wedding until he was self-supporting, but with deft doses of wheedling charm, he was able to gain her blessing—and two hundred dollars monthly support. Caroline's father, the genteel millionaire lumberman, offered his Canadian son-in-law a cushy job as a corporate doctor in Spokane, a position that could set him up for life. But he refused; Jack FitzGerald had much to prove, and he would prove it on his own.

After the honeymoon, the newlyweds lived in Baltimore. One night, Jack introduced his new bride to Dr. X; as it turned out, his ardour for his Hopkins teacher had not entirely faded, for he had secretly phoned her on the night before the wedding. When the internship ended in December, the couple moved to Toronto and lived with Mrs. FitzGerald at Balmoral. Jack wrote Dr. X several letters and when she responded, he told her to direct future letters to his fraternity house. But before long, Caroline discovered one of the letters and from their steamy content she assumed that the affair had continued after the marriage. When my father denied it, she reluctantly believed him.

Jack and his first wife, Caroline Leuthold, Spokane, Washington, September 1942

Months later, in February 1943, Jack enlisted in the Royal Canadian Navy, although he did not travel overseas until the following year. The navy was an odd choice given his near-phobic aversion to water. He was posted for training in Port Arthur–Fort William on the northern shore of Lake Superior, a frozen outback that could not fail to brutalize Caroline's sensibilities. Balking at the prospect of moving north with a man she realized she no longer trusted or loved, Caroline, now pregnant, left Canada on New Year's Eve 1943, and returned to the protection of her family home, known as Deer Park, a wooded, five-acre tract of land outside Spokane. That March, weeks away from giving birth, she invited Jack to Spokane in a bid to revive the marriage, but he failed to endear himself to his in-laws when he drank all day and night and slept in till noon.

On May 1, 1944, a hazel-eyed, red-haired boy was born in Spokane, the fifth, first-born FitzGerald son in five generations to be named John. But his father was now half a world away, serving overseas as a surgeon-lieutenant on a newly commissioned Canadian frigate. In his beige, toggled duffle coat, Jack stood night watch on the conning tower, scanning the storm-tossed North Sea with a pair of binoculars. Anticipating the dreaded torpedo attack, he always slept in his clothes, joking with his buddies to allay the terror he felt. In its only action, in October 1944, the HMCS *Annan* engaged a surfaced, wounded Nazi U-boat in a running gun battle, then sank the sitting duck with a barrage of depth charges. As German sailors floundered in the frigid, oil-slicked waves, a crew-cut, trigger-happy, young machine gunner from Winnipeg raked the sea with bullets, killing and wounding several of the defenceless enemy. The *Toronto Telegram* saluted the victory on its front pages. My father typed up an unpublished first-person account of how he dosed the survivors with morphine, transfused bags of blood plasma, and "probed and picked steel out of Gerries for many hours." A fellow doctor on board, Stuart Macdonald, was a son of Lucy Maud Montgomery, author of the beloved *Anne of Green Gables* books, who had killed herself only months earlier. The dark family secret would not be made public until 2008—such was the silence, shame, and stigma attached to depression and suicide, especially among the socially prominent.

In the spring of 1945, my father, the returning veteran, made a second visit to Spokane and once again drank excessively. He paid little attention to his son whose recent first birthday he had forgotten. Jack wanted the family to live in Toronto; Caroline preferred to stay in the States. Caroline filed for divorce and on their third wedding anniversary in September 1945, it was finalized. Not only did Jack agree to let Caroline retain sole custody of their son, he waived all visitation rights. The divorce was legal in the US but not Canada, where the law required proof of adultery. Jack tried to persuade Caroline to testify to a trumped-up claim of adultery to restore his single status, as he had recently fallen for my mother, but Caroline refused. All in all, it was a royal mess, and it wasn't over yet.

Months afterwards, at Christmas of 1946, Caroline remarried, this time a Spokane pediatrician. Six weeks later, she died suddenly, ostensibly of infected fallopian tubes, but some whispered suspicions of a botched abortion. She was twenty-eight. Devastated by the loss, little Johnny withdrew into a nervous shell. His stepfather hastily relinquished the boy to Caroline's wealthy brother, Sam, who was assuming the reins of Deer Park Lumber, the lucrative family business.

This strange turn of events lit a fire in my father's belly. With Caroline's death, he was now legally free to remarry. Repeating the pattern he had followed with Caroline, he hectored Janet with the passion of a small boy, urging her to marry him within weeks, but that was not all. Racing over to his fiancée's house, he stood in the living room of 21 Delisle Avenue and cried: "I'm going to the States to get Johnny!"

Jack was now passionately determined to reclaim custody of a boy he had abandoned to convenience; my mother, understandably, was unnerved by the prospect of bringing up a three-year-old stepson—especially when her betrothed, aflame with Irish impetuosity, neglected to consult her first. When Janet asked for time to think about it, he stormed out, slamming the door. Years later, she told me she regretted she did not stand up to my father and refuse to marry him if she was expected to mother a small boy. Or was it two small boys?

Taking a leave of absence from his new job—he was working as a chest clinician at Hamilton's Mountain Sanitorium, a tuberculosis hospice outside Toronto—my father and his sister, Molly, travelled three thousand miles by train to the state of Washington, checked into a Spokane hotel under assumed names, and filed a temporary restraining order. When they brusquely demanded that the Leutholds, one of the plutocratic families of Spokane, hand over the child on the spot, the reply was short and swift: "See you in court."

In a highly publicized society trial, the American lawyers depicted Canada as a "backward country," the traffic intersection of Avenue Road and Balmoral as "suicide corner"; noxious fumes allegedly oozed from the sewer grates—a laughable fiction given that "Toronto the Good" boasted one of the best sanitation and public-health records in the world. My father was portrayed, not entirely inaccurately, as a feckless lounge lizard, addicted to drink and the unsavoury jazz underworld, indifferent to the welfare of his baby son. Clinching evidence of his moral failings came with the revelation of his premarital fling with Dr. X at Johns Hopkins.

Eminent character witnesses were stacked up in my father's defence; if the Leutholds had "pull," so did the FitzGeralds. In Toronto, glowing depositions were given by an Ontario high court justice, the medical director of the Royal Canadian Navy, distinguished doctors who had been friends of his father's, and Dan Lang, an Upper Canada College classmate and navy shipmate destined to become a Liberal senator. Another UCC friend bent the truth more than a little when he testified that he had never seen Jack drunk. The seventy-eight-year-old Reverend Dr. Henry John Cody, former president of the University of Toronto and a pillar of the Anglican establishment, the man who had presided over my parents' wedding only weeks earlier, testified that Jack was "the son of the most outstanding medical man on the staff of the University of Toronto. Everybody knew him as his father's son."

In the end, the Spokane judge—who happened to be a regular golfing partner of the Leutholds—ruled that Jack FitzGerald's sudden, new-found attachment to his discarded son was "pecuniary, not parental." My own intuition is that my father wasn't after the money, rather that his small son needed rescuing—and suddenly reminded him of himself. In any event, Jack appealed the case and lost. Little Johnny, heir to a $300,000 trust fund, was adopted by his maternal uncle and aunt, a childless couple, and the boy's name was changed from FitzGerald to Leuthold. The judge noted, in passing, that my mother was the only fully honest witness to testify, as she had openly admitted in court that she was not keen to raise a three-year-old boy not her own. As the two-week trial adjourned, a rouged, fur-draped grande dame sidled up to my mother—married for only six weeks—and hissed the prescient words: "My dear, you'll be very unhappy."

Wedding of Jack and Janet, Toronto, April 1947

The case of a father losing custody of his own biological son attracted the press on both sides of the border (including a satirical squib in *The New Yorker*); the same Toronto newspaper that had championed a hometown doctor as a naval war hero now ran headlines proclaiming that an American court had judged him an "unfit father." To cover the legal expenses, Jack burned through ten thousand dollars of his mother's, Edna's, stock portfolio, a small fortune in those days. By coincidence, the trial ended on the day my father turned thirty, no doubt reinforcing his lifelong aversion to the celebration of birthdays—his and anyone else's.

The loss of his son was an event of which my father would never speak—just as he never spoke of his dead father, his years in boarding school, the war, the wild alcoholic binging of his sister, and the suicide of his close friend Peter Spohn—a chain of batterings that must have ripped pieces from his soul. On top of it all, he still carried inside him the cutting words of Duncan Graham—a bloodless, cantankerous Scot who was chief of staff of Toronto General Hospital and a close colleague of his father—words that Graham had dropped when Jack was still a doodling student in medical school, lost in the clouds of Ellingtonian swing: "You'll never be as good as your father."

My father already burned with fierce ambition; Graham had simply thrown gas on the fire. Jack FitzGerald would show *him*—and any other of his father's cronies who dared think the same. In this he was not so different from Sigmund Freud who believed that all politics could be reduced to the primal conflict of father and son. In *The Interpretation of Dreams*, Freud recalled a searing moment in his childhood when his father scolded him for brazenly showing off his penis as he pissed into an adult chamber pot: "The boy will come to nothing!" The father of "the talking cure" would have something to say about that.

═══

The autumn after the custody trial, my father started a nine-month research fellowship in immunology and pathology at Roosevelt Hospital in Manhattan. He interned under the pioneering American allergy specialist, Robert "Pops" Cooke, an exacting, sober-minded, fourth-generation doctor who sported an ugly Frankenstein-like scar across his forehead, the result of an operation for severe sinusitis. Cooke became my father's first mentor.

Allergies had been discovered accidentally in 1906 when laboratory pigs injected with diphtheria antitoxin became asthmatic. Coined by a German scientist, "allergy" was derived from a Greek word meaning "other energy." But for years the reaction was dismissed as a mere idiosyncracy. As a young intern in that same year of 1906, Cooke was assigned to ride horse-drawn ambulances through the streets of New York. After each call, he was left gasping and choking, needing a shot of adrenalin. Finally, someone deduced that his asthma was caused by the dandruff of the horses.

In 1908, after exposure to a patient suffering from diphtheria, Cooke was treated with a shot of antitoxin produced by the blood serum of horses. Before the needle was out, he was in a bad way; again, an emergency shot of adrenalin saved him from death by anaphylactic shock. Medical opinion held that the offending substance was poisonous. Yet a mystery remained: why were some people sensitive and others not?

Driven to understand the roots of his affliction, the thirty-eight-year-old Cooke opened the Institute of Allergy, the world's first asthma and hay fever clinic, in 1918. To neutralize allergic reactions, he used tedious desensitization methods, injecting patients with tiny amounts of the offending allergen—ragweed, grass, dust, hay, etc.—in incremental doses; as he lived on a farm in New Jersey, he desensitized himself to horses. Three decades later, there were still only a few hundred allergy specialists in the world, more than half in the US, and one of them was my father.

With his trophy wife on his arm, my father rented a brownstone walk-up on 101st Street near Riverside Drive, where they listened to the woman in the upstairs apartment routinely beating her kids. One day in the spring of 1948, Janet rendezvoused with "Fitz"—for this is what she called him—after work in his favourite haunt, a jazz nightclub on 52nd Street. Newly pregnant, she suddenly fainted and slid off the bar stool.

That December, my sister, Shelagh, was born in Kingston, Ontario, as her father was now sweating out an FRCP, the elite fellowship degree, in the labs of Queen's University. He was studying the nuances of an embryonic new science—the immunology and pathology of tissue sensitivity in allergic disease—and he was dead set on dominating the field. If Dr. Cooke's reference letter to the National Research Council was to be believed, his future was a promising one: *"Dr. FitzGerald impresses me as a man of great intelligence and high ideals. He possesses an unusual flair for medical research; he is thorough, painstaking and extremely critical. In other words, he has a natural aptitude for medical research, which in my experience is unusual."*

In the fall of 1949, my parents returned to Toronto, where they paid Edna fifty dollars monthly rent for the third-floor apartment at Balmoral. Nearly penniless at thirty-two, Jack FitzGerald had passed through the universities and hospitals of McGill, Cambridge, Toronto, Johns Hopkins, Manhattan, and Queen's, putting a gruelling postgraduate education, degree by degree, behind him; before him stretched a life designed not to fail.

Renting an office in the elegant Medical Arts building at the corner of Bloor and St. George, where six snakes of Aesculapius hung over the art deco portico, he launched a private allergy practice that grew in leaps and bounds. In January 1951, as I cradled a stuffed bunny in my infant arms, my father co-authored an article in an American pathology journal, entitled: "Diffuse glomerulonephritis induced in rabbits by small intravenous injections of horse serum."

The time had come to live up to his dead father's formidable reputation—and, ideally, surpass him.

═

Every day after work, without fail, my father escaped to the only corner of the world where he was truly happy. Passing under the porte cochère and through the front door of Balmoral, he headed straight for the living room. Thumbing through his voluminous collection of vinyl records, he pulled a disc from its sleeve, placed it on the turntable, and dropped the needle in the groove. For my father, jazz was the coolest drug that soothed the restless blood coursing in his veins.

My father was a hipster. Being hip, of course, held long associations with drugs—opium smokers lay in their dens, smoking their pipes "on the hip"—and my father fit the profile perfectly. Since the war, he prescribed himself hefty doses of Wigraine to melt the migraine headaches that stabbed his brain like a burning knife. Many of his heroes, both medical and musical, used opiates and their blissful extracts; for my father, the medical, musical, and drug scenes fused as one.

In his lonely boarding-school days in the 1930s, my father's stacks of seventy-eights rose higher than his textbooks; during his arduous medical training, he trailed big bands across two continents, as they played clubs in Toronto, New York, London, and Paris. Over the years, he would worship the virtuoso piano players—Art Tatum, Erroll Garner, Oscar Peterson, Thelonious Monk—moaning softly as they ran their quicksilver fingers up and down the ebony and ivory keys. Swinging high, swinging low, Jack snapped his fingers to their seductive rhythms, revelling in the golden age of jazz that strangely coincided with the rise and fall of the Third Reich. He loved mimicking the slang esoterica of the coal-skinned jazzmen, his body bursting into jags of convulsive joy.

When Duke Ellington's band played in Toronto in the days before my birth—often in the art deco lakeside dance hall, the Palais Royale—my father would liberate many of the musicians from their downtown fleabag hotels and lead the way up to Balmoral (his friends called it "Club 186") for wild, all-night bacchanals. The Duke's signature catchphrase at live concerts—"We love you madly!"—was repaid in kind. Belting back the shots of booze, they'd rivet themselves to the phonograph player, blacks and whites together, digging the latest, hottest cut spinning on the turntable. They'd rarely dance; reverent listening was the sign of the true aficionado. Once my father gave an impoverished Ellington trombonist a second-hand pair of pants. "We don't get treated this well anywhere in the world—except Paris," the musician gushed in thanks. Their gratitude intensified when the obliging Canadian doctor "prescribed" soothing narcotics for whatever ailed them. A cautious few of Jack's private-school buddies warned that he was risking career suicide by consorting with negroes, let alone serving as their drug pusher, but he didn't care. And so Balmoral took on a Jekyll-and-Hyde hue: On sunny afternoons, my grandmother Edna, the elegant, Victorian chatelaine, held formal tea parties with her polite society friends; under the cover of night, she pretended not to notice her son, Jack, the hipster doctor, sink into the same settee and pass rolled up leaves of "tea" to unsavoury jazzmen with faux-aristocratic names like Count Basie and Duke Ellington, as they idly tinkled the "eighty-eights" of her baby grand piano.

One night in the early 1950s, as I was dreaming in my crib, Count Basie stood regally by the crackling fireplace downstairs in the living room, intoxicated with marijuana chased by shots of

Crown Royal, as divine jazz sprang off the turntable. He was holding forth to my father when suddenly, in mid-sentence, he crashed unconscious to the carpet—Basie down for the count. Some time later, I sat on my father's cherished Django Reinhardt record, *Sweet Chorus*, and broke it—another family story played back over the years in tones of rueful loss. I might as well have shit in the lap of the great man himself; I might as well have committed an unpardonable sin.

═══

Our backyard fence abutted the playground of Brown Public School, vast and cinder strewn like the surface of a dead planet. In September 1954, I was sent there to junior kindergarten. Each morning, the entire school body assembled to drone "God Save the Queen" and I puzzled over the line: "Long may she rain over us." I did not know that a generation earlier, both my parents had sat in these same wooden desks in these same drafty, high-ceilinged classrooms. Not until the 1960s, pulling out a tattered 1923 class photo, did they discover they had attended the same kindergarten class, ages five and six.

I was a veteran nail-biter and thumb-sucker and, like my sister, I obsessively twirled my hair into knots. In my report cards, my teachers observed I was "easily upset, readily reduced to tears, but can be aggressive and noisy. . . . Jamie is a very sensitive child who is anxious to succeed. . . . he is sensitive about mistakes and accepting criticism." My marks show the early signs of a schizoid personality: straight A's in writing and straight C's in oral expression. Some invisible authority had swung an invisible axe and cleft my brain in twain, as if separating church and state. Semi-mute, I was one-quarter Helen Keller, teetering on the lip of autism. I craved attention, yet recoiled when I got it. Among groups, silence became my second language.

By now, I had cultivated a knack for dissociation that in retrospect probably saved my sanity. On my first day of grade one, the teacher read out the roll call, an unfamiliar ritual. When our names were called, we were required to say, "Present." When I heard "*Jamie FitzGerald*," I did not answer, for I was *not* present in any real sense. My classmates swivelled their heads as one and stared at me; their attention only deepened my paralysis. I felt stripped naked, accused, antlike. I did not yet know who I was—nor did I know how to know.

Each breakfast before school, my father hid his face behind *The Globe and Mail*; although he rarely lowered it to speak, his self-immersion indirectly imparted to me the importance of words. Perhaps printed language could make sense of my world; if I could not speak, I could read and write. And so I took to turning the pages of children's books, hand-me-down, English literature from the 1920s, redolent of prams and nannies and Little Lord Fauntleroy, reflecting a topsy-turvy world where children were treated like adults; when they grew into adults, they were treated like children.

A Fifties Family: James, Janet, Shelagh, and Michael; Jack "out of the picture"

Family photograph albums survive as vivid yet unreliable witnesses to my "picture perfect" childhood. Before the days of the ruthlessly truthful video camera, emotionally doctored images sliced us into precise sections of time, sometimes candid but mostly posed, forever voiceless and still, models of familial formality and normality; no prints exist of my sister being drugged and tied to her bed. In one portrait taken by Ashley and Crippen, a firm catering to the carriage trade, I sit perched on the arm of a settee, my crewcut head backlit like a religious corona. My mother, incandescent as a Hollywood movie star, wears a dress with plunging décolletage and a pearl necklace as white as her teeth; her jet black hair is pulled back tightly in a bun. Within the gilded frame she gazes up at me, together with my brother and sister, all smiles, all eyes on me. My father is conspicuous in his absence, literally "out of the picture"; the photographer had unconsciously cast me, a six-year-old boy in neatly pressed shirt and tie, in the timeless role of Prince Oedipus-in-waiting.

Although we were not allowed to miss Sunday school, my father never set foot in a church himself. On infrequent Sunday afternoons, he would drive the family down Russel Hill Road, originally a twisting Indian trail, to the Toronto Western Hospital, his place of work, my place of birth, for lunch in the cafeteria. He owned a convertible, which my gap-toothed brother, angelic with chubby cheeks and curly blond hair, called the "broken glass." As if in childish imitation, my father called the hospital, "the doctor place."

The serious, white-coated men pushing trays in the cafeteria line seemed busy and important, and so did my father. We spooned up cylinders of vanilla ice cream inexplicably wrapped in strips of wax paper. After lunch, in our father's spare and clinical office, my sister stood on a chair behind the smooth black plate of his X-ray machine, munching an arrowroot cookie; transfixed, we watched a black blob travel south, down her esophagus, like a rabbit coursing through a boa constrictor. We sat on the crinkling paper rolled over the examining table and fiddled with the thermometer and tongue depressors, saying "ahhh"; oblivious of our shenanigans, our father toiled at his desk, head bent as if in prayer, mumbling odd words into a dictaphone. All work and no play was making Jack a dull boy.

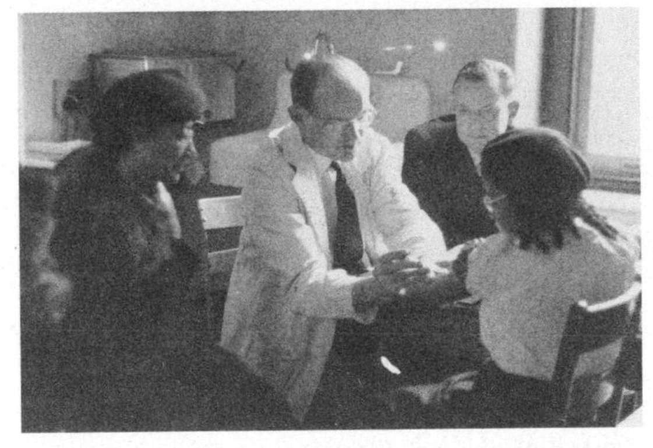

Jack, with Dr. John Hamilton, immunizing a young patient

During my own childhood years at Balmoral, an ethereal figure named "Granny" drifted cloudlike in and out of my life. My grandmother Edna Leonard FitzGerald had lived in the house through two world wars, watching the city of Toronto quadruple in size. She loved the old place dearly; naturally she wanted her children and grandchildren to love it as dearly as she. Sadly, none of us could oblige.

Edna's inherited wealth stemmed from the industry of her iron-willed great-grandfather who had built a foundry in London, Ontario, in 1834. Her grandfather, Elijah Leonard, was appointed a Senator in 1867, the year of Canada's Confederation. Born in 1882, Granny grew up in the family home, Trewiston, surrounded by the high-collared, aristocratic trappings of the local anglophile Establishment; alcoholism ran like a quiet stream through branches of the family tree. Her mother died when she was two and her father, Charles, remarried. An Upper Canada College old boy, he was a retiring character who whiled away the hours at the London Hunt and Country Club. In the late 1890s, Edna was sent to Toronto to board at Havergal College on fashionable Jarvis Street, the newly opened private school designed to fashion agreeable young ladies of taste and breeding.

Not surprisingly, my grandmother was a snob, one of the "Leonards of London." Frequenting the Toronto Ladies Club at Yonge and Bloor, she gracefully played out her predestined role of chatelaine—shopping at Britnell's bookshop, taking cabs to the ballet and opera, playing rounds of bridge, contributing to charity, sailing back and forth to Europe on vacation. When first introduced to my mother, Edna inquired snootily: "And what does your father do, dear?" Happily, my mother's social bloodlines made the cut, and soon they formed a bond of affection, likely based on the mutual recognition of a certain toughness.

In 1948, the widowed matriarch had rented the third-floor apartment of Balmoral to her daughter, my aunt Molly, who was raising two small kids, my cousins Anna and Patrick. Her husband, Major Tom Whitley, my godfather as well as my uncle, was a hard-headed war hero and rising young executive in the Royal Bank of Canada. Only a few years earlier, in the summer of 1944, Tom had commanded a company of infantry in a bloody battle with SS panzer tanks in an apple orchard outside Caen; in the end, the Tommies beat the Gerries, but not before an enemy bullet ricocheted off the silver cigarette case my uncle carried in his breast pocket. Postwar, he remained a tight-lipped man of action; whatever scars he bore under his impeccable suits and cut-glass Oxford accent, he pressed bravely on, winning even more battles in the trenches of Corporate Canada.

Molly was a thin, stylish redhead whose dry wit matched her epic consumption of dry gin, her deadpan drolleries echoing the English actress Maggie Smith, the masterful parodist of snobbish affectation. No stranger to scenes of alcoholic excess, my mother once told me that until she met Molly, she had never seen anyone so floridly drunk; she was shocked, too, to catch her sister-in-law one night, lit up, her husband out of town, lolling on the sofa in the living room of Balmoral, flirting shamelessly with an old flame, a handsome Irish cad named Gerry O'Sullivan, as the children slept upstairs.

In 1949, the Whitleys moved out of Balmoral. It was now my father's turn to reoccupy the place of his birth as the new third-floor tenant. With my birth a year later, Granny ceded the bottom two floors to my parents and took the upper apartment herself. When my brother arrived in 1952, Granny quit the house entirely, but she did not stray far, moving down the street to a rented apartment at number 133. Balmoral, the gothic dream castle of his childhood, now belonged to my father; or rather, he belonged to it.

And so did I.

Wisely, Granny saw her five grandchildren one at a time; each week, she spent an afternoon alone with me at 133 Balmoral, taking tea at precisely four o'clock. As her Bakelite bracelets jangled on her wrists, she taught me how to hold a knife and cut the tops off shelled, soft-boiled eggs, then dip narrow slivers of toast "soldiers" into the yellow yolk. With classical music playing on an old Victrola, she'd park me at a polished antique table and hand me a string bag of ivory mah-jongg tiles; I was never sure what to do with them. I felt frightfully grown up to be served coffee ice cream, dished into porcelain bowls decorated with scenes of Chinese pagodas and intricate garden paths, where my mind wandered lonely as a cloud. Later, my mother told me I was Granny's favourite, as I looked like a perfect replica of Jack, her only son.

Yet none of us ever heard my grandmother speak of her husband, Gerry, dead for over fifteen years.

═

In 1955, Hank and Haans Besselar, a newlywed Dutch couple, moved into the self-contained apartment on the third floor of Balmoral, the first in a series of live-in "help." Both of my parents had been raised under the ministrations of full-time nannies, and we too would submit to this "upstairs-downstairs" world. It was up to the third-floor spare bedroom that our mother would often send us, mostly one at a time, where my bad dreams far outnumbered the good. Perhaps her motives were benevolent; perhaps she thought we might receive something from the young couple she herself could not give.

Hank was a blond, wiry carpenter who carved and hammered my parents a new double bed; Haans made us meals in the kitchen and bathed us afterwards. One night when a dish of pears raised the gag reflex in my throat, I dropped my compliant persona and refused to choke them down. Haans calmly turned out the lights and left me to stew alone in the dark. Around this time, my brother began running away from home. Mike was only three when he first slipped out the front gate of Balmoral and dashed a full mile down the Avenue Road hill. Eventually, a stout Irish cop picked him up and delivered him back home, berating my mother for her negligence.

"What are ya doin'," he fumed, "lettin' a wee lad with a fine

Irish name like Michael FitzGerald wander the streets alone?"

Mike's escape attempts persisted; a second cop rescued him, and again my mother received a stern lecture on parenting. Whenever my mother repeated the story, she said she was appalled—not by her own maternal disconnect, but by "what the neighbours might think." Born in captivity, the youngest and smallest member of the family had looked up and seen two preoccupied adults, a hellraising sister, and an inert brother, all nervous as cats; my little brother voted with his feet and tunnelled the hell out of Stalag Luft Balmoral. But such experiences failed to inhibit his growth, for by his late teens, Mike would sprout to the height of six feet six inches.

By 1957, our time in our grandparents' house was nearing its end. That winter, my sister contracted a case of infectious arthritis that wracked her right leg with excruciating pain. At first, our father dismissed her complaints as growing pains, but after she suffered a night of agony, she was taken seriously and driven to Sick Kids Hospital where she spent the next six weeks. I retain the strong image of my father carrying Shelagh in his arms out the front door, an unsettling sight, not only for her distress, but for the fact that my father was actually embracing one of us. To this day, my sister loves hospitals, for at Sick Kids she received a steady diet of kind attention.

Family Christmas at Balmoral, 1956—Standing: Jack, Molly, Anna, Patrick, Edna, Janet; Sitting: James, Shelagh, Michael

I was six and, aptly enough, reading the British books *Now We Are Six* and *When We Were Very Young* by A.A. Milne. One poem included this singsong line, which resonated deeply—"James James / Morrison

Morrison/ Weatherby George Dupree/ Took great/ Care of his mother,/ Though he was only three." "Waiting at the Window" featured an ink drawing of a mooning, solitary boy and his teddy bear, sitting together in a window seat, gazing out on a rainy day. I knew my own name was James; perhaps I hadn't yet grasped that John Desmond and John Gerald were the names of my father and grandfather.

These are my two drops of rain
Waiting on the window-pane.

I am waiting here to see
Which the winning one will be.

Both of them have different names.
One is John and one is James.

All the best and all the worst
Comes from which of them is first.

James has just begun to ooze.
He's the one I want to lose.

John is waiting to begin.
He's the one I want to win.

James is going slowly on.
Something sort of sticks to John.

John is moving off at last.
James is going pretty fast.

John is rushing down the pane.
James is going slow again.

James has met a sort of smear.
John is getting very near.

Is he going fast enough?
(James has found a piece of fluff.)

John has hurried quickly by.
(James was talking to a fly.)

John is there, and John has won!
Look! I told you! Here's the sun!

Life, it seemed, involved a lonely, grievous contest between fathers and sons—a contest about which no one dared breathe a word.

At Balmoral, of course, I was too young to voice the questions that would come much later: Why did my father shun his own three children as if we were lepers? Why did he refuse to share meals with us, banishing us to the pantry, hapless among "the help"? Was he competing with his own children for the caress of our overtaxed mother? Was he unwittingly replaying his own starved Balmoral childhood with the precision of a sleepwalker?

Years would pass before I would understand that my father had a childhood, too; he, too, had parents; he, too, harboured and hid contrary feelings. It was simple yet astonishing: My father had been conceived in the same bedroom where I had been conceived; his eyes had traced this same wallpaper pattern on the same nursery wall; he had been washed in the same claw-footed bathtub, sat in the same stifling church pew, the same cavernous Brown School classroom; he had dreamt and tossed behind the same wooden slats of the same wooden crib, wearing the same pyjamas made of lead. He had been picked up by the same giant fingers, dipped in the same dish of liquid porcelain, and set out to dry and harden on the same windowsill.

Years would pass before I would understand that *both* my parents were rebels, each in their own way, which helped account for their initial mutual attraction. Both were the youngest children in their families; both defied their parents' expectations and married Americans during the war; both surrendered to the boozy allure of the smoky house party and the forbidden, primal rhythms of the jazz nightclub, the only club they ever joined. Yet in the end, each for their own reasons, my parents played the Establishment game, honouring the straight and narrow path.

Like my father before me, Balmoral taught me how to make a virtue of necessity and fortify my capacity for solitude. But then, unexpectedly, some sharp shaft of feeling would rise up within me and puncture my calm. I remember my mother used to "bleed" the black, cast iron radiator in our nursery, turning a knob to let off steam. Sometimes, as I wandered alone through the rooms at dusk, I felt sure that some daunting, suppressed secrets were poised to burst out of the rad, flooding the hardwood floor.

Sometimes, like the boy in the sad English poem, I would part the curtains and gaze out the nursery window for hours on end. I would never bring my intuitions to words, but I felt I was living someone else's life, like an actor following a script. But whose life? And whose script? And who, beyond the curtains, was that persistent old man in baggy pants who trundled up and down the sidewalks as I napped on somnolent summer afternoons? Why did he lug a grindstone on wheels, spun by a foot pedal? Why did he seem so indispensable to the story? Did my boy-father hear this same old man, year after year, ring this same eerie brass bell? Did he hear him calling us out to the sidewalk, awakening him from his own slumbers? Did he, as entranced as I, run out and offer up our dinner knives to be sharpened?

When Carl Jung writes that a grandparent can influence a child as much, or more, than a parent, I am inclined to agree. At Balmoral, the citadel of repression that ringed the ghost of my grandfather

Gerry served only to connect me to him in uncanny ways; the three-storey house has three crowded generations of stories to tell, and mine is only one of them.

One of the single most powerful memories of my early childhood remains indelible: Walking home from kindergarten one day, I turn the corner of Balmoral—*Step on a crack, break your mother's back*—and open the front door. I climb the stairs to my parents' bedroom, where I startle my mother with the solemn words: "I want to change my name from Jamie to Gerry."

I had never heard my grandfather's name spoken; my mother's look of amazement made me sense that I had, as curious children do, unlocked some long-neglected closet door.

Some time later, the family drove past an impressive medical building on College Street on a Sunday outing. My young ear caught my parents whispering references to its builder, my grandfather—still a taboo subject, not unlike sex.

"Where does he live?" I piped up from the back seat of our 1955 Studebaker.

"Oh, he's dead now," my mother explained.

Brow knitted, the budding journalist in me retorted: "Who shot him?"

Even then, my parents' evasive, nervous laughter seemed a kind of clue, another tenuous lead to something hidden.

Perhaps, one day, I would find out who Gerry was—and who I was.

TWO

Room at the Top

The point of imperfection which we occupy—is it on the way up or down?

RALPH WALDO EMERSON

I left Balmoral, but Balmoral would never leave me.

In October 1957, my father sold his father's house to an Italian diplomat named Nuti. He told my mother that he was keen on "cutting the umbilical cord" and abandoning the gloomy, stuccoed house and desolate childhood years he had quietly hated. Yet he did not escape nearly far enough.

Only three blocks north of Balmoral, our new street once again echoed the name of a dank medieval Scottish castle transplanted to the forests of Canada. Dunvegan Road tilted northward on a slight incline, as if reaching for the promised land of the upper-middle-class. (Aptly, "dun" is the Irish word for fortress.) Hundreds of grand mansions huddled together, shoulder to shoulder, mirroring each others' subdued opulence. Like Rosedale, its older cousin to the east, Forest Hill and its well-groomed, well-heeled gentry exuded a concentrated force field of affluence and influence; the envy the neighbourhood inspired in outsiders glowed as green as the manicured lawns. Here, the wealthiest residential enclave in Canada stood a world apart, riding the postwar economic boom, sealed off from "the lesser breeds."

Just up Dunvegan lived the faceless figures we knew to be

multi-millionaires—Jack Kent Cooke, the blustery, freewheeling sports entrepreneur whose mansion would later be occupied by the cable TV magnate, Ted Rogers; the Bassetts, the owners of Canada's first private TV station; the Molson beer barons; and the Eatons, the department store family of Northern Irish blood whose insouciant fourth-generation sons would ultimately fritter away their corporate dynasty. The sprawling properties of the plutocracy seemed to dwarf our modest lot; their ultra-rich proprietors seemed to hover so much further up the ladder than us. In contrast to the gated castles, 75 Dunvegan Road seemed like a cramped tarpaper shack, our lot in life puny and second-rate—a feeling echoed by the near pathologically status-conscious writer F. Scott Fitzgerald who once observed that he grew up in "a below-average house on an above-average street."

Three nondescript storeys of beige brick, our new home felt less stark and sinister than Balmoral for warm broadloom carpets covered the hardwood floors from wall to wall. A two-car coach house included a second-floor snooker room, an essential adolescent refuge, but for the time being, I had to share a bedroom with my brother on the un-air-conditioned third floor of the house where, during the stifling summer months, I rolled sweating and sleepless on the bedsheets. For successive Julys, we were bused two hours north into the mosquito-infested Ontario wilderness to Camp Waseosa, near Huntsville, for the entire month; the first time, my brother was only five. Night after night, I wet my bunk in our tent, too ashamed to tell the ill-tempered counsellor. Inevitably, the stink grew so fierce that my bunk mate ratted on me and I was made to wash the piss-streaked bedsheets in the lake—the first airing of my laundry in public.

The same month we moved out of Balmoral, the University of Toronto appointed my father an assistant professor of medicine; now forty, Jack FitzGerald's professional star was rising in concert with an unprecedented, postwar economic boom. Three months later, he flew to a medical conference in Philadelphia; the day before, he quarrelled with his mother over her insistence that he dutifully report his every last move. That frigid January night, our seventy-five-year-old granny, Edna, took a taxi with friends to the symphony, then returned to her newly acquired apartment in an elegant, horseshoe-shaped, four-storey building at 400 Avenue Road, a one-hundred-yard walk down the hill from Balmoral. As she did every day of her life, she spoke with her daughter, Molly, on the telephone; from her bay window, decorated with stone angels, she was fond of gazing across the road at the massive, Gothic revival mansion of yellow brick, Oaklands, now a Catholic boys' school. The next morning, Mabel, the Irish housekeeper, nudged open Edna's bedroom door, bearing a breakfast tray. My grandmother was propped bolt upright in bed, cold dead, her pearl necklace still round her neck. A single word—"Seconal"—was scribbled on a wall calendar. As part of the protocol for suspicious deaths, a policeman was posted at her apartment door.

The official version was a heart attack. When our mother delivered the news, my sister, Shelagh, ever the emotional release valve of family repressions, burst into tears; even though I was deemed Granny's favourite, I felt nothing, as if there were no room for my own reaction. In time, my mother would say that the selling of 186 Balmoral had broken Edna FitzGerald's heart.

My father cut short his medical meeting and flew home. On the Saturday afternoon of my grandmother's funeral, I was engrossed in a special episode of *The Lone Ranger* that promised to reveal how the virtuous, wounded hero made a blood-soaked mask from his torn tunic, forged his silver bullets, and assumed a secret identity. My father offered a choice: I could stay home and watch the TV show or come to the funeral with everyone else. In the end, my innocent eyes followed the polished mahogany coffin as it slid smoothly up the church aisle, my widowed grandmother's Victorian secrets lost to the ages, and I pondered in my pew how the Lone Ranger became lone.

Even though Edna had loyally attended Grace Church on-the-Hill for decades, the Reverend Canon Craig gracelessly forgot her name during the eulogy. Miffed, my mother pulled up stakes and redirected the family down to Timothy Eaton Memorial Church, a severe grey stone edifice at the corner of Dunvegan and St. Clair. In those days, switching allegiance from a High Anglican to a United church seemed a brazen act of apostasy: In fact, we simply migrated a few hundred yards south, substituting one blandly sanctimonious WASP enclave for another.

For months, a family stalemate kept Edna's ashes in a shoebox in my parents' bedroom closet. Her cousin Ibbotson Leonard demanded that Edna's cremated remains be committed to the Leonard family crypt in London; my father finally agreed, but on the condition that his father, Gerry—whose ashes had been stashed inside his bust at the University of Toronto since 1940—be reunited with his wife. For some unexplained reason, Jack's idea was simply *not on.*

The family plot quietly thickened. A decorated World War I hero notorious for driving his car down the dead centre of the streets of London as if leading *The Charge of the Light Brigade*, Colonel "Ib" was an inveterate bully whose arrogance would have shamed a bewigged British aristocrat; in 1901, he had donated a scholarship to Upper Canada College on the proviso that its candidates could establish that white, British, Protestant blood flowed in their veins. In the Leonards' ultra-snobbish universe—they claimed, with straight faces, to have traced their lineage back to Noah's Ark—the likes of low-born Gerry FitzGerald, despite his medical eminence, did not cut it. In the end, my father dug in his heels; when he defiantly placed his parents' urns together in the Doric-columned mausoleum in Toronto's Mount Pleasant Cemetery, he and the Leonards were no longer speaking. The names of my grandparents were inscribed on a small, simple panel of Carrara marble, in an almost anonymous niche dwarfed by the grandiose Eaton crypt outside, guarded by a surly pair of bronze lions. Whether consciously or not, Jack chose a panel so small that it did not leave room for the names of his sister or himself.

We inherited Granny's live-in Irish housekeeper. A prim, soft-spoken, middle-aged spinster with a sharply pointed nose and chin,

Mabel Dunlop might as well have stepped out of the plot of *Jane Eyre*. She had lost her mother in early childhood and grown up in a Protestant orphanage in Northern Ireland during The Troubles in the 1920s. Now she occupied a bedroom on the third floor, directly opposite the room my brother and I shared. I remember the kelly green linen tea towel, embroidered with a map of Eire, that hung on her door; from here grew my first dim awareness of an obscure island across the sea, a "land-of-ire" ringed with a mystic aura, its jutting, daggerlike peninsulas thrusting westward.

"The Irish" I once heard Mabel remark in her lilting Donegal accent, "can soar up into the heavens and plunge down into hell in a split second." From her, I first heard the eerie legend of the banshee, the supernatural harbinger of death that would fire my boyish reveries. Her eyes sunken blood red by centuries of mourning, the solitary, silver-haired wraith was said to circle the hills overlooking the homes of Irish families of ancient lineage, her keening wails portending some unspeakable doom. Another memory of Mabel left an equally deep imprint: On random nights, as I washed myself in the claw-footed, porcelain bathtub, she'd scurry past me in a blur, vomit violently into the toilet, then leave without a word. Each time it happened, I pretended it hadn't—for such was the custom in our world. Only once did she actually say: "Don't tell your mother." And I didn't.

But Mabel did tell my mother something else: She was bothered by a recurring dream of lying flat on an ironing board and floating about the room. My mother suggested it was the influence of Edna's ashes, still brooding in the bedroom closet, a dark space that smelled of fur coats and mothballs. As for me, a recurring dream stalked my own sleep, one I never told my mother, my father, or anyone else: I am frantically digging under the concrete floor in our basement, burying a body I had murdered, terrified of being caught and condemned. When I awoke, I felt scalding stabs of guilt, believing the dream to be true.

═

In September 1958, I walked the two blocks north to Upper Canada College Preparatory School. The sidewalks of Forest Hill might as well have borne me along like the belt of a Ford assembly line, for my name had been placed on the entrance list at birth. Wearing knitted ties and crested blazers, we prep boys deferred to the masters as "sir"; they, in turn, addressed us by our last names only. And suddenly, all the girls vanished, like lepers. I was just turning eight.

UCC was best captured by the Latin phrase *in loco parentis*, a concept that I would, in time, come to disdain as just plain loco. My father had wanted my brother and me to board at the school, as he had done, but my mother, to her credit, refused. Only in a father-starved institution like UCC would you find ranks of tow-headed boys, ranging from seven to eighteen years old, confined to cramped, darkened dormitory cubicles, flashlights glowing under their blankets, devouring books like *The Rise and Fall of the Third Reich*, struggling to decode the distant, silent world of the heroic war generation, and thus, perhaps, find their fathers, and themselves.

In my first year, I was a keen speller, working hard on a perfect record, until the day that my form master summoned me to the front of the class. He pointed to a red X on my test paper and I froze: I had misspelled the word "docter," my first and last spelling mistake that year. The master, himself the son of a doctor and, like my father, a Cambridge grad named Jack, laughed gently at the Freudian slippage, as tears burned my eyes.

The clockwork, lockstep world of UCC conferred a sense of historical continuity, alternately comforting and suffocating. I was only dimly conscious that long lines of sons, fathers, grandfathers, and great-grandfathers stretched back to the founding year of 1829; or that two of my own maternal ancestors numbered among the head boys in the 1830s; or that this special place, this ancient orphanage of abandoned sons and absent fathers, trained something called "leaders"; that there was some refined ideal we strove to inhabit—the image of the perfect man, a superman who learns to fly before he learns to walk.

On my first day of grade eight, I ran my fingernails over the initials "JDLF 1929" that an "old boy" had carved on the rough surface of my wooden desk. Suddenly a feeling stole through me like a mild electric shock: these were *my father's* initials. My own father had shifted and squirmed in this very same seat, gazing out these same windows latticed with lead, as snowflakes melted on the iron black radiators. Yet my father, the man, never spoke of my father, the boy, and his time in this place. Why was I here? Why was I being thrust inside his own skin?

As the years rolled on, I absorbed the institution's compelling icons and classical rituals: the moonlike clock tower, named the Four-Faced Liar, standing like Father Time high on the Avenue Road hill; the gilt-lettered rolls of the glorious war dead; the sonorous recitations of Latin verse; the Gilbert and Sullivan comic operas where boys dressed as girls; the compulsory boxing tournaments where skinny eight-year-olds, paired off by height and weight, were induced to punch each other's faces into puddles of blood and tears; the arcane English game of cricket, which boys in white ducks played on "Commons" and "Lord's" over long, soporific June afternoons, breaking for tea as the clock tower struck sharply four o'clock.

On my ten-year odyssey through the Scylla and Charybdis of UCC, I soaked up its fiercely competitive culture. On winter weekends, we played shinny for hours on end, wielding hockey sticks like sabres, sports as sublimated war. I proudly wore the ragged burgundy hockey shirt my father wore for the Cambridge Eskimos in the 1930s; it made me feel "big." I loved the strutting exhilaration of winning; I hated the bitter, acid feeling that rose in my throat when I lost. Gradually, I despised myself for feeling either. For a time I had loved parading up the Avenue Road hill from St. Paul's Anglican Church on Bloor Street, the site of my parents' wedding, wearing the same horsehair cadet uniforms our fathers had worn, and lugging ancient phallic rifles. But this, too, would lose its cachet. Both my father and grandfather had served as medical officers in their world wars, conflicts won at unspeakable cost; but what war was I destined to fight? Would I win or lose? And what would it cost me?

I rose to the challenge and the rigour of the place, but eventually I was worn down by the sudden, unpredictable moments of humiliation and perversity; the purple welts raised on my flesh by a master's cane burned my psyche the deepest, echoing the experience of my father a generation earlier. I remember a father-son cricket match where a burly old boy bowled a red, stitched ball so fiercely that it bloodied his young son's face; I was quietly appalled at the way the violence was laughed off as just something that happens naturally between the generations. At UCC, I joined the hundreds of boys scanning the hard faces of father figures, gowned and chalk fingered, imagining love where far too little, or none, was given. I grew obsessed with a book on the curriculum, *Norse Gods and Heroes*, not yet aware that the Norman Irish were descended from the Norsemen. I can still picture the book's hard crimson cover, its etchings of pagan lore, its wild, mythological characters of Loki, Thor, and Odin, a fantastic world that seemed far more real than my own. My fixation so alarmed my mother that she took the book away from me.

75 Dunvegan Road, 1961: Michael, James, Janet and Shelagh, with lurking plaster statues of Jack and Molly

From there, I devoured the Arthurian legend of the Holy Grail, which likewise disturbed something deep in my blood. Why was I so drawn to the Fisher King—my father? my grandfather?—who was maimed by a sword in his side and his country—our family?—reduced to a sterile wasteland? Why can he only be restored to fertility by a questing knight who perseveres through many ordeals, answering ritual questions about the Grail?

I was being forced inside my head; it turned out to be as good a place as any for a knight errant to live.

═

My father's career was now thriving. At the Toronto Western Hospital, he was earning a reputation as a well-organized pragmatist, an "ideas man" who used his mind incisively, like a surgeon. In the late 1950s, he served as president of the Canadian Academy of Allergy, delivering addresses to his peers at the annual convention. He published journal articles about his research on assorted subjects: drug reactions, grass pollen, the seasonal incidence of atmospheric mould spores, blood transfusion reactions, penicillin reactions, and bone marrow morphology. He diligently collected bags of pollen, plant mould, house dust, and dander—the paradoxical ingredients of his medicines. Ontario housewives streamed through his office, extending their raw hands, complaining of rashes caused by endless washing with detergents and floor wax; he treated countless cases of poison ivy, hives, and assorted food allergies. I was impressed that he once saw Jacques Plante, the Montreal Canadiens goaltender as famously masked as the Lone Ranger. On a road trip, Plante complained of an allergy after spending a night at the King Edward Hotel; my father's detective work uncovered a feather pillow as the culprit.

He started up an allergy clinic at Toronto Western, modelled after Robert Cooke's New York clinic. Soon, other practising allergists arrived, and it became a university-based training centre. Every morning, the corridors were lined with men, women, and children awaiting their latest, graduated dose of allergy extract designed to desensitize their symptoms. My father sat beside a tray of fifty vials of stock materials—standardized "Cooke units" of cat and dog antigen, ragweed, and tree extracts broken down by dozens of species. He would ask, "How did you tolerate your last shot? Did you react? Did you swell up?" If not, he did the mental arithmetic and mixed the next dose on the spot. He'd pull out the precise amount of aspergillus or timothy for that particular patient that particular day. After injecting his patients, he told them to sit down for two minutes to see if they had a reaction.

The pace was hectic. Every morning, my father worked in the increasingly crowded clinic, then dashed off to his private practice at the Medical Arts building, a mile north on Bloor Street. He knew that many other clinician-scientists were doing the same, and they were all tired of it. Then my father had a brainwave. He approached fifty teaching-research consulting specialists—radiologists, hematologists, biochemists, and allergists. He convinced them to collectively invest over one million dollars in a limited company with a stock option plan. They bought a patch of land next door to the hospital, erected an office building, and stocked it with specialized lab equipment.

The Toronto Western Medical Building opened in 1960, just east of the hospital, at 25 Leonard Avenue; auspiciously, the street bore the same name as his mother's family. The first medical building of its kind in Canada proved an aesthetic nonentity—a squat, four-storey red-brick box, prosaic and utilitarian. But my father was not interested in winning beauty contests and moved into an office on the fourth floor. Then he did something radical: He opened a private allergy extract lab in the basement. Unlike other products, allergy extracts had no set of standards. It was difficult to make a reproducible standard with biologically active material, as so many ingredients were subject to change; if one batch turned out too strong, you'd risk triggering a violent reaction. The extracts produced by the existing small commercial labs were simply not of consistent enough quality for him to recommend to his patients. So he hired a technician, Martha, a blond woman with a European accent, who helped him make mass quantities of extracts and serums for the treatment of asthma and hay fever.

Selling allergy extracts to doctors for personal profit was a dicey proposition, yet no one openly accused my father of a conflict of interest, for he truly believed that he could not promote in good conscience the products of the commercial labs. However, Dr. Hurst Brown, the hospital's Physician-in-Chief, quietly disapproved of the private lab—at its height, it grossed $85,000 annually—as unethical. The progressive socialist leader

Tommy Douglas was now pushing the idea of universal medicare in Canada that would become law in the late sixties. Yet, like many of his peers, my father remained a vehement defender of private medicine—a stance that, I did not realize until much later, flew violently in the face of the public service ideals of his dead father.

My father was determined to disprove a common misconception—that many allergic reactions, even when the chemical mechanism could be proved, were due to nerves. Asthmatics were typically stressed and depressed by their illness; it was an effect, he said, not a cause. He believed the symptoms were simply exacerbated when patients were emotionally upset, a vicious cycle that pushed them into chronic illness. In the 1960s, the "hygiene hypothesis"—that overly bathed children growing up in overly scrubbed urban environments (much like my own) paradoxically upset their immune systems—had yet to be conceived. Preventing kids from rolling in the mud denied natural exposure to friendly bacteria; perhaps this was the genesis of my own fierce allergy to cats. In my father's era, the puzzle lay unsolved: Why did the allergic immune system, like a raving paranoid, seem to sense external danger when none existed, and become irrationally enraged at innocuous substances like dust or peanuts or kittens?

To practise my father's slippery profession was like trying to eat soup with a knife. Patients spontaneously acquired or lost a capacity to react to allergens. My father's inability to predict the course of allergic illnesses over a lifetime added to the problem of treatment; short of herding livestock into his patients' bedrooms, all he could do was inoculate them with carefully graded doses of the offending substance, such as the house dust he so diligently collected from under our beds. Because there was no cure for allergies, prevention remained the best medicine. Much, therefore, hinged on eliminating the environmental sources: down pillows, pets, foods, drugs, cosmetics. His retired American mentor, Dr. Cooke, remained optimistic about the future of the specialty: He predicted that one day, my father's generation of doctors would find a way to biologically switch off allergic reactions altogether.

My father's specialty, elusive and indefinable and strange as he, had not yet earned scientific respectability. And so, for over twenty years, he never ceased lobbying the Royal College of Physicians and Surgeons to recognize clinical immunology and allergy as a separate subspecialty, like orthopedics or rheumatology. Finally, they gave in.

All this, of course, sailed over my head. My childhood visits to my father's lab were as few and far between as the words we exchanged. I remember most the bland sterility of the place where he buried himself in his work: the pervasive antiseptic smell; the sickly fluorescent lights; the slick, mop-washed linoleum floors; the minimalist chrome furniture; the trim rows of test tubes; and grey metal cabinets stuffed with patient files. It was here, like a secular priest attired in bow tie, horn-rimmed glasses, and white lab coat, that my solemn-faced father performed his daily sacred rituals. I could get more laughs in church than in the house that Jack built. A first-hand experience of his brusque bedside manner was driven home on the day he gave me a hepatitis shot after the school swimming pool had been infected; tersely instructing me to drop my pants, he jabbed a needle in my buttocks and pushed down the plunger with an alarming insensitivity to my mounting anxiety. I wanted to believe that speeding things up was his way of being kind.

A few years after my father's death, I looked up several of his former protegés. A startling portrait emerged, dramatically different from the self-absorbed man I knew. While it was unlikely his colleagues would badmouth my father to my face, I was bowled over by their sincere expressions of respect and affection.

A former Polish underground resistance fighter, Roman Bladek had emigrated to Canada after the war, struggled to learn English, and put himself through medical school. When he met my father in the early 1950s, he was debt ridden, and so my father arranged for him to receive a five-thousand-dollar FitzGerald Memorial Research Fellowship, named after his father, from the University of Toronto. "In conservative, prejudiced, 1950s Toronto," Bladek told me, "your father was an outstanding example of nondiscrimination." In the anti-Semitic Toronto of the 1950s, he also vigorously promoted the careers of "bright Jewish boys" traditionally denied faculty positions. There was nothing self-congratulatory about his efforts to break down the quota system—he simply believed in hiring the best people for the job. And he had to fight hard for them.

"Fitz bristled with creative ideas," Bladek continued. "He was such a tremendous, prolific idea man that if there was a hint of a trip to the moon, he would promote it. His energy pulled other people along with him; his enthusiasm was simply infectious. He had the incredible ability to somehow disregard the obstacles, the negative and pessimistic aspects of life. He could see the light and get other people to go towards it with him. After Jack FitzGerald came on the scene at the University of Toronto, no one dared belittle or disregard the concept of allergy. That was a great accomplishment because he was going against the mainstream."

One former medical student recalled my father's quirky humour. One day on clinical rounds, he pointed to a rash on the wrist of a chubby female allergy patient. "MacPherson, what is that?" he asked. "Fat," MacPherson replied. "None of my patients are fat," came the deadpan retort. "They have *adipose tissue*." Another colleague, Alice Briggs, the first female pathologist at the Western Hospital, remembered a similar incident after finishing patient rounds. "It was a Sunday and the light was streaming through the windows, gleaming on your father's car keys, freckles, and red hair. We were standing at the foot of the stairs and suddenly he said, 'Well, I guess we'll leave 'em lay where Jesus flang 'em.' I killed myself laughing. It was as Irish a phrase as I'd ever heard, and I've used it ever since."

But the most effusive of my father's former colleagues was Stan Epstein, the son of a Russian-Jewish immigrant whom my father took on as a research fellow in clinical immunology. "He was quick, quick, quick," Stan recalled. "Very pragmatic and decisive. He could diagnose a patient before he said hello. . . . He was my God."

As we spoke, Stan sat behind my father's old desk, which he had inherited. (When he retired not long afterwards, he offered me the desk, but I declined.) As I rose to leave, he stepped forward and hugged me. It was an act performed by my father exactly once in my life and I could only stand as still as a stone statue; but on this day, decades later, I returned a doctor's embrace. As I descended the elevator of the Western Hospital, the place of my birth, I struggled to digest Dr. Epstein's astonishing parting words: "You know, we're brothers—we had the same father!"

═

Returning home from work each day, my father wore a different face. As a family, we watched *The Ed Sullivan Show* together on Sunday nights, huddled round the electronic campfire. A dark coat of Brylcreem residue—a mark of his countless hours of stargazing—stained the corduroy headrest of my father's brown wingback chair. Hogging the remote—a plastic, chocolate brown box with rows of buttons and a long cord that plugged into the set like an umbilical cord—he nervously surfed up and down the channels to avoid the ads, the captain controlling the ship, indifferent to the preferences of the second-class passengers. I found myself watching my father watching TV, occupying a silent parallel universe, as if we were each cryogenically frozen in separate glass boxes.

Like allergies, my father presented a Chinese puzzle: His passions were palpable, yet buried, his thoughts withheld from us—all we had to go on was his outward behaviour. As my mother once memorably characterized him, Jack FitzGerald was forever "uncomfortable in his own skin." He loved watching high-performance athletes, leaning anxiously forward in his chair at critical junctures in the game, nibbling peanuts and piling the shells high in a wooden bowl. As a Toronto Argonaut tight end pulled in a silky spiral pass, then made an end run down the sidelines, evading the tacklers, my father would rocket out of his chair like a jack-in-the-box, as if squirting through the defensive line of his own internal adversaries, screaming, "Go-Go-Go-Go-Go-Go-Go-Go! Go, baaaaaaby, go!" As the player triumphantly crossed the goal line, my father would jump up and down like a kid on a sugar high; plaster-cracking thumps resounded through the house, sending the neurotic miniature poodle—named Ella Fitzgerald—scurrying in terror behind the couch. Every Saturday night in winter, when the *Hockey Night in Canada* signature tune came on at eight o'clock, it acted like a Pavlovian cue, once again setting the dog all aquiver. Never once did it occur to my father that he could step outside and pass a puck or lob a ball to *us*.

His world was clearly split into winners and losers. The people on TV seemed to be performing *for him*; in those stolen hours, he didn't have to perform himself. All was vicariousness. He never once took his sons to the temple of hockey, Maple Leaf Gardens, to see a live game; we'd go by ourselves, alone on the subway, downtown to the bustle of Carlton Street, pressing our way through the wall of surging bodies. In his absence, we sprouted into sports fanatics, cutting out newspaper clippings, compiling reams of statistics, inheriting a Protestant love of facts. I remember going so far as to kneel at the side of my bed, folding my hands in prayer to God to let the Toronto Maple Leafs win the Stanley Cup. Typical of my father's casual cruelty was the time when our heroes stood poised to hoist the Holy Grail in the spring of 1962. I was eleven and Mike ten, both of us mad with excitement. During the deciding game, the Leafs were leading at the end of the second period; inexplicably, he sent us upstairs to bed, denying us the ecstatic release of victory.

As with sports, so with music. One day my father bought a state-of-the-art stereo, complete with enormous mahogany speakers and a reel-to-reel tape deck. Stretching nearly six feet long, the system looked more like an elegant liquor cabinet—or high-end casket. He leaned over it for hours, lifting and lowering the lid, lifting and lowering one holy disc after another. His instructions were clear: "Hands off!" He repeatedly played Duke Ellington's double album, *Ella and Duke at the Cote D'Azur (Live)*, which included a killer version of "Mack the Knife." The aristocracy of jazz seemed like a magic inner circle where mere children were neither hip nor welcome; The Quintet of the Hot Club of France cleanly trumped the quintet of my own family. I sensed he was much happier before we were born.

Only once did our father take us to a nightclub, the Colonial Tavern at Yonge and Queen. As kids, we'd never seen a black-skinned person—not in our schools or streets—until that night. There sat Earl "Fatha" Hines, as creased as the leather furniture, bent over his piano like a praying mantis, humming and swaying, reminiscent of my brother soothing himself to sleep. During a break, Hines sidled over to our table to greet my father like a long-lost war buddy; they murmured in intimate tones beyond my reach. Long before Motown, my father had something called soul.

But it was the Duke of Ellington who was king of them all. Whenever his band appeared on TV, a cry of "Duke's on!" resounded through the house. Once I summoned the courage to ask my father what it was like to know Ellington personally. With typical terseness, he simply said: "A hard guy to know." I felt like saying, *just like you.*

Here was a man who could resist scrutiny in his sleep; like the music he loved, Jack FitzGerald blew hot and cold. It was as if he was wearing a permanent "Do Not Disturb" sign on his forehead, but of course, the disturbance had long since crept inside. On weekends, he slept off his hangovers, often till noon. "I'm suffering from well-deserved toxicity," he declared one morning with a classic deadpan delivery that drew no wisecracks from the peanut gallery. Higher wisdom was found in the "talk show" and the sophisticated aggressions of the stand-up comedy routine, and so we listened attentively to the albums and the TV spots of sad-eyed satirists and comedians—Jonathan Winters, Lenny Bruce, Nichols and May, Woody Allen, Jack Paar, Jackie Gleason, Jack E. Leonard, Art Carney, Oscar Levant, Bill Cosby, Flip Wilson, Allan Sherman, Shelley Berman, Bob Newhart, Norm Crosby—the "characters," the cut-ups, the Irish and black and Jewish outsiders, the hip out-foxing the square. When my father cracked up, like monkeys so did we; we laughed even when we didn't get the sly allusions, the slippery shots, the double takes, the hair-trigger jibes. "Getting it" was everything; you were either "with it"—or not.

When he did speak, our father spritzed the air with his own terse one-liners, catchphrases, and wry asides. I suppose his glibness was the closest he could ever come to being tender. After a song by one of his beloved sultry songstresses, Lena Horne or Peggy Lee, he'd quip: "Wrap her up!" He passed sporadic judgments on athletes—"He's not worth a pinch of coonshit" or "He skates like his jock's on backwards" or "What a performance!" If a player choked in the clutch, or fumbled the ball, he "got the yips." If he was too flamboyant, he was dismissed as a "showboater." When offering encouragement, he said: "Keep your pecker up." Disasters brought out "Jesus wept!" or "He went up like a rocket and down like a burnt stick." A burst of joy delivered: "Hot damn!"

Often it seemed he was quoting some quirky old aphorism, like, "He's strictly from Missouri" or "Cheaters never prosper" or "If I had to do it all over again, I'd do it all over you." Once he came home from work and told our mother: "I'm so smart I make myself sick." Another time, he astonished me by suddenly screwing up his face and adopting the mock accent of a stage Irishman: "Faith an' Begorrah, I'll slap ya in the belly with the guts of a hot cod!" It was the only time I ever heard my father even hint at his Irish blood.

But of all his minimalist maxims, one entrenched itself the most deeply on my memory: "There's always room at the top." I sensed that he had heard it repeatedly, like a mantra, back when he was a boy like me. Now it was my turn to absorb this message of seemingly incontrovertible truth, bearing the inexorable force of a paternal, and grand-paternal, command.

═

My parents never owned a summer cottage; instead, we visited my father's doctor friends in Muskoka, the Forest Hill of the Ontario lake district. It was a privilege lovely in itself, yet it only reinforced that subtle sense of outsiderdom, our status as perpetual guests. I could never shake the insistent feeling that our family really didn't belong in this magical world of open sky, fresh water, and earthy woods; that we were imposters, unrooted in the flesh of nature. Summer was the only time I ventured within range of girls my age; I retain lasting images of the enchanting doctors' daughters water-skiing over the silky surface of Lake Rosseau, vivid and giggling like storybook nymphs. Our conversations were too brief and awkward to forge a tie that might survive into the dead winter months; in September, as I resumed my well-trodden path to school up Dunvegan, I might spy one of those same alluring girls of July, the pedigreed, pink-skinned sirens of Bishop Strachan, clad in their pleated tunics of crimson and grey, coming towards me. I was so consumed with shyness, I averted my gaze. Sometimes I'd even flee to the other side of the street.

It was in August 1962, the summer that the Hollywood sexpot Marilyn Monroe died, that I felt my blood stir with the first tinglings of lust. I was staring at Sally Wodehouse, the long-legged daughter of a doctor who ran the University of Toronto student health service, as she was suntanning on a dock at De Grassi Point on Lake Simcoe. Her tomboy, stick-figure body had grown fleshy and curvaceous over the winter. I thought: So *this* is what everyone is talking about. Within weeks, as if on cue, my father made an unannounced appearance on the third floor of Dunvegan, the rarest of events. I was twelve, and the school had asked parents to help impart "the facts of life" to their pubescent sons. In the bathtub, I was preparing to set a world record holding my breath under water when I heard his open-heeled slippers flapping up the stairs; he swept past me, flipped down the lid of the toilet and sat down. I sensed he was nervous.

"Do you know how to give a girl a squeeze?"

I nodded, saving him the agony of elaboration. My father, the expert physician, a guy who knew his way around the human body, stood up, lifted the toilet seat back up and walked back downstairs. Back down, I did not suspect at the time, to his own squeezeless marriage, his own private tragedy of the bedroom. Ever so briefly, and with a dab of spare poetry, the cold, hard facts of sex and death brought father and son together.

═

It was around this time, I later learned, that my mother finally threatened my father with a divorce.

The last straw came when my thirteen-year-old sister, Shelagh, stood on the edge of the living room, wordless, teetering, watching our father sipping a pre-dinner "smash" and unburdening his heavy day into our mother's lap. Shelagh was implicitly seeking paternal permission to cross the threshold, but she knew the rules. Not only were we forbidden to share meals with him, we must *never* bother him when he came home from work. My mother read the riot act: In future, my father would eat with the children and we would behave like a "normal" family. To placate my mother—she radiated the energy of two men—he gave her a part-time job as the accountant for his growing allergy lab and talk of divorce subsided.

But the trade-offs inherent in every marriage had long ceased to work for my parents. Janet resented Jack's lack of support in the child rearing, his shunning of her parents, his chronic self-absorption; he went so far as to ban all talk of children, housework, or domesticity when they were together. I saw no kissing, no hand holding, no gestures of affection; I heard no terms of endearment—no honeys, no darlings—flutter through the empty spaces between them. In those prefeminist days, I remember being puzzled by the salutations of letters, addressed to "Mrs. John D.L. FitzGerald," that tumbled through the brass mail slot in the front door. She used to say that a large part of her wifely role was to submerge her identity and "bask in my husband's glory." But the clouds of glory were fast melting away.

Every February, our parents flew south for a two-week vacation on the beaches of the Caribbean, each year a different island: Tobago, Barbados, Antigua, Nassau, or Guadeloupe. Mabel had moved on, so we were left in the clutches of Mrs. McCurry, a crabby, chain-smoking Irish harpy who felt it her job to carp at us day and night. It remained a mystery why our parents invariably hired witchlike, kid-phobic emotional gangsters to "take care of us." Mrs. McCurry would force us to write our parents letters; there was no choice. So our index fingers pecked away on a typewriter, reporting dire winter weather conditions to my mother and sports scores to my father; my brother's standard exit line, part fear, part wish, was: "P.S. Don't forget to get killed on the way home."

Annual Caribbean vacations fail to calm Jack's sea of troubles

When our parents did come home, unkilled, we ritualistically greeted them at the door, my mother tanned and glamorous in the winter night, my father glum and preoccupied, unloading their baggage from the trunk of the airport limousine. They brought us trinkets, as if to appease the restless natives of Dunvegan; once my mother dropped a blowfish, big as a softball, in my outstretched palm, its sharp needles pricking my flesh. Janet, the artist *manqué*, enjoyed showing us her artfully composed colour slides of the soft, distant world of the tropics. In the darkness of our living room, the clicking projector draped random, foreign images on the white screen: palm trees, bronzed flesh, sunglasses, iguanas, my father sprawled on a deck chair gazing at the eternal sea, the brilliant tropical rays failing to redeem a heart receding into the darkness of a deepening affluenza. As ever, an Irish proverb had it nailed: "Some people reach the top of the ladder only to find it is leaning against the wrong wall."

My world seemed a chain of static, fragmented visions, like the pieces of my dreams I dare not reveal to a soul. More and more, I was retreating into the cool detachment of the schizoid pathologist, a neutral observer locked in an emotional Switzerland, collecting bits of evidence to solve some unspecified puzzle, reveal some classified information. The evidence was piling up fast: trouble in paradise.

==

In those juvenile days, I coveted a jackknife with a gnarled wooden handle. One summer day, I was perched on the front steps of 75 Dunvegan, whittling a stick, when I accidentally gashed the heel of my palm. The cut ran deep and bloody. Just then, my father's silver Oldsmobile pulled up the driveway. I slipped the knife in my pocket, afraid he'd take it away from me. When he saw my wound, I made up a story about running past the fence along the side of the house and catching my hand on a protruding nail. Genuinely concerned, he asked me to show him the nail, for I might need a tetanus shot. As we trolled up and down the fence, vainly feeling for the fictional culprit, he must have realized I was telling a white lie, for the head of a nail could not have caused such a clean cut.

But I kept the jackknife.

I was now showing early promise at track and field at school and my success seemed to awaken something inside my father. At twelve, I was still growing, my legs as long and spindly as a colt's. It was the only time in my life my father showed direct interest in my ceaseless strivings—I was winning races, setting records, and gleaning a raft of medals—and I found his sudden attention confusing and unsettling.

In preparation for our annual June track meet, he trained me relentlessly over the winter and spring months. He bought me a dense, bewildering book by the über-trainer Lloyd Percival; I was bidden to emulate Canada's reigning track star Bruce Kidd. Week after week, I tore around the cinder track at UCC and up and down the steep hills overlooking Lord's cricket pitch; I pounded around the neighbourhood pavements, across Heath Street, up Forest Hill, along Kilbarry, up and down Warren, Russell Hill, and finally, home on Dunvegan. Sometimes, in a hyperventilating sweat, I'd race past a gaggle of giggling BSS girls in a blur, imagining their heads swivelling to admire my whirling legs, all sweat and bone and sinew. Each night, my father stood at the foot of our driveway, stopwatch and clipboard in hand, waiting, shouting, urging me to sprint the last fifty yards as hard as I could, to go all-out, hard, hard, hard; each night, I was expected to shave a few seconds off my time, achieving, by increments, some imaginary, cosmic state of perfection.

For the hundred-yard dash, he trained me to burst out of the starting blocks quickly to gain a jump on my rivals, taking short, choppy steps before breaking into my full stride, though my long legs best suited me for middle distances, not sprints. On the day of the track meet, he stood among the crowd of parents at the finish line. I was so nervous, I botched the start that I had rehearsed with him so many times; within a full second of the pistol shot, my four rivals had left me in the dust. I raced forward, tears flowing, my body pulled by the distant sound of my father's rising voice, spurred by my own panic and the terror of an impending loss. *No guts, no glory.* The twelve-second race felt like twelve minutes; miraculously, I cut the tape a step ahead of the rest. I won the medal, but I felt a terrible burning inside, as if the tape had encircled my throat, choking me with the bitter taste of a victory I felt I did not deserve.

The following year, my last at the lower school, I was poised to win the Duke Somerville Cup, a trophy named after a nasty British-born UCC headmaster of pedophiliac tendencies who had caned my father in the 1920s. But sudden, stabbing pains in my heels stopped me dead in my tracks, and I finished out of the running. Unconsciously, I had dug in my heels. My father called a surgeon friend who, in two separate operations, cut open my heels and shaved bone spurs that were so small as to almost not exist. The transient glory of my track career vanished under the knife, leaving me to contemplate equal feelings of relief and disappointment.

═══

For family Christmases, we'd visit my cousins, the Whitleys, at their elegant home at 6 Highland Avenue in Rosedale. My aunt Molly reposed regally in her chair—her children dubbed her "the Duchess"—while her husband, Tom, the war hero and Royal Bank executive with the impeccable Oxford accent, served flutes of champagne cocktails and a platter of smoked salmon in his silk ascot and smoking jacket. They were not unkind to me, just polite and slightly aloof; my cousins exuded class—certainly more class, I thought, than clung to me.

In our rare moments of contact, Molly seemed as cool (in both senses of the word) as her brother, my father. Over time, vignettes of Molly's past filtered through stray bits of family conversation. How, game for worldly adventure, she volunteered for the Canadian Women's Army Corps stationed in England during the war. How one time she was being served cocktails in the lounge of Toronto's Windsor Arms Hotel by a gay waiter, in the preliberation days of the closet; as he left to fetch her drink, she puffed on her Matinée cigarette and offhandedly remarked, "Isn't it nice that they're giving those people something to do?" When a robust dinner guest once leaned back on a valuable antique chair, causing a leg to break, Molly famously sighed, "Don't worry, dear—it's old."

But I had no idea my aunt Molly was terribly troubled, bored to distraction by the clockwork Royal Bank parties, gliding about town in a black limousine driven by a uniformed chauffeur, playing the role of the chatelaine, and the slavish keeping up of appearances that her mother had down to a science. In those days, the wives of Royal Bank executives were forbidden to work and female employees were fired when they married. So she performed her volunteer work—Havergal old girls, National Ballet—with a desultory style all her own. I had no clue she drank too much, at times blurting something racy or *outré*; that she ranted and swore like a banshee; that she'd answer the front door in the nude; that some kind of blind fury burned deep inside her. That she was an irredeemable alcoholic, sliding down an ever-darkening path. But on me at least, the facade worked like a charm.

Strange cravings swirled inside me, too, although nothing to compare to my father's or aunt's. I'd wander over to Maxie's, the drugstore in Forest Hill Village, and shoplift hockey cards or slip a copy of *MAD Magazine* under my windbreaker. Sometimes I'd sneak into my parents' bedroom and sift through the pile of loose change my father left on the mirrored dresser that had once belonged to his father, pocketing a few quarters.

My maternal grandmother paid for much of our expensive private education. Every birthday, she gave us each a dollar for each year we had passed; when I received ten dollars for my tenth birthday, I remember quipping that I wished I was a million years old. My father, too, never gave us gifts, only cheques at Christmas; he consistently forgot our birthdays. Later, I took comfort from the words of Sigmund Freud, who said money never brought true happiness—true happiness stemmed from the adult fulfillment of a childhood wish, and children, like me, rarely wished for money. We wished for something else.

Each Christmas, our spinster great-aunt made an appearance. Like her niece, Molly, Hazel Kathleen FitzGerald was slender and stylish; on formal occasions, she wore black Chanel suits, a foxhead stole and puffed du Maurier cigarettes, although she rarely inhaled. I remember most her thin, Virginia Woolf–ish face and Irish aquiline nose, the liver spots dotting her hands, and an impish laugh that made me like her. On occasional Sundays, we drove the thirty miles west down the Queen Elizabeth Way to Burlington, where she lived on the edge of the lake in a cottage stuffed with Victorian bric-a-brac. Out back, she tended her rambling garden in a wide-brimmed hat and rested under a weeping willow tree. As the Studebaker pulled up her driveway, we passed a wooden sign painted with the single word "FitzGerald," and in my young mind,

I puzzled: Who was this sweet, gnomish old lady who shared my last name? How was she connected to us?

Only much later did I learn that in the 1930s she had opened a couturier shop, Gerald's, on a second floor above a Coles bookstore at the corner of Charles and Yonge Streets. Catering exclusively to the Rosedale and Forest Hill carriage trade, she competed with the Russian-Jewish furrier Eddie Creed. (She worked the WASP side of town, Creed the Jewish.) Hazel lived with another single woman, Eileen Robertson, known locally as the "Queen of Rathnelly," in a house at number 30 near the foot of the Avenue Road hill. Each summer, they travelled north to the rocky wilds of Tobermory overlooking Lake Huron, where they roughed it in a cabin unburdened by plumbing or electricity, basking in the glorious pink sunsets. Perhaps they were lovers, in the manner of Gertrude Stein and Alice B. Toklas; perhaps not. But I do know that Hazel treated us kids kindly, like equals, with a twinkle in her eye and a buoyant sense of fun.

Like most women of her time, she had been expected to sacrifice her own needs for others. In the nursery at Balmoral, she had been enlisted as a nanny for young Jack and Molly, filling in for their absent parents who travelled much of the year. On weekends in the 1920s, Hazel would walk up to the UCC prep boarding school and stand on the curb on Lonsdale Road; my father, the abandoned nine-year-old, forlornly poked his head out the barred window of the gothic Peacock Building, and they talked. She called him "Dinny," an Irish-ization of Desmond, his middle name. She was forbidden to enter the dorm to console her nephew, or so went the family story.

Hazel's loyalty was not repaid in kind: As the nanny, she became the habitual target of Molly's and Jack's displaced tantrums and hatreds, for she was the adult closest to hand. Many years later, Molly's "other side" was dramatically revealed to me for the first time when my teenaged brother told how he was dispatched by my mother to pilot Molly down to Burlington in the family's canary-yellow Firebird to visit our great aunt. Now eighty and nudging senility, Hazel was dragging her heels about trading in her beloved cottage for an institution. Mike was shocked to see Molly's dignified public persona—the one we saw unfurled every Christmas Eve—explode in his face: With an intensity uncalled for by the situation, she berated Hazel as a *fucking selfish bitch*, pouring a hot stream of bile on her grey head. Under the barrage of invective, Hazel turned to Mike, smiling sweetly, and said, "Isn't she being silly?"

At this point in my young life, I knew Hazel had three older brothers: Gerry the doctor, Sidney the lawyer, and Bill the banker. The first two were long gone. I had a dim sense there was a stooped, reclusive, thorny old man named Bill FitzGerald living in an old house up north in Thornhill, but I never asked after him for he, too, seemed enshrouded by an aura of taboo.

═

It is not easy to plot precisely when my father started to crack up. But an American president may have played a part.

Like the press corps, my father worshipped the public image of John Fitzgerald Kennedy. He'd watch the televised press conferences—the charm, the Chiclet grin, the off-the-cuff quips, the charismatic air of Camelot—and like the obsequious reporters, eat out of the hand of the chief executive. When Kennedy visited Ireland in the summer of 1963, he famously told a rapturous working-class audience: "If my great-great-grandfather hadn't left Ireland, I'd be down there with you on the factory floor." Perhaps this was the mystique that spoke to my father: the ceaseless striving for something bigger, better, brighter, happier; something beyond himself; the outsider turning insider. Besides his birthday, he shared much with JFK: both had struggled through a series of elite schools, fought in the navy, and married women admired for their physical beauty. Both laboured under the heavy expectations of powerful fathers; both suffered assorted illnesses, which they did their best to conceal; both embodied a high promise that would never fully play out. The family codes were similar, too: "Kennedys don't cry."

A code I broke often enough.

On the late afternoon of Friday, November 22, 1963, my father opened the front door of 75 Dunvegan Road, his downcast face as grey as a battleship. He looked as if his own father had just been murdered. I mumbled something like, "Isn't it terrible?" but he said nothing, brushing past me as if I were a hat rack.

Like millions round the world, we sat trancelike in a tribal circle, fixated on the TV screen all weekend long, balancing our meals on our knees. I was thirteen; perhaps this was the moment I began the mental habit of conflating outward historical events with the inner family drama raging behind the drawn curtains of our home. We didn't know it then, but Kennedy would have survived if his back brace had not held him erect after the first shot hit. In Texas, Oswald's bullets had splattered the brains of a sickly Irish-American president across his wife's pink suit, and no one knew why; in Toronto, an Irish-Canadian doctor, born on the same day as the martyred chief executive, sank precipitously into a wordless funk, as if poisoned by some secret, conspiratorial knowledge—knowledge that fell to his eldest son, in the years to come, to decode.

Studio portrait taken November 23, 1963, the day after the JFK assassination and the beginning of Jack's descent

THREE

Daddy Ded

When the shark bites with its teeth, dear,
Scarlet billows start to spread.
Fancy gloves though wears Macheath, dear,
So there's never, never a trace of red.

BERTOLT BRECHT,
"Mack the Knife"

As my father began his prolonged descent, I was passing through puberty and from the Prep to the Upper School. After reading H.G. Wells's *The Time Machine*, I was much given to daydreaming of fantastic escapes into the past and the future, creating mental scenes as wondrous as my nightly dreams. A rare family outing to see a film—*Dr. Strangelove*—only confirmed our ever-widening generation gap; I loved it, my parents hated it. After years of earnest effort, I was now hitting the wall at school; random acts of humiliation of students by teachers—and students by students—seemed a near-daily experience. When I was unjustly caned, I did not dream of telling anyone, out of shame; when my mother learned of it through the grapevine, she merely laughed. A turning point came when a supercilious, lemon-lipped master, himself a UCC old boy from the 1950s, dressed me down in a grade ten English class for earning a perfect mark on a piece of Shakespearean memory work.

"FitzGerald! Stand up."

I stood up.

"You got ten out of ten. What happened? This is not *down* to your usual standard. . . ."

Enraged by his perverse sarcasm, I silently vowed I would no longer play by their rules. For years, the school made strong demands on me, which gave me a conscience; and yes, decency and kindness did exist. Yet why did so many decent and kind people turn a blind eye to the ceaseless acts of derision, cruelty, and intimidation? I had not yet figured out that "winners" needed to create "losers" in order to feel like . . . winners. Or allow that I could dish out the nastiness as sharply as anyone. I chose to lay low, hiding in the weeds of mediocrity. No longer able to take these men seriously, I retreated into the consoling clouds of adolescent iconoclasm.

In 1965, my father's latest mentor, Hurst Brown, died, only in his sixties; both men had worked hard at building up the allergy clinic at the Toronto Western Hospital. My father had already absorbed the shock of the suicide of Peter Spohn, his American roommate at UCC, in 1960; and the murder of JFK felt like a family tragedy. In losing Brown, he lost a sympathetic friend and professional ally, and his idea for founding an allergy research institute at the university soon crashed on the shoals of political infighting. Smart, younger doctors were coming up behind him, nipping at his heels. He felt desperately alone.

In the winter of 1966, my parents returned days early from their annual Caribbean vacation; as usual, we met them at the front door. Once more, the tropical breezes had failed to soften the furrows on my father's hairless forehead. Like a sleepwalker, he walked right past us, up the stairs, and shut the bedroom door. The rage in his heart was explosive enough to murder us all, but on this day he found the strength to turn it inward. On the door, he taped a scrawled note for all to see, for all to never speak about, for all to never forget:

"Daddy ded."

Over the next six months, my father took more and more time off work. He was having a nervous breakdown, but we pretended not to notice. He cried to my mother that he was all washed up, that he could not handle his job any longer, and cancelled appointments with his patients. Only during my archival searchings decades later did I discover that he had been invited in June 1966 to attend the official opening of FitzGerald Building, a new laboratory facility on a three-hundred-acre property north of the city, honouring the memory of his illustrious father. Engulfed by a merciless malaise, he declined the invitation.

A week later, on July 1, 1966, a honeycombed, twelve-storey block of concrete opened its doors on 250 College Street. The Clarke Institute of Psychiatry was considered the country's cutting-edge institution for the treatment of people deemed "mentally ill." Soon afterwards, a figure from the Clarke fatefully crossed the threshold of 75 Dunvegan on the first of a long string of house calls. He had come to help my father; I would come to cordially hate his guts.

Dr. Don McCulloch was a University of Toronto professor of psychiatry whose craggy, melancholic Scottish face resembled R.D. Laing's—minus the charisma. Bright, serious, well-liked, he had a puritanical streak, the kind Laing himself spent a lifetime trying to shed. Dressed in turtleneck sweaters, McCulloch embraced the antimedication, "let it all hang out," Esalen-styled liberation movements of the day. Bent on exploding rigid social conventions and scything through the bullshit, he led sensitivity-training sessions in an old farmhouse in rural Ontario, breaking for bouts of skinny-dipping in the lake. Besides house calls, he even moved into his patients' homes, if granted permission, to suss out the larger family dynamics. When a Clarke psychiatrist reported that a psychotic patient thought he was in direct communication with God, McCulloch dryly replied, "So do three-quarters of the human race." This hip doctor, it would seem, was a good fit for my quirky, offbeat father. But soon it became clear he was in over his head.

One night, the entire family assembled for group therapy in our living room. Sitting next to my father, McCulloch turned to the four of us and announced matter-of-factly: "You know what this family's problem is? You don't love your father enough."

My mother burst into tears and a hammerlike silence fell over us. Then McCulloch pitched another dart into our chests: "See, your mother is in distress and no one went over to comfort her."

To her everlasting credit, my feisty, volatile sister hit the ceiling and gave McCulloch hell. I sat there, stunned as a toad on rock; I couldn't find the words to utter the obvious. Yes, the FitzGeralds were a collection of emotional Neanderthals; but if our father didn't know how to love, how could we? How do you give back something you have not received?

Portrait of a marriage, July 1, 1967: As Janet impersonates Britannia for a party of admirers on Canada's one-hundredth birthday, Jack votes with his feet

Family therapy ended then and there, although my father continued to see McCulloch in weekly sessions, one on one, in the TV room at our house. I remember him running downstairs, stirred up in midsession, pouring two scotches to the brim of each glass, then bounding back upstairs to the sound of clinking ice cubes. The pair of them jabbered away up there for what seemed like hours on end, more manic frat house rap sessions than therapy. Once McCulloch invited my parents to recline on their queen-size bed; then he lay down between them—his version of "free association." (My mother later remarked, "I was sure he was going to roll over on top of me at any second.") Later I heard McCulloch was experimenting with psychotropic drugs, dropping tabs of LSD with his students. For all these misguided, "bash-through-the-defences" stabs at therapy, no one seemed to be speaking of fathers, of doctors, of doctors who had fathers, of my grandfather's unacknowledged ghost hovering at the door. Neither was anyone inviting me to say what I felt, and I felt much.

A rare uninhibited Christmas, with cousins Anna and Patrick, 1967

My father stacked up piles of pop-psychology books with titles like *The Art of Loving*; he read a tome on relaxation techniques to try to quell his obsessiveness; he used an electric vibrator for his migraines, pressing its rubber tip to his forehead. Sometimes, I'd sneak into his bedroom, plug it in, and guiltily rub my groin; my father, I thought to myself, was focusing on the wrong organ. For a spell, he took up the buzzword "authenticity," complaining that he felt false, inauthentic, unreal. I remember thinking: *He's onto something there*. For, of course, I felt that way, too. Maybe we all did.

McCulloch urged Jack to take up hobbies to distract himself from his worries, no doubt heeding the advice of Canada's greatest doctor, Sir William Osler, who once opined: "No man is really happy or safe without a hobby . . . anything will do so long as he straddles a hobby and rides it hard." And so my father dutifully repaired to the basement where he joylessly stripped wooden chairs and tables, or concocted homemade wine in bulbous glass vats, bottling and labelling it "Fitz's Finest Fizz." I imagine the stuff did not help his migraines. Habitually drawn to cutting-edge technologies, he bought an "instant" sixty-second camera, bidding us to pose in toothy grins that instantly dropped after the sound of the clicking shutter. As a family, we passed a winter playing a fiercely competitive round robin cribbage tournament, a last-ditch attempt to create a feeling of togetherness. I revelled in beating my father, jamming matchsticks triumphantly into the peg holes; at the same time, I felt as if we were propping up a mannequin in a store window. Sometimes I would dare, under my breath, to mock his solipsistic immersion in jazz and sports and TV at our expense, waiting for the right moment to slip in my needle: "It don't mean a thing if it ain't got that mood swing." One time, punning on a list of movie titles in the *TV Guide*, I quipped, "How green is my Valium?"; he had, I recall, the grace to laugh.

Soon he was distracted from his distractions and began to shuffle down Dunvegan on Sunday mornings to sit alone in the balcony pews of Timothy Eaton Memorial Church to listen to the droning sermonizing. But that hollow ritual ended soon enough. The United Church was well known for its social conscience, its Christian charity, its helping of the poor. But no one seemed to know how to reach inside the spiritually impoverished doctor from up the street.

One day my sister borrowed a spider monkey from a friend and brought it into the house. Ella, the pink-bowed poodle, took one whiff and went berserk, yapping madly, her dormant killer instinct aroused. My father was slumped in his chair before the altar of the TV. Mischievously, Shelagh let the tiny beast out of its cage and it made a beeline for Jack, perching on his shoulder like a parrot. I will never forget the look of terror that seized my father's face—nor the trickle of yellow monkey piss that ran down the front of his white shirt.

For anyone who cared to see, Jack FitzGerald had a monkey on his back.

I escaped, as millions of insufferably horny adolescents still do, into narcissistic fantasies of rock 'n' roll stardom. In our coach house out back, I windmilled my imaginary electric guitar, conjuring waves of tartan-skirted private-school girls swooning at my feet. Jimi Hendrix's album *Are You Experienced?*—I wasn't—fell into my hands as if dropped by angels; I cranked up "Manic Depression" to near ear-bleeding decibel levels. I was loathe to admit it, but I was just like my father, bathing indulgently in cathartic rhythms—only my black genius was topping his black genius. I might as well have screamed to my father across a vast void, "Are . . . you . . . listening?" One day I was bold enough to place a set of headphones on my father's head, gripping his skull like jumper cables. I played him Hendrix's "All Along the Watchtower," its guitar lines boring through his brain, shooting back and forth from ear to ear. I saw my father's eyes roll in amazement at the groundbreaking effect, but he said nothing. He wouldn't give me the satisfaction. The message was unspoken: If I thought the 1960s were wild, they couldn't hold an incense stick to the 1920s. Or, as our very own Marshall McLuhan liked to say: "In one era and out the other."

Dr. McCulloch's boozy rap sessions were opening up my father like a watermelon, but I was never sure to what end. One spring day in 1967, he invited me on a walk through the leafy canopy of Forest Hill. Such extended contact was unusual. We headed north, ending up circling the quarter-mile cinder track at UCC, the scene of my past athletic triumphs and failures, washed up at thirteen, the surgeon's knife leaving pink scars on my Achilles heels. We paused to sit on a bench overlooking the playing fields, silent, alone together. We both stared down on the freshly mown grass marked with white lines. After a tense silence, he turned to me and in a trembling voice, said, "I've ruined my life. I don't know what to do. . . ."

My throat tightened. What could I possibly say that might lift my father out of his hell? I listened as he horsewhipped himself as a loser, a failure, a nothing. Out flowed the jagged, black soliloquy: In his inherited, unforgiving world of zero tolerance for being human, you were either a Hero or a Zero. Then his words melted away and like a swallowed narcotic, the deadening silence returned.

I mumbled something lame, something to the effect that he had maybe another twenty-five years to live; why not make the best of it? If only *once* my father had told me, during one of my *own* states of anxious dread, "I know how you feel," he might have moved mountains. But something held him fast. As we headed home, side by side, heads bowed, I counted the concrete slabs of pavement, burying the ashes of my impotence six feet under our feet.

═

A year later, lifted by the electric June breezes of 1968, I walked off the grounds of UCC, confident that I would never return, except in my dreams. *Free at last. Thank God almighty, free at last.* I remember a billowing wave of relief, together with a sad, reluctant suspicion, confirmed by the years to come, that my schooling had done more harm than good; there were bodies buried there, too.

A few days before graduation, I came downstairs for breakfast to find the headlines blazing yet another horror: Robert Kennedy, poised to sweep into the White House riding a radical agenda of social justice, had been gunned down in Los Angeles. Martin Luther King had been murdered only weeks before, and now here lay RFK across the front page, his iconic bushy head, like his older brother's, shattered by a madman's bullets. At the table, my eyes averted from my father's face, I struggled to hold back my tears, puzzled by the intensity of my feelings. Pierre Trudeau had been elected our prime minister only weeks earlier, a dynamic, Kennedy-esque, social liberal of youthful promise; I was caught up, like most, with the excitement of the times, the energy of change, that just might migrate from the streets into my father's living room, that just might cut loose the iron weights in my body and let me sail free. Now that promise, too, seemed doomed.

That summer, my maternal grandmother paid to send me on a trip to Europe with twenty-five other teenagers from across Canada, half of whom were those alluring, elusive creatures called girls. "The Odyssey"—museum tours by day, bacchanal by night—was, I suppose, my unspoken reward for surviving a decade of UCC. Or my own family.

In Europe, I fell for the quick-witted daughter of a U of T doctor, a colleague of my father's. A kilted Branksome Hall girl, Sally Wodehouse had been the object of my first rush of erotic longings six years earlier on the dock on Lake Simcoe, and I hadn't seen her since. We had some catching up to do. Over six delirious, liberating weeks, we laughed and drank, tearing through the ancient cobblestoned alleys of Venice, Florence, and Rome. In my naiveté, the student riots of Paris 1968 meant nothing to me; hedonism trumped politics every time.

Then down came the anvil: On August 13, Sally was killed in a freak traffic accident in the Rhine Valley.

If I had been fundamentally religious, I would have believed we were being punished for our sins. When my parents picked me up at the airport two weeks later, they seemed incapable of extending any palpable gestures of comfort. I can still hear my sister playing *Moonlight Sonata* on the upright Heintzman piano. Perhaps it was the music that triggered me; I fled to the bathroom on the third floor, locked the door, and sat on the toilet seat where my ghost of a father had tersely imparted the facts of life. As I wailed, I prayed to God, if there was one, that no one was listening.

═

Within days, I was enrolled in Queen's University, a limestone ivy-league enclave on the shores of Lake Ontario that I was naive enough to think might offer a refuge from home, from strife, from feelings. My brother was escaping into ink-black forests of baseball statistics, devoted to his "lovable losers," the Chicago Cubs and Boston Red Sox, both teams afflicted with some ancient, mysterious curse. Like the Irish, my sister, Shelagh, was mad about horses; and boys were mad about her. Having left BSS, she had struggled at the coed Oakwood Collegiate, puzzled as to why her quicksilver mind couldn't digest information off the blackboard in the conventional manner. She didn't know it yet, but her brain hemispheres were perfectly balanced; ambidextrous, she could write forwards and backwards with both hands simultaneously, with equal speed and facility. (I am now ready to confess that sometimes I'd sneak a read of her backwards-written diary by holding it up to the mirror.)

At Queen's, I got busy flunking most of my first-year courses, caught in the boozed thrall of a mad passion for a Havergal girl that grew so intense I could do nothing but sabotage it. I was doubly shocked when my mother smiled approvingly at my news of the break-up. Mired in self-pity for months, I remember sitting alone with my father at home one weekend, installed in front of the flickering blue light of the TV screen as if hooked up to an IV bag. We were watching the final scene of the William Inge–scripted film, *Splendor in the Grass*, a poignant story of adolescent first love and lust. When Inge killed himself only a few years later, somehow it seemed to fit.

As the credits rolled, my throat burned and eyes misted; my father astonished me by asking, in a near-whisper, "Did you ever feel like that?"

I murmured, "Yes"; then the old silence enveloped us like a cloud of ether. He hated small talk, but feared big talk—real talk? deep talk?—even more.

In that year of 1969, the mayhem of the world funnelled week by week through the cathode ray feeding tube, pouring lavalike onto our beige wall-to-wall broadloom: another Kennedy tragedy at Chappaquiddick, the moon landing, the Manson murders, all juxtaposed against the mass love-in of Woodstock. Most of the World War II generation were aghast: Why were their selfish, ungrateful teenagers pissing on everything they had fought for? Why were millions of middle-class white kids like me growing our hair as long as Norman knights and bleating sappy songs of the Age of Aquarius, falling back on our pagan past, smeared with mud and body paint, back to the garden? Most parents were baffled, but not my worldly father; nothing we did ever shocked him.

Likewise, nothing could heal the pain raging in the lining of my guts, a pain so agonizing that it summoned images of knife-wielding samurai committing ritual seppuku. I dreaded telling my father, the doctor, about my condition, reflexively keeping my "inner life" to myself. Finally, after months of unbending stoicism, my brother heard my muffled moans in the dead of night and turned me in. I anticipated the cold steel of a clinical response, and I got one: When my father drove me down to the Western Hospital, he offered no word of warning of what ugly indignities lay in wait. My lanky six-foot-three-inch frame was hoisted up, butt exposed, on a specially designed apparatus. Then I was cornered by a horseshoe of grim, white-robed men resembling Roman assassins. One of them thrust a sigmoidoscope into my bowels like a sword. I gritted my teeth and sweat poured from my forehead; I nearly passed out as shards of pain shot through my belly—the seat of the soul, some say. But I refused to scream.

No disease was found. Over the months, I endured rounds of barium swallows and enemas, the results still maddeningly inconclusive; maybe it was simply a case of my family being too much to stomach. Finally, in the fall of 1969, a new set of X-rays at last revealed evidence of lesions rooted in my inglorious guts: I was diagnosed with Crohn's disease, a mysterious autoimmune disorder of lifelong duration, cause unknown, that inflamed the ileum, the juncture of my small and large intestines. ("Autoimmune" means the immune system, unable to distinguish self from not-self, misfires and attacks its own body, a process analogous to a kind of subtle suicide.) I was prescribed a drug of only moderate help; not once did a single doctor concede that the unruly emotions I was so successfully squashing played a part in my condition. The irony hung in the air: My father, the reluctant immunologist, was unable to redeem his immunity-deficient son.

That fall term at Queen's—I had been readmitted after failing all of my courses but English in my first year—I was struggling to make sense of John Stuart Mill's essay, *Utilitarianism*, unaware that the philosopher had suffered a breakdown under the oppressive weight of his father's expectations. Extracurricularly, I was also dipping into Freud's *Interpretation of Dreams*, an iconic tome conspicuously cleansed from my "rat-ified," behaviourist-oriented psychology courses. When I read random passages aloud, I never failed to provoke derisive barbs from a skeptical flatmate. I was coming to suspect that the sport of Freud-bashing deserved its own set of interpretations.

One Sunday afternoon, like a dream, my father unexpectedly appeared at my lopsided hovel of an apartment in Kingston, the very town where he had sweated out his elite FRCP degree twenty years earlier. He climbed the stairs and stood at the threshold of my bedroom, ragged and pathetic; he looked like a whipped dog. I was playing *Beggars Banquet*, loudly, and I turned to lift the needle off the vinyl grooves. As he moved hesitantly towards me, I was shocked to see tears streaking his face. Then he lifted his arms and, like a dream, embraced me.

I stood as still as a statue, encased in a familiar sheath of nothingness. I felt as if a bum on the street, an utter stranger, had lost his way and stumbled into the wrong house. If I could have found the words, I might have bellowed in his face: "*Where the fuck have you been all my life? What do you want from me?*"

But you don't kick a man when he's down. And so my father turned around, descended the stairs, and slipped back into his Oldsmobile. Then he drove the 160 miles back to Toronto—mad Lear on the edge of the heath, and me feeling the fool.

═

It was the end of the '60s; I didn't know it yet, but it was also the end of our family. Exhausted by my father's escalating abuse of drugs, intercut with bursts of weeping, my mother was reaching the end of her rope. One by one, my father was now dusting off his old friends; and one by one, they were letting him—all except Jack Kee, a fellow jazz fiend from the days of the UCC boarding house. My father was touched that Jack so often tried to reach out to him, but he could not respond. Once he struggled to thank him, but he had to do it in a letter that Kee's daughter, Janet, gave me decades later. "I couldn't tell you to your face," my father wrote, "or I'd come all unglued."

When I bused home on weekends from university—as infrequently as possible—I felt as if I were watching a relative stranger implode inside a glass box. Plagued by migraines since the war, my father had grown recklessly addicted to a staggering succession of narcotics—Wigraine, 222s, alcohol, Valium, amphetamines, Seconal, and morphine—often knocking himself out cold. Under the pretense of walking the dog, or buying a gift for my mother, he'd head over to the pharmacy in Forest Hill Village, scribble himself a prescription, and load up on drugs—drugs that were, by now, only partially suppressing the sudden jags of panic that burst like crazy ghosts out of his gloom. He was ensnared in a classic Catch-22: No one dared suspend his prescribing privileges, for his livelihood depended on it. More often than not, a bizarre drama was now playing out at the dinner table: At warp speed, my father would wolf down his food, then erupt into an inevitable fit of choking, his face turning beet red. Each time, I thought he had breathed his last; each time, no one knew how to stop the cycle.

For reasons I did not yet grasp, my recent reading of the grandiose Greek myth of Helios, the sun god who drives a horse-drawn chariot of fire across the sky, was strangely resonating inside me. When his son Phaethon, born of a mortal mother, learns his father's divine identity, he seeks him out and begs to let him take the reins of the chariot. "In your ignorance, you aspire to do that which not even the gods themselves may do," Helios thunders. "None but myself may drive the flaming car of day."

Yet Phaethon persists and tricks his father into letting him drive the chariot. Sensing a weak hand at the reins, the horses bolt, careering wildly, blackening all the inhabitants of Africa, scorching all living things in their path. Phaethon would have set the whole world ablaze if Helios had not exploded the chariot with a mighty thunderbolt. As the horses break free, the mortal Phaethon plunges to his death.

My father, thankfully, had never pressured me to become a "docter," the word I had misspelled as an aspiring eight-year-old "writer." I felt no urge to drive his flaming car of day; the pressure I felt came from somewhere else.

That Christmas of 1969, the five of us flew south, along with Aunt Molly and her family, for a vacation in a Jamaican resort fittingly called Blue Hole. It rained incessantly. I watched American tourists in plaid shorts throw handfuls of coins off a bridge so that naked black boys would dive down deep into the sea to retrieve them. I envied the strength of their lungs. My parents had at last included us in their winter escapes, but paradise could not relieve the heavy lethargy coiled round my body like a boa constrictor. When I fell severely ill with an acute bout of Crohn's and anemia, a side effect of the drug I was taking, my parents and I flew home early. As we drove to the airport, my mother turned to me in the back seat and said: "You're just like your father."

First at the Toronto Western, and then when I returned to Kingston, I endured another battery of invasive tests. That same week in early January marked the start of my father's winter term of teaching. As she did most mornings, my mother gave my father pep talks to rouse him out of bed. Dreading the prospect of facing rows of medical students, he phoned in sick. As he lay alone on the queen-size bed, he filled a syringe with morphine, then jabbed the point into a vein in his arm.

He was lost but he wanted to be found; and sure enough, my mother found him. She called an allergy colleague for help and together they pulled him through with cups of coffee. Later, the doctor, struck by my mother's sang-froid, told her: "You're a good actress." Then and there, my mother silently vowed not to tell her children what had happened, at least not now. Nor could she admit that there was no worse feeling than being pulled into the sad swamp of my father's needy manipulations.

At Queen's, oblivious of my father's death wish, I was cutting classes, more out of fear than indifference. Sometimes, as I approached the door of a seminar room, I was so overcome by anxiety that I stopped dead, as if a child reliving the intense, threshold-crossing taboos of Balmoral. Turning tail, I'd regress into my pop-culture cocoon, devouring my latest copy of *Rolling Stone* and collecting rock albums with every spare dollar I could muster. I was intrigued by the dropout theatrics of John Lennon and Yoko Ono, spending a week in bed in a Montreal hotel room. The primal screams, the stabs of brutal honesty: This was it, this was the answer. *Make love, not war*. Dead simple, really.

In April 1970, the Beatles broke up. In my shabby Kingston apartment, I held a memorial concert, stacking up the complete oeuvre on my turntable, thirteen straight hours of beloved music melding elegiacally with the light of dawn. Back in Toronto, Dr. McCulloch was drawing up a battle plan. In May, he sent my father to a "Human Relations Lab" at Talisman, a resort north of the city. There he joined a week-long encounter T group run by Vello Sermat, an Estonian psychologist from York University. Dozens of people sprawled on lawns, played games, and confronted each other and their deeper selves. McCulloch hoped the experience would pull my father out of his downward spiral, but it backfired. A man who dreaded all forms of confrontation, Jack flew high as a kite, then crashed in a heap.

In June 1970, my twenty-one-year-old sister, Shelagh, now beginning a career in the travel industry, tracked down our long-lost half-brother Johnny Leuthold in San Francisco. When she first laid eyes on our twenty-six-year-old sibling, she freaked. Even though he had been raised on the other side of the continent, John was a dead ringer for our father—the same facial features and receding hairline, the same mannerisms and idiosyncracies, the same hand gestures, the same quirky turns of phrase, the same passion for music; he even played competent jazz piano. When Mike and I met him two summers later, we felt the same as Shelagh. John seemed more like our father than any of us; even if it wasn't true, the thought felt comforting.

My sister's search for a big brother echoed my own inarticulate longings for a loving father. In my universe, fathers and sons seemed to move like orbiting planets, oblivious of the profound mutual influence that gravitated through the empty space between them. Surely, I wondered, my father had once been a son and understood the natural feelings of sons? Why was he leaving me in the dark to make sense of it all? I had next to no clues to go on, just the dim awareness through my teenage years that my father's father, in his day, had been a medical star of international reputation and influence. All my mother had told me was that Gerry had founded a laboratory, known as Connaught, that manufactured vaccines free for the poor and that he ran a training institute for public-health doctors called the School of Hygiene. The questions quietly whirled in my head: Why was my father self-destructing before our eyes, trying to live up to the legend of some ethereal, long-dead, godlike figure none of us knew or remembered? What was he reaching for? What was he trying to prove?

Edging out of the fog of youth, my imagination took random stabs in the dark, filling the silent vacuum with fanciful scenarios of my grandfather's unspoken fate. Had he been cut down by some gruesome infectious disease? Shot by IRA terrorists? Kidnapped by wailing banshees? It was an act of sheer adolescent intuition that one day compelled me to ask my mother point-blank: Why the silence enveloping my grandfather? Did he kill himself?

Startled, she shifted her eyes and nodded wordlessly. Then she murmured something about a drug overdose, although she wasn't entirely sure. No further information was forthcoming, but at least I had a clue. As I studied the ticking time bomb that was my father, I wondered why everyone was colluding in dodging the obvious fact, the elephant in the bathtub: the unspoken impact of my grandfather on my father's life. And, of course, my own.

It is always worst in the early morning, the spectral hours that echo memories of birth. From his loveless bed my father arises before dawn, agitated, anxious, lashed by the insomnia, laced with the traces of nightmares that crawl under his skin like an army of black ants. Thoughts of the impending day, the call to perform, tumble and swarm over his brain like the unbroken hum of an electric lawn mower he cannot turn off. He cries, he wrings his hands, he paces up and down, he despairs of life. Filled with equal parts fear and rage and grief, his burned-out wife can no longer bear his tears, his whining, his weakness. How could she, when she has long ceased to love him, if she had ever loved him at all?

75 Dunvegan Road, fall 1970

It is Wednesday, September 9, 1970, a week before my twentieth birthday. Once again, my father is scheduled to begin a new term of teaching; once again, he can't face those unsmiling tiers of medical students arrayed before him. My mother has decided to spend a few days at their recently built split-level country property forty-five miles away in the pacifying hills of Caledon. My parents had designed and built "The Place" together and the project had, for a spell, revived Jack's creative energy; but on the stroke of its completion, he receded into his cave of mute indifference.

My mother takes Michael, poised to enter grade thirteen at UCC, with her up north to help plant trees. I have left for Kingston for my third year of university the previous day; years will pass before I will fully absorb what happened next. My mother knows that leaving Jack alone in the house poses a risk, for the conditions are ripe for a repetition of the January scene nine months past. The cleaning lady is away on vacation and so Janet asks Shelagh, now living in an apartment on Charles Street, to spend the night at Dunvegan. Shelagh is instructed to assume the wifely role of meal maker and cheerleader, making sure Jack keeps his appointments at the hospital in the morning and to lecture in the afternoon. Then, before she steps out the door, our mother drops the bomb: *You should know something—your father is suicidal.*

The next morning, Jack summons the will to drive down to the hospital but returns home for lunch. To blot out the shock of our mother's revelation, Shelagh had caroused with friends at the

Chez Moi pub the night before and is sleeping off her hangover. Standing outside her bedroom door, Jack announces he is going upstairs for a nap; Shelagh wonders why he has decided on the third floor and not his own bed on the second. She knows he must be at the university by two in the afternoon, but he insists she must not disturb him—no arguments.

And so she watches our father climb the broadloomed stairs to the third storey, a place he has rarely gone. He could have turned left or right; he chooses right, lying down on my brother's bed, not mine. That narrow bed, that timeless rectangle of sex and birth and death, where dustballs gather below for the immunologist to collect; the coffinlike bed of the dreaming pyjama-clad son, which the father never chose to visit in life—but he goes there now, strangely, silently, symbolically, in search of the dust and ashes promised by eternity. "We act out until we remember," the sage Freud once said; but here there is only a sleep and a forgetting.

As he had asked thousands of his own patients to do, year in, year out, our father rolls up his sleeve. This time, he will do it right. This time, he jabs twelve grains of the lethal narcotic into his veins, then hides the needle behind a row of books on baseball.

As the minutes tick by, Shelagh silently wills him to come down on his own. As two o'clock approaches, she climbs the staircase, heart thumping, the air turning white, the dread of what she has imagined suddenly made flesh: our father lying on the bed on his side, his breathing slow, harsh, and shallow, his face a horrific shade of blue.

She stands still for a moment, uncertain. She speaks—"Dad? Dad?"—but he does not hear; she shakes his shoulder, but he does not wake. If she had not learned only hours before of our father's dire intent, she might have turned back down the stairs and left him to his fate. Instead, my sister strides trancelike to the phone. She calls Dr. McCulloch; he tells her to call for an ambulance and take Jack to the ER at the Western Hospital. Her trembling finger dials zero, raising the bored voice of the operator; then comes the agonizing wait, the pacing and the panic and the frenzy slashing her soul; the converging ambulance, police cars, fire trucks, the red lights swooning in crazy circles; the limp body strapped to the gurney, tilting down the two flights of stairs and out into the late summer air, the neighbours parting curtains or lingering at the foot of the driveway, at once curious and ashamed of their curiosity.

The driver does not head for Toronto Western, but for Toronto General, as it is closer. As the ambulance tears down Avenue Road, down past Balmoral Avenue, down the hill, past the University of Toronto and the packed classroom of medical students our father will never teach again, the poison in his blood hounds the dreamscape of his fifty-three years. The doctors of TGH, the place of his birth and the place of his imminent death, await his delivery.

"What did your father take?" comes the voice of the ER doctor. "He's dying. We need to know so we can give him the antidote." But how could Shelagh possibly know? And so, as the father hovers on the edge of endless night, the trapped daughter calls the mother at the country retreat; and as the tall white coats and hanging stethoscopes ring her in a circle of urgent expectation, from the far end of the line comes the familiar voice—the voice of the long-disaffected housewife locked in the state of unholy acrimony, the failed artist, the mother who had calmly orchestrated her first-born child to face the crazy music in her stead—and the cold, flat note of a single word, in all of its inevitability: "Morphine."

As the doctors rush away, a young intern slips into the glassed-in waiting room that feels like the vacuum of outer space and sits down beside Shelagh, who is numb with shock and rage and guilt and shame and humiliation. "You don't seem to be too upset about your father," he observes with McCulloch-esque tactlessness, a blade thrust she can only receive with stunned incomprehension. She stares at the wall where, on the other side, our father lies. He wants to die and he wants to be saved; yet even if his body survives, nothing in life can redeem his starving Irish spirit now. Not the best of homes in the best of neighbourhoods in the best of countries; not the most prestigious boarding school money could buy, not the gentle nanny-gestures of his sweet Aunt Hazel, compensating for his absent parents; not the respect of his colleagues or the affection of his friends; not the melodic elation of Ella Fitzgerald, his mad love of the Duke, nor the divine Django's virtuoso fretwork, ringed with halos of marijuana smoke; not the cobalt blue waters of his heated swimming pool, infused by the sensuous light of the summer moon; not the warm beige broadloom nor the whitewall tires of his gleaming Oldsmobile; not the reflected sheen of the silver Stanley Cup nor the blue bottles of narcotics arrayed like toy soldiers behind the mirror of the cabinet door. Not the brisk dog walks on the green grounds of Upper Canada College and the tranquil rolling hills of the countryside; not the sweet-scented flower garden of blooming lillies, nor the sleek-necked bottles of fine red wine; not the sugar brown sands of Caribbean paradise, the spontaneous riffs of witty repartee, the dreamy promise of perfect orgasmic sex with a perfect, soft, smiling, loving, giving, all-surrendering wife; not the smooth hypnotic eye of the television screen, the clockwork flow of *Laugh-In*s and *Twilight Zone*s and *Mission Impossible*s. Not the headlong flailings of Dr. Donald McCulloch, paving the road to hell with his good intentions.

None of these things, together or apart, can save the skin of Jack FitzGerald, MD, FRCP. Somehow, somewhere, deep down, he knows the score: He has not been true to himself. And there is a price to pay. By us all.

FOUR

Dr. FitzGerald Must Learn to Cover Up His Feelings

Don't try to be master in everything. What you once won and held did not stay with you all your life long.

SOPHOCLES,
Oedipus Rex

My mother tried to shield my brother and me from the drama of the third floor, but truth, especially the suppressed variety, has a way of leaking out. In time, I came to understand and forgive the family's protection racket, but the decision of my mother to withhold the grisly details would prove detrimental to our collective health. Tragically, Janet's cold-blooded set-up of Shelagh as the designated saviour remained a bottomless source of unresolved grief and pain and guilt for them both—a squalid beast squatting in the living rooms and kitchens and gardens, the birthdays and Christmases and Mother's Days, a ubiquitous, Medusa-headed monster demanding to be fed, spewing scenes of addictive spite that bound mother to daughter and daughter to mother with an unspeakable force.

Only when I got my hands on my father's psychiatric files twenty-five years later—and found the courage to start my book and ask my mother and sister some hard questions—could I finally begin to digest some incremental doses of "the truth." But back in 1970,

I could find no person or place to receive my scrambled emotions. That same week of September, my guitar hero, Jimi Hendrix died of an accidental drug overdose at age twenty-seven, choking on his own vomit in his sleep. I remember tears leaking reluctantly out of my eyes—as with the assassinations of the Kennedy brothers, I was puzzled as to why it seemed easier to feel grief for the disembodied images of public figures I had never met.

My sister was thrust into the front lines of my father's madness. Dr. McCulloch failed to show up to the hospital emergency room, and the next day, Shelagh lit into him with the intensity of a Hendrix solo. Had McCulloch deliberately chosen not to come to the side of his suicidal patient? Was he trained to thwart the selfish manipulations of the borderline personality? Who knew for sure; but when he did finally see Jack alone, he berated him for "his pessimism."

Within weeks, my father staggered back to his feet. At our mother's urging, he meekly apologized to Shelagh and thanked her, puppetlike, for saving his life. Thereafter, no one spoke of the events of September 9; the day was wiped from the calendar like raindrops from a windshield. Jack returned to his practice, part-time, bored senseless by the rote nature of his work. Rushing manically through his patient load, doling out prescriptions or injecting sufferers of asthma and hay fever, he'd careen back home, bashing parked cars in a chain of fender-benders on downtown side streets. Alarmed at the sight of their mentor unravelling, his closest colleagues felt powerless to help. "Is there anything I can do?" asked Stan Epstein. "I wish you could, Stan. I wish you could," my father replied. One day, minutes before walking into an auditorium to lecture two hundred medical students, he froze; on the spot, he asked his colleague Roman Bladek to take over, then fled home to bed. More than once over the years, he had told Bladek that he was convinced he was going to die by suicide in his fifties.

With the family income radically diminished, my embittered mother sold the house on Dunvegan Road on April 11, 1971, the day of their twenty-fourth wedding anniversary; she would soon unload the weekend country retreat that had failed to spark my father's lost vitality. In her diary she wrote: "I wouldn't recommend anyone marry a doctor, unless he's a broad, Renaissance kind of man who reads literature. And don't ever cross him!" She rented a duplex at 100 Rosedale Heights, then a smaller one at 14 Highbourne Road, as if clinging to addresses with elevated names.

In the summer of 1971, my father crossed the threshold of the Clarke Institute of Psychiatry at 250 College Street; in a moment, he exchanged the trim white coat of the highly reputable doctor for the baggy white gown of the vaguely disreputable patient. Whatever he was before, he was now officially branded "mentally ill." The narrow, concrete-encased windows of his room resembled the arrow-proof slits of a medieval battlement; if he happened to peer eastward, he would have seen the roof of the University of Toronto School of Hygiene, the epicentre of his father's storied empire, at 150 College. Perhaps a further irony stung his thoughts: "The Clarke" was named after one of his father's mentors and heroes.

My father unfurled his history to the admitting psychiatrist. When I first read his file, procured from the Clarke archives in the 1990s, I discovered much that I already knew and much that I didn't. Throughout, I was intrigued by my father's perception of his own history. Jack proudly described his mother as "the belle of London, Ontario, who went to the best girls' school." His father was "an influential physician in Toronto and abroad, a remarkably effective, energetic, thrusting man who suffered from frequent depressions and ultimately committed suicide in a paranoid state." Jack described his father as rigid and obsessive, but also as a human being and a good father who tried to leave him free to make his own choices.

He recalled his childhood as "retreating" in nature and said he frequently went to the movies alone. Much of the time, as his parents travelled abroad, he was brought up by his spinster Aunt Hazel, but suffered "no loss of love." In his teenage years, he ran away from home frequently. He was popular, renowned for making people laugh and had several close friends. At fifteen, he began stealing his mother's sodium amytal tablets to sedate his own anxieties and tensions. He started dating at sixteen, often girls older than himself. He was sent to Europe twice and described a precocious worldliness.

One by one, my father unfurled the travails of his life—the premature death of his mentor Hurst Brown, the crushing burdens of work, his sister's worsening alcoholism, his marriage on the rocks. His sex life had long since dried up; when pushed, he admitted it had never really been satisfying for either of them. Then he revealed to the psychiatrist something that had never reached my own ears: That as we three began moving out of the nest, he felt increasing regret for not having devoted time to his children as we were growing up. He confessed that he was either asleep or at work most of the time: "I can count on my hands the number of times I have taken them out."

He had to work extremely hard to succeed, applying himself completely to medicine and dropping other fields of interest. He felt an intense need to prove himself and gain the approval of others. He struggled to earn his medical fellowship to prove he "could get on the faculty without the help of my father's name or any of his cronies. I was under constant stress to beat the gun, but I liked it." He took on an extended workload, doing a great deal of teaching; he did it rather well, although he found it an increasing strain.

His struggles to establish the Western Medical Building, opposed by the University of Toronto, "ate his guts out." Attempts to expand the Toronto Western Hospital were similarly frustrated by the university; he worked on several feasibility committees but his efforts proved futile. Younger men whom he had trained were taking over positions and doing the things he wanted to do. Teaching classes of medical students precipitated episodes of acute anxiety, eroding his confidence.

He recounted the troubles of his alcoholic sister, Molly, and his own fierce drug addiction. "The old Fitz people knew—the conversationalist, the comedy man, the happy Irishman who got the party off the ground—was a boozed Fitz. It made me forget my worries." He admitted that he was often shabby, unable to take care of himself, and that he cried repeatedly. Panicked by thoughts of suicide, he resorted to taking pills. It seemed to him only a matter of time before his father's fate would overtake him.

The psychiatrists diagnosed my father as manic-depressive, a label that concealed as much as it revealed. In the months ahead, the names of doctors—Hunter, Stancer, Lamon—mixed inextricably with the names of drugs—Pertofrane, Seconal, Mellaril, Haldol, etrofan D, Nozinan, meprobamate, Cogentin—sifting back to the ears of the family in fragments. No matter how many pills my father swallowed, the waves of anxiety broke through the chemical haze into moments of full-blown terror. "He denies it, but there is no doubt self-pity is present," observed Robin Hunter, the psychiatrist-in-chief. "For instance, he wept briefly on two occasions when discussing people who have been good to him. My feeling was that he, like the crocodile, weeps over those he devours and defends himself against this realization by self-pity."

On October 13, 1971, a white-coated doctor opened a drawer full of pieces of rubber and inserted one into my father's mouth. He attached two shiny metal discs the size of silver dollars to my father's temples and then turned the knob on the black box to "treat." The ritual was repeated ten times over six weeks. In the end, the doctors admitted the bursts of electricity, deliberately inducing grand mal seizures in his brain, did no good. In fact, his condition was now worse than on admission. A crisis is a terrible thing to waste; and wasted this one surely was.

1621-71
DR. H. C. STANCER 5TH FLOOR CIU
10/11/71

CLARKE INSTITUTE OF PSYCHIATRY
CONSENT FOR SPECIAL TREATMENT OR INVESTIGATION
23285

I, the undersigned, being John Fitzgerald (RELATIONSHIP) do hereby give full consent to have performed on me my ______ (RELATIONSHIP)
DR JOHN FITZGERALD (NAME)
a E.C.T (PROCEDURE)
having had explained to me by Dr. Lamon, the nature and risk of this procedure. If any conditions are discovered at the time of this procedure which were not recognized before and which, in the judgment of the attending physician call for other procedures than those contemplated above, I authorize him to do whatever under the circumstances he may deem necessary in the interests of

John FitzGerald (NAME)
Nov. 24, 1971 (DATE) ______ (SIGNATURE OF PATIENT OR PERSON AUTHORIZED TO ACT ON BEHALF OF)
L Bara R.N (WITNESS)
Clarke Institute (ADDRESS) ______ (ADDRESS)

Jack's consent form for electroshock treatments, 1971

When Shelagh visited our father after the shock treatments, she found an unrecognizable, prematurely aged man in an open-backed white gown; a vacant, shuffling, trembling, dishevelled mess cast into the blasted heath, his memories stripped clean as bark from a tree. I repressed the memory of my sole visit with my brother, as if my own brain had been washed clean with ECT. Years later, when Mike recounted the pathetic details of our visit, I drew a blank. Maybe we were all skating round the guilt of what we really felt: that maybe our father *was* better off dead.

At the time, I was still ensconced in the ivory tower of Queen's University three hours east of Toronto, safe from the intrusions of the disturbed, or so I thought. I was cutting classes, spinning my discs of electric music, and growing my hair down to my shoulders. My roommate, Sean, was a stocky, quick-witted character whom I had met in English class in our first year, and we had become fast friends. Visiting our home in Toronto one day—he had gotten to know my father a bit, forbiddingly remote as he was—Sean startled me with the offhand remark: "Your father is one of the most charming yet tortured people I've ever met."

By strange coincidence, the autumn of my father's shock treatments, Sean would rock my world like never before. At first I dismissed the early warning signs; then came the day, en route to psych class, that he started conversing with the flashing traffic lights on Division Street, believing he was receiving coded messages. The same with the sonorous voices of newscasters on the car radio: he thought that political events—the India-Pakistan war was currently raging—were fraught with personal significance. We were taking Abnormal Psychology lectures together and the pages of our textbook vibrated with clinically precise descriptions of paranoid schizophrenia behaviour uncannily like his own, although the link seemed lost on him. Day by day, Sean's hallucinations intensified, yet in his periods of lucidity, he persuaded me that I was the one losing my mind, and he was half right. Weeks later, I persuaded him to check into the psychiatric wing of the city hospital; he insisted on driving. When we arrived at the ER in

"Wolfgang," the beat-up black 1962 Volkswagen we co-owned, my hands and voice were shaking. The nurse looked at us and asked, "Who's being admitted?" Simultaneously, we pointed at each other. I envisioned a Kafkaesque nightmare of being dragged off kicking and screaming as Sean drove away, smiling and waving, with everyone none the wiser. As it turned out, "sanity" prevailed; like my father, my friend was fed into the ranks of the drug-and-shock troops. I felt both relieved and guilty as hell. In a raucous student beer hall, I cried on the shoulder of a girl I barely knew. There was something noble about Sean's Randall McMurphy–like swagger, his Irish knack for squeezing a laugh out of catastrophes large and small, that I loved. I thought of the old joke that, in better days, had raised a knowing smile on my friend's face: *A paranoid schizophrenic is someone who has finally figured everything out.*

Walking out of an exam hall that spring, I bumped into Jane, an attractive blond cheerleader whom I had flirted with at parties. The Havergal-educated daughter of a war hero, she wore a ceaseless, perfect smile that matched her ceaseless, perfect cartwheels. I was girlfriendless, and Jane seemed the kind of iconic, tartan-skirted beauty my social class had programmed me to court and marry; at the time, I wasn't averse to getting with the program. She flashed the smile and cheerfully invited me over for a beer that night. I agreed, but my shyness intervened and I never showed up. Weeks later, Jane sealed herself in her parents' garage in Toronto and asphyxiated herself with car exhaust. She thought no one liked her.

My father's doctors, meanwhile, were concentrating on getting him out of the hospital and back to work, even though he was failing to improve. Diagnosing and treating Jack FitzGerald was like nailing Jell-O to a passing ambulance. He spent the mornings in the ward as a "day care" patient, worked in his office in the afternoons, then slept at home overnight. The mornings were the worst; he felt "empty, utterly incapable of doing anything. The only thing I can think of is to escape to bed." He still suffered from terrible headaches and overwhelming nightmares filled with scenes of catastrophe. He'd take up to eight tabs of chloral hydrate to kill the chronic insomnia. As he backed his car out of the driveway on his first day of work, he hit a tree.

The return to work proved a mistake; in reaction to the stress, he was now mixing chloral hydrate, Elavil, Haldol, 222s, and chlorpromazine. "He's in a bad way—flushed, bug-eyed, sweaty, agitated, and slurred speech," noted one of the doctors. "My approach has been disapproval—intellectually—but I don't want to discourage him from reporting his drug intake." By this point, no one had a clear idea where the drugs left off and his disturbance began. "Wife not especially complimentary toward the patient," another psychiatrist noted dryly. For his part, Jack dismissed his wife as "big, beautiful, and aggressive, obsessed with minutiae and detail . . . she's strictly from Missouri." Between the magnetized poles of my battling parents, I swung from sympathy to indifference to contempt and back again; one week, *she* was to blame for it all, the next week, *him*.

In the spring of 1972, my father returned to the Clarke as an in-patient. "Recent evidence," wrote Dr. Harvey Stancer with studied understatement, "indicates that we have not been completely successful . . . I do not now feel that this patient can benefit from any extended psychotherapy as he was receiving previously from Dr. McCulloch. It is my opinion that medication, environmental manipulation, and behaviour therapy are our only hope."

As the doctors started to wean him from the toxic stew of self-medication—intending to eventually get him on the "right" drugs—he slid into a state of withdrawal. He was boosted by a constant stream of talk with another patient, but the doctors now feared he was becoming dependent on *people*, as if this were a sin. One morning, he greeted the psychiatrist with, "If one more person says I look better today, I'll kick them in the nuts!" Whatever models of normality the doctors were selling, my father wasn't buying.

After much agonizing, he decided to close down his allergy practice and retire from teaching; at age fifty-five, my father was out of work. Three of his protegés, Drs. Briggs, Bladek, and Epstein, proposed a dinner in his honour, but my mother rightly said he could not possibly handle it. Over five straight days at the Clarke, a flood of tears, agitation, and suicidal despair mounted with irresistible force. My father found the overriding sensation of inner hollowness intolerable. "I feel," he confessed, "as bad as I felt before trying to knock myself off."

The Clarke had done its best and its worst, but the big-shot doctor proved stubbornly "treatment-resistant." It was time to pass Jack FitzGerald, the hot Irish potato, on to the Donwood Institute.

═

At the time, I remember hearing great things about the first public hospital in North America specifically designed to help addicts. If the guys at the Donwood couldn't separate my father from his cravings, no one could. Born into a Scottish Presbyterian family in small-town Ontario, the founder, Gordon Bell, had heroically fought the stigma doctors traditionally imposed on alcoholics, whom they scorned as the agents of their own downfall. Before World War II, research on addiction in Canada was virtually non-existent and Bell was keen to redress the situation, working out of his house starting in 1946. Hailed as the best anywhere, Bell's clinics evolved into a national and international model, notably imitated by the Betty Ford Center in California.

On the day of my father's admission, the examining psychiatrist declared he had enough drugs in his system to kill a horse. "Dr. FitzGerald," wrote Dr. Neil Maharaj, "is an obsessive man who has tended to overwork, narrow his field of increasingly intense activity, and require constant approval. He is a passive-dependent person who is almost constantly preoccupied with himself. He is highly manipulative in terms of seeking out other persons to walk and talk with him. His early morning agitation drives him to search out other persons to be with him during this highly anxious time of the day."

My mother was growing increasingly frustrated by the revolving door of ineffectual shrinks and the Donwood proved no different. "She feels," noted Maharaj, "that she has accepted the complete masculine role and detests this in every way possible and John is not the man he used to be. She feels that separation would be ideal, but at the same time ambivalent in that she is afraid that on leaving John, he might contemplate suicide, which he has tried on two previous occasions. The therapist then confronted her as to whether she likes this masochistic role and does she ever realize that she might want him sick. With this, she stormed out of the room."

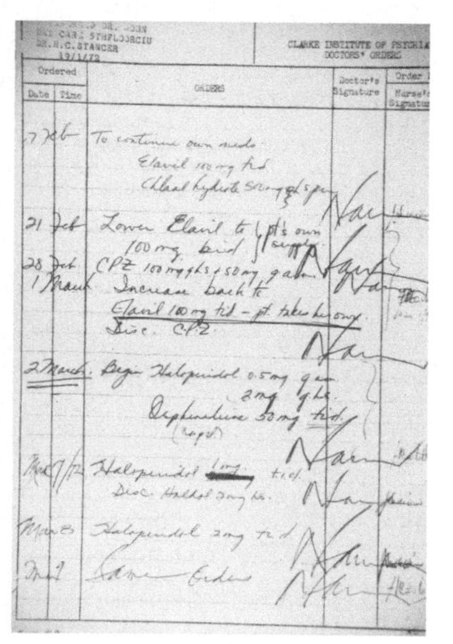

DR. H.C. STANCER
19/1/72

CLARKE INSTITUTE OF PSYCHIA[TRY]
DOCTORS' ORDERS

Ordered: Date	Time	ORDERS	Doctor's Signature	Order: Nurse's Signature
7 Feb		To continue own meds Elavil 100 mg tid Chloral hydrate		
21 Feb		Lower Elavil to 100 mg bid (pt's own supply)		
28 Feb		CPZ 100 mg hs + 50 mg q a.m.		
1 March		Increase back to Elavil 100 mg tid – pt. takes his own. Disc. CPZ		
2 March		Begin Haloperidol 0.5 mg q a.m. 2 mg q h.s. Orphenadrine 50 mg tid.		
Mar 7/72		Haloperidol [illegible] tid. Disc. Haldol 2 mg hs.		
Mar 8		Haloperidol 2 mg tid		
Mar 9		Same Order		

Jack's psychiatric drug regimen, Clarke Institute

A louche, long-haired twenty-one-year-old, freshly graduated from Queen's, I visited my father in the Donwood one balmy day in June—the one time I did—only to encounter the usual prickly silences. "Keep your pecker up!"—my father's pet phrase to encourage others in distress—came to mind, but I resisted the impulse to parrot it back to him. I handed him a heavily underlined copy of Leonard Cohen's novel *Beautiful Losers*, naively hoping my newfound intellectual excitements might rub off on him; maybe timeless art could trump temporary science. But my father rarely read non-medical books; only once did I catch him leafing through a biography, aptly enough, of Ernest Hemingway, the Nobel Prize–winning writer who, despairing that his memory had been fried by electroshock, blew his head off with a shotgun, exactly as his physician father had done a generation earlier.

My father's ten-week stint at Donwood would prove fruitless. The experts threw up their hands, admitting that he was "the type of individual who is going to probably require medication for the rest of his life, hopefully at a supervised level. . . . Unfortunately, the prognosis is poor."

In July, he was sent home and my mother was entrusted with doling out the drugs. The old patterns re-emerged; by November, he had reached the point where he was demanding tablets every half hour and Janet was at the end of her tether. That same month, she managed to sell the allergy lab for fifty thousand dollars, and I helped her clean out the place. The sale was no easy task; practising allergists saw the private enterprise as too controversial to take on. The Connaught Labs also declined, as they were embroiled in their own public controversy, which, at the time, I was only vaguely aware: Its owner, the University of Toronto, was in the process of selling off my grandfather's monument to public service—"a national treasure," the media called it—to private interests. Acting on the belief that most infectious diseases had been successfully controlled (AIDS and SARS as yet unimaginable), the university then broke up the various departments within the lab's allied School of Hygiene on College Street and folded them into the medical faculty. I did not realize it until years later, but the collapse of my father's medical career coincided with the destruction of my grandfather's groundbreaking vision of public health and preventive medicine, first conceived in the years before World War I.

One weekend that fall, my father became frantically suicidal; a psychiatrist laid on doses of Nozinan and Cogentin, to no avail. This time, Jack was admitted to the Addiction Research Foundation, a newly erected concrete building tucked behind the Clarke Institute, falling within the shadow of the old Spadina Division of the Connaught Labs. No matter where in the city he landed, the ghost of his father hovered in the wings.

My father joined a therapy group of fellow addicts. Speaking freely of his drug abuse, he sailed off on tangents, relating his own stress and anxiety to the rise and fall of successful American businessmen. He feared his marriage would break up: Only his illness kept his wife committed to him, and tenuously so. He sensed that she would not stay with him if he improved—another Catch-22 that might be causing him to perpetuate his "ill" behaviour.

Each evening, he paced the halls with the nurses in a state of high-pitched agitation. He talked about a "jumble of feelings" that were "life threatening," but he was unable to name his fears, other than a constant dread of "losing control." He wanted to discharge himself because of the noise and swearing in the ward. Bothered by the age difference between himself and the younger patients, he felt unable to relate to them and envied their vitality. In the group, a nurse noted that his forehead wrinkled with worry as he took an "intimacy test." People suggested ways of relieving his intense agitation; instead of the negative act of pacing, perhaps he could play a game of cribbage or checkers with a staff member or patient. Everyone called him "John," a name his friends and family never used; it was either Jack or Fitz. Now that he was unemployed, the nurses tried to gauge his aptitude for a job outside medicine. On all 280 questions of the *Ohio Vocational Interest Survey*, he answered either "dislike" or "neutral." The test, my father likely figured, was strictly from Ohio. After ten days, he was gone: On November 17, 1972, my father was transferred to Homewood, a private sanitarium in the town of Guelph. There he would stay for the next year.

Institutions, like people, have histories; and some, like people, might even have souls. A stately stone mansion reposing on forty-seven wooded acres, Homewood was Ontario's first private sanitarium. Overlooking the Speed River, one that would claim its share of suicides, the twin-towered red-brick edifice was built in 1883 by several affluent Toronto businessmen as a middle-class reaction to the horrors of the crowded public asylums. The first superintendent, an Irish-born graduate of Upper Canada College and an expert on opium addiction, would eventually die in his own hospital as a patient; his successor would notoriously perform gynecological mutilations on streams of depressed upper-middle-class women. But by my father's time, Homewood had garnered a worthy reputation, drawing a host of celebrities, including Lucille Ball and Bing Crosby, and a steady intake of addicted professionals, many of whom, like my father, were doctors. At the time, a quiet irony was lost on me, and likely my father: Homewood was once run by Clarence B. Farrar, Canada's pre-eminent psychiatrist and my grandfather's close friend and confidant.

Dr. C.F. Story, the Homewood psychiatrist-in-chief, inherited my father's ever-thickening case file. "The informant, Mrs. FitzGerald, is a tall, grey-haired, immaculately dressed, fifty-three-year-old individual," he wrote. "During the interview, she came through very much as a professional woman and initially related with very little feeling attachment, with an undercurrent of guilt, in relation to her husband. Since the interview was brief, the information obtained was sketchy. The wife felt that the central difficulty with her husband was that he was continually 'fighting his father's image.'

"There seems to be no real communication. Even early in their marriage, in 1963, when their eldest child was 14, there was much conflict between the mother and father in the area of discipline and upbringing of the children. At that time, the wife stated that she brought up the possibility of separating from her husband and that with this obvious manoeuvre was able to coerce her husband to her way of thinking. She stated that at that time, the father wished to utilize an 'English style' of upbringing in that he wanted to send his children to boarding school.

"The wife noted that it was significant that in 1969, their son James, then 19, was diagnosed with Crohn's Disease. The father avoided talking about it and there was an element of denial and avoidance. The wife herself seems to come through as an individual who tends to keep feelings to herself . . . she is a very domineering, aggressive individual who seemed emotionally highly constricted and over-controlled."

For her own part, my mother memorably diagnosed Dr. Story as "flat as piss on a plate."

Story characterized my father as a victim of "obsessive-compulsive thinking in that he will feel that he should remember the name of a certain football player and other meaningless entities which go round and round in his head and give him no peace." He started him on individual psychotherapy and group psychotherapy sessions three times a week: "Although often nothing very dramatic came out of the individual therapy sessions, the patient would obviously seem much better as if the ritual of having the appointment was in itself the main feature to him. He tried to manipulate the staff into providing him with more medication but it has been necessary to be firm with him."

Then Story ordered a round of subcoma insulin. The injections "made a marked difference in his agitation, but unfortunately, towards the peak of his tolerance, he had some unpleasant side effects and we had to cut back rather rapidly in the dosage. He remains a very dependent and manipulative individual. Every effort is being made to involve him in occupational therapy, recreational opportunities, as well as group psychotherapy. Nevertheless, up to the last, he is still dependent on other people to talk to him, walk with him, or play games with him. He seems totally incapable of putting in an hour or so completely on his own. He not only pressures the staff for medication and for extra time with them, which the latter was felt to be legitimate, but he also puts a good deal of pressure on other patients to provide him with their time."

At this point, my father took a personality test that Story interpreted to represent "the current personality dynamics of a gentleman in a borderline state whose neurotic defences are being used to cover up and control frankly psychotic behaviour. . . . Dr. FitzGerald appears to be a somewhat shy, moody, indecisive person whose self-esteem is considerably lower than one would expect of a man of his socio-economic status and educational background. His current defence system is very inadequate, especially here in the hospital where others from his own profession with no apparent problems may be reinforcing his own views about his loss of status and deviation from socio-economic mores . . . he may have already gone beyond a neurotic type of illness and is in a developing schizophrenic-like or borderline psychotic state. If it has not already not occurred, one can expect a severely disrupted home and professional life, since open hostility and aggression will more frequently occur towards family, friends, etc, as time goes by."

One day in December, my father precipitously discharged himself, against the doctors' wishes, taking a cab the sixty miles back to Toronto. Being a voluntary patient, he was free to do as he liked. Again, the familiar drama played out at home. Two days before Christmas, he overdosed yet again, fell out of bed, and smashed his forehead. My beseiged mother drove him down to emergency at Toronto Western. A physician phoned Homewood to say that Dr. FitzGerald could not be managed in their—his—hospital and should be sent back "certified."

Back at Homewood, he talked of the hopelessness of his situation; there was no point in living. When Shelagh visited him the next day, she was appalled by the sight of his gashed head, his terrified weeping, his begging her to help him and take him home on the spot. Wracked with guilt and shame and helplessness, she extricated herself from his lunging grasp, turned back to the front door, only to collapse, wailing, on the snow-laden lawns of Homewood.

The doctors set to work, ordering hefty doses of phenothiazines, along with Haldol and Cogentin. The drugs dramatically lowered his blood pressure, causing ataxia, the breaking down of his muscular movements; he became confused, then delirious. Then came the crumbling of the last line of his long-battered defences, the torrid rush of delusions and hallucinations, the tipping over the border into stark madness. The doctors' punishing treatments were that and only that—punishing. After all, what did medical science have to do with nurturing relationships between human beings? What masqueraded as a helping hand he could only experience as the final insult in a lifelong line of insults. If my father was mad, doctors had everything to do with it.

═

As the cruel winter months wore on, my father continued to wallow in a tar pit of despair, seeing life as completely meaningless. "We are trying," wrote Story, "to get across to Dr. FitzGerald that he must learn to cover up his feelings to some extent and to carry on regardless."

Even though my father had a bad reaction at the Clarke, Dr. Story decided to retry electroshock. After his first two sessions, Jack appeared "much more relaxed and less demanding"; in fact, he stopped taking tranquillizers because he was forgetting to ask for them. "Some of this forgetfulness is of course due to the bilateral ECT treatments," noted Story, "but there does seem to be some general improvement." After further bouts of ECT, he had his house privileges suspended "because of memory disturbance and some vacant staring. He is still rushing through work which he is given to do in occupational therapy and we have asked that he try to concentrate on slowing down and pacing himself."

In the spring, my father was transferred to a long-term back ward; the ECTs and drug regimen continued. As his symptoms shifted, so did the multiple diagnoses; like a bar of Irish Spring, my ever-elusive father kept slipping through the doctors' fingers.

Then a consultant psychiatrist, Thomas Male, assessed my father: "This patient did not get the expected improvement with the bilateral ECTs and has continued to be quite anxious. Dr. FitzGerald presents features of a 'mixed depression,' showing characteristics of both psychotic and neurotic types. Despite his dependent characteristics, I feel that a full-scale trial of medication and ECT should be made. His paradoxical reaction to phenothiazines tends to influence use in the direction of either large doses of Valium or one of the newer medications. As there is a serious breakdown in communication with his wife, I would suggest a long series of joint sessions for them with counsellors.

"Finally, if there is no appreciable change, we should consider a total staff conference to look at all possibilities—*even a leucotomy*."

My father's madness peaked; then the tide slowly turned and the hovering blade of psychosurgery was withdrawn. To date, he had been treated by a succession of sixteen different psychiatrists; in May, he started seeing a psychologist, Ed Staples, in daily sessions of intensive psychotherapy. After a relentless battering, he now seemed ready to sustain face-to-face contact with a single, receptive, dependable human being—who also happened to be unburdened by a medical degree. By the summer, Jack's routine included occupational therapy, calisthenics, walks, medical reading, listening to jazz, and taking trial weekends at home. Staples routinely "challenged, admonished, and encouraged" him to try to regain a foothold in the real world. As the fog lifted slightly over the battlefield, the dire prospect of confinement to a long-term ward spurred my father to take a stab at normality. What did he want to do with the rest of his life? The choice was his.

═

As my father paced the corridors of Homewood, I was passing through another kind of institution: the postgraduate journalism school at the University of Western Ontario, an hour's drive west, in London. The birthplace of my grandmother wore a monochromatic WASP complexion for which I fast lost my tolerance; listless and uninspired, I dreamt of dropping out. Skirting the edges of my father's madness, I made a single, token visit to Homewood, which, in my memory, I have reduced to a single image: a solitary stranger shuffling, head-down, across the smooth green lawns, past tennis courts and terraced gardens and empty gazebos, moving slowly towards me, his pale, puffy, bug-eyed face silent as the drifting clouds above. Whatever stilted monosyllables passed between us have long since melted into nothingness; all I knew was that onto this pleasant wooded Eden, my father had mapped his personal hell.

In those days I was reading R.D. Laing and the radical crusaders of the budding antipsychiatry movement. Crazy as he had become, Laing had intriguing things to say: "Madness need not be all breakdown. It may also be a breakthrough. It is potentially liberation and renewal as well as enslavement and existential death." It seemed like a noble theory, but even in the cushioned rooms of my imagination, I could not picture a place on earth, nor the open minds of enlightened men, where such a miracle might unfold. (As he lay dying on a tennis court after being stricken by a heart attack, Laing uttered the memorable last words: "No bloody doctors!")

Back home in Toronto, my father became an outpatient at the mental hospital at 999 Queen Street West, his fourth institution in three years. Again, the ironies were breeding like lab rats, for it was here, decades earlier, under the soot-encrusted dome of the old lunatic asylum, that his father had once worked as a neuropathologist, cutting open the dead brains of mad Irish syphilitics. Jack saw a chaplain with the unlikely name of Reverend Bert Massiah, who practised something called reality therapy—the idea of concentrating on getting out of bed each morning and putting on one pant leg at a time. My father was floating in a kind of purgatory; neither husband nor doctor, brother nor teacher, father nor son nor friend, a man without qualities—the past and the future did not bear thinking about. The overdosing continued sporadically; I retain a gauzy image of my wobbly father leaning on my shoulder as I shepherded him down the stairs to a cab en route to the ER. One day, my brother watched him pull a Duke Ellington record from its album sleeve and place it on the turntable; he lowered the needle into the grooves of a favourite cut "Autumn Leaves," lay on the sofa and listened one last time. The ritual had the quality of a funeral elegy, a final parting with the enthusiasms of youth; over the last eighteen years of his life, my father would never again lift the lid of his stereo and tempt the gods of memory. Perhaps his inertia, his giving up of the ghost if not yet the body, was the price he felt he must pay—together with the daily doses of lithium—to imitate a semblance of what doctors called normal.

═══

That summer of 1974, I was transfixed by the Watergate scandal, caught up in the culture of lies, denial, and cover-up in high places. I was working as a cub reporter for a weekly newspaper, living in a dingy basement apartment across the street from Scarborough General Hospital. Comforting the afflicted and afflicting the comfortable—it seemed like a lofty ideal, even if I wasn't fully up to it. I was fast losing my stomach for "the straight life" that had trampled my father. At night, a marijuana-dealing Upper Canada College friend and I would drive down to the foot of the budding CN Tower, the tallest concrete phallus in the world, rising foot by foot into the cosmos, and pretend to worship the round-the-clock erection of Canada's own secular Stonehenge. My irreverent friend joked that the syringelike telecommunications tower was "man's puny attempt to fuck God." (Within a few years, at age twenty-seven, he would be found dead in his bed, some believed of a drug overdose, the same age as our hero, Jimi Hendrix.) Other nights, I'd lie spread-eagled on the cricket field at UCC, alone on the grass, an insider-outsider staring up into the black eternity of outer space. The American space station, SkyLab, was orbiting the earth, a tiny dot of light, and I thought of the bland, bubble-headed astronauts performing airless experiments, stripping away all variables, sacrificing their imperfect humanity on the altar of absolute truth.

After nine months, I quit my newspaper job. I was itching to get the hell out of squeaky-clean Toronto, out of Canada, out of myself. To get down off the hill. To get dirty. To get real. My father was gone, a writeoff, drifting among the living dead, the drug-coated silence holding hostage some unspeakable violence. Jack FitzGerald was like a soldier in Stalin's Red Army: He had retreated and retreated and retreated, and like the pursuing Germans, we were freezing to death.

I was gone, too, never to return, or so I hoped. I set no time limit on my travels, imagining backpacking as far as Tierra del Fuego. For over a year, I drove and bicycled through the back roads of Nova Scotia, Yorkshire, and Wales, picking up manual jobs along the way, cycling through the Western world as my father had rotated through the madhouses of Toronto. As I skimmed silently over the ancient lawns of Trinity Hall, Cambridge, I could only vaguely intuit that this grand place was where my young father had struggled to live up to his father's great expectations. The fingers of the jagged island that adorned Mabel's kelly green tea towel beckoned me in my dreams; Ireland, after all, was a land kind to dreamers and artists. Here, the Celtic twilight held me in its thrall, and so I delayed my plans to push further afield into western Europe and beyond. I cycled through village after village, cutting through the sheets of mist, as if expecting a rake-thin, redheaded FitzGerald to spontaneously burst out of a thatched cottage, pour me a jar and magically explain the meaning of it all. I twisted through the Ring of Kerry, sleeping in hostels and haystacks, blithely unaware of the trail of my own bloodlines, oblivious of the ghosts of medieval FitzGeralds who trod the desolate, stony fields and Norman ruins. No matter that there was no such thing as "pure Irish," that we were all wandering mongrels, mere tenants of time, chasing the illusion of permanence, longing to belong. One day as I coasted down a steep hill, a fierce west wind blew off my hat—my father's crumpled, 1940s brown fedora—and I let it go, taking it as a sign. The same wind blew me back across the Atlantic; I hated to admit it, but for all my disdain for the phony Forest Hill air-kissers, I missed my middle-class comforts.

To my relief, when I got home I found my parents had at last separated, without dire consequence. I found a job as a mild-mannered reporter for the *Port Hope Evening Guide*, the smallest daily newspaper in Canada, an hour east of Toronto on the shores of Lake Ontario. At the time, the quaint Victorian town of ten thousand was enmeshed in a public-health scandal that was attracting international attention: Eldorado, the local uranium refinery (a partner in the making of the atom bombs that destroyed Hiroshima and Nagasaki), had for decades been dumping nuclear waste into the harbour and landfill sites, spewing radon gas that forced the closing of schools. I was not sure what underground magnetic force guided me to Port Hope, but my nine-month stint proved a happy one, due in no small part to a robust sexual relationship with a kind kindergarten teacher who lived, believe it or not, in a nearby village called Welcome. All the while, I didn't know that 150 years earlier, my grandfather's grandfather had landed in his immigrant ship in Port Hope's harbour, then cut bravely inland into the Upper Canadian wilderness, in quest of his own private Eldorado.

History—and especially family history—did not, of course, interest my father. By now, he had confined himself to a musty, cramped, high-rise apartment in midtown Toronto, married to his TV set; borne by her tough Scottish genes, my "phallic mother"or so the Freudians might have painted her—had rented a duplex and taken an alumni administrative job at the University of Toronto, carving out the professional independence she had thrown over as a young woman in deference to the preordained roles of wife and mother. The role reversal was complete.

On the day after Christmas 1976, my great-aunt Hazel died at age eighty-eight, having fallen into spells of dementia in a retirement home, spitting racist epithets at the black-skinned nurses. Near the end, she had told my mother, "I'm nothing but a toad on a rock." Obeying Hazel's instructions, my sister and my cousin Anna, Molly's daughter, hurled her ashes off the Burlington escarpment into Lake Ontario. Hazel had made only a single admonition: "Don't throw me downwind or I'll end up in the neighbours' vegetable patch." Two years earlier, when I was cycling through England and Ireland, her last surviving brother, my great-uncle Bill, the eccentric Thornhill recluse, had died, also at age eighty-eight. My father did not attend his funeral.

With the deaths of our mysterious great-uncle and our sweet-tempered great-aunt, I had no clue—callow, incurious twenty-six-year-old that I was—that I had let something invaluable slip away: the last living familial links to their brother, my grandfather Gerry, the big-shot doctor whose name no one chose to repeat.

FIVE

In Dreams

The struggle inside oneself is the only history that matters.

D.H. LAWRENCE

Back in Toronto, I spent a one-year stint as an assistant editor of a medical journal. Pathologically shy, I'd phone up busy doctors, many of whom could not spare the time to give me the time of day—a masochistic replay of my emotionally starved childhood. Sometimes, paralyzed by the sudden upsurge of a panic attack, I'd stare at the phone for hours, as if the promise of some toxic doom oozed from the mouthpiece.

By age twenty-nine, teetering on the threshold of a long-delayed maturity, I switched from journalism into educational book publishing, then back to journalism a decade later, restless and distracted in both worlds. My father had lost everything a man can lose—his job, his lab, his house, his wife, his friends, his dignity, his mind—and the pathetic image of him, languishing alone in a threadbare bachelor apartment, zoned on lithium and TV, haunted me even as I struggled to erase him from my thoughts. I felt no passion for focused career building, no desire to buy a house, start a family, or amass the material wealth and status that defined masculinity. I did not know it then, but I unconsciously equated being "distinguished" with being "extinguished." Notions of success and failure, winning and losing, pride and shame, doctors-as-patients,

patients-as-doctors—all seemed fused, confused, interchangeable, a surreal, topsy-turvy miasma, devoid of meaning. What kind of person did I want to be? Was it even possible to be a person? Sleepwalking through the winter streets of Toronto, I felt the ice-encrusted concrete shift capriciously beneath my feet, as if a single slab of sidewalk might suddenly drop open like the trap door of a gallows. In my trancelike state, I could not shake a nagging feeling of tentativeness; that I was a perpetual tenant, holding on, holding back the best—and worst—parts of myself, living a vaguely fraudulent "as if" life; as if I was neglecting some ill-defined, unfinished business. In the pockets of my nervous isolation, all I carried were my wits, my words, my mind—none of which I intended to lose.

By 1983, I had reached a crossroads in my life. I had just quit a troubled romance of two years when I met Ann—a lively, charismatic palliative-care nurse with straight, jet black hair that fell to her waist. On our third date, we slouched on a park bench on the amnesiac streets of downtown Toronto as a cloud of blue cigarette smoke ringed my darkening thoughts. It was an unforgivingly cold autumnal night and as I squeezed her hand, Ann looked me dead straight in the eyes and blurted: "I think you're depressed. I think you're suicidal."

She caught me off guard and I laughed nervously. I thought she was being melodramatic, a novice caregiver honing her diagnostic skills. Yet, decades on, I'm still grateful to Ann. I *was* depressed. I *was* flirting with thoughts of self-annihilation. And I was the last to know.

When our relationship collapsed a few weeks later, the loss pitched me into the roiling depths of psychotherapy—a prospect I had contemplated, but dodged, for years. But I had reached a crisis and thus an opportunity; long ago I had stowed my body, like a fur coat, in cold storage and it was high time to air it out. I was thirty-three.

As it turned out, I was lucky where my father wasn't. That fall of 1983, a friend of Irish-Catholic blood steered me to a benign, burly psychotherapist working out of a basement office in a three-storey house in the leafy Annex neighbourhood of Toronto. I paid a modest fee, which I viewed as a long-term investment in my health. Peter, the rock, was aptly named; I found him disarmingly honest and unsentimental, yet empathically and intuitively attuned, an astringent bullshit detector leavened by a subversive sense of humour. He was a complete stranger, yet strangely I sensed I could grow to trust him. Crucially, he was a layman, a former high school teacher trained by a community of peers, steeped in the humanistic traditions of literature, film, music, and art. After all, who better to speak of madness than Shakespeare?

In our first session, Peter latched onto my Irish surname, suggesting that the legacy of my ethnicity was something I should take seriously. Everything began with my identity—and my search for it; in this I was a typical English Canadian. In our face-to-face hourly sessions, he acted as a Virgilian guide to my sealed-off underworld, an emotional midwife restoring myself, piecemeal, to myself. While most of my upwardly mobile peers thrashed about the golf course or squash court, I plunged downward, sifting through the tangled vines of my psychic undergrowth. With Peter I encountered no labels or pills, no verbal psychosurgery, no fostering of dependence or kowtowing to victimhood—just the power of simple words massaging the hollow, cadaverous silence I carried deep inside me. When I learned that the word "psychotherapy" means "soul healing," I was suddenly reminded that I had such a thing as a soul. In stark contrast to the sad, outrageous example of my father, battered from without by needle, pill, and electrode, I would eventually realize it was humanly possible to change the system—the nervous system—from within.

In my sixth session, I brought in a dream that would prove uncannily prescient of my future searchings: A doctor in a white lab coat picks up a scalpel and makes a deep, vertical incision down the middle of my face, forehead to chin, releasing a violent, Niagara-like torrent of water. Dreams, I learned, do not yield to a single interpretation; yet this one seemed to symbolize the accumulated generations of untapped grief, of which I was the contemporary carrier. And it hinted at my psychic split, the precise, ultra-rational, surgical hand of schizoid Western civilization that cut my head in half at the bloody moment of my birth: an agreeable, polite, deferential Canadian persona masking my wild animal nature, my exiled Irish madman.

Month after month, dreams spilled onto my pillow like burnt offerings, subtle, furious, cryptic, cosmic, comic, cathartic. Through musiclike leaps of words, feelings, images, and associations, I connected patterns and formed meanings; in the molehills of my psyche I hid my mountains. Emotions erupted with sudden force, the soothing after-rush of endorphins dabbing the dark holes of half-numb pain. Soon I realized I was unwittingly mapping the narrow habits of my past on the tough-minded but compassionate human screen sitting and listening—always listening—but five feet away; it is called "transference," a universal psychological phenomenon that Freud called "of undreamt of importance." Parked in my leather armchair, I had turned a very accessible Peter into my remote father, among others in my past; I began to understand how the hands of history had so indelibly formed me, like a solitary plaster statuette of a boy set on a mantelpiece. Yet if I could work through the transference, nothing was preordained, nothing written in stone; the clay statue was free to move.

It was in this basement sanctuary that I was jolted into an epiphanic discovery: I actually possessed an unconscious mind. Many compare the electric feeling of revelation to Columbus and his first glimpse of the dark, unexplored continent, and such was my own experience. No longer just an abstract concept I read about in books, the unconscious was a timeless, oceanic cauldron of desire and terror, love and hate, sex and creativity, where courage was a necessary prerequisite to entry. Out in the "real world," the intellect, or a debased form of it, was valued above all else, indeed at the expense of all else. Here, I was learning how to understand my intellectual defences, how to tolerate the essential ambiguities of life, how to respect the stored, buried power of childhood experience, how to listen to the stirrings of my flesh and blood. Here I was learning that my heart was as smart, or smarter, than my brain. And even as I did, I realized, incredulously at first, that deeply hidden forces animated my motives and actions; that a separate intelligence, an Oz-like man-behind-the-curtain, was calling the shots. Like a protoplasmic puppet-master, the "Un," as I came to call it, pulled the strings; but now, at last, I could pull back.

Soon I joined a group, run by Peter and his wife, composed of fifteen or so men and women not unlike myself—teachers, journalists, truck drivers, actors—and there, in our Aboriginal-like tribal circle, the transferences multiplied and the education deepened. At first, my terror of the group threw me back into a frozen silence that uncannily recapitulated the isolation of my early childhood. Then, in one memorable session, a charismatic, blue-eyed young woman reported a dream that ended with the oracular line, "We weep between the museum and home," and somehow I sensed I was in the right place. Gradually, with help, I started to find the buried voice inside me, an experience subtle yet radically transformative, profound—ironically beyond words. Along the way, I learned for certain that all is uncertain: There are no guarantees, no panaceas, no cures. How can anyone be "cured" of the mystery of the human condition? I was simply one of the lucky ones who learned, through disciplined self-scrutiny, how to live in relative accord with the ceaseless riot of my inner life. More than I knew, I had been taught to aspire to the superhuman toughness of my Victorian grandfathers, to shun help, and to isolate myself from others, to ruthlessly kill off the "weakness" inside. Yet here I was, absorbing, inch by inch, a stunning paradox—I could draw strength from my own sensitivity and vulnerability. For me, psychotherapy was ultimately a spiritual quest—an art, not a science; and, yes, the truth, stranger than any fiction, held the power to set me free. The questions would always outrace the answers, but what is an answer but a kind of death?

I realized that my father hadn't suffered from an overdose of Freud, but from no Freud at all. The founder of the talking cure had predicted that the medical profession—"the man with the syringe"—would blunt the radical edge of his psychological discoveries, and time proved him right. Challenging the dogmatic model of biological determinism, Freud insisted that most troubles of the mind could not be traced to diseases of the brain; for Freud, a doctor, a knowledge of anatomy and physiology was next to useless for an effective therapy. Therefore educated, gifted laymen steeped in the humanities generally made for the best emotional healers—so long as they pursued a rigorous therapy of their own.

My own experience of therapy ran against the grain of mainstream notions of "mental illness," an unfortunate phrase that assumed only doctors were qualified to "cure" it—usually with invasive physical treatments. The medical model—the idea that human beings who were suffering profound sadness, grief, and anguish could be categorically reduced to spongy brainpans upset by a "biochemical imbalance"—struck me, as a teenager, as an example of dangerously "unbalanced" thinking. Chaotic relationships, emotional trauma, and spiritual crisis could scramble one's body chemistry—just as health could be restored by the chemistry of words. Granted, not everyone is fitted for psychotherapy, but nor is everyone suited for a chemical straitjacket.

Peter and I worked light years away from Freud's dusty Viennese couch, yet we owed everything to his legacy and the creative, fractious, revisionist discourse of his antecedents. Freud bequeathed us the foundation of a healing vision of profound subtlety that, a century on, Western culture still struggles to grasp, expand, and reinvent. He established beyond doubt the timeless axiom that human beings are wired to deny and repress and dissociate from disturbing truths (or displace them onto convenient scapegoats). Nothing disturbs the unimpeachable authority of biological psychiatry more than the suggestion that certain human truths will forever elude the objectifying, stainless-steel instruments of science. As long as I laboured in the trenches of honest talk and sustained listening, as long as I struggled to shoulder my own disturbing truths, I knew I was not doomed to re-enact a family script of self-destruction. In the steady presence of an alert, attentive, empathic human being, I had a fighting chance.

═

Within a month of entering therapy, my dramatic knife dream thrust itself into my waking life—not with a flood of water, but blood. For Christmas night of 1983, our annual festival of Keeping-Up-Appearances, we visited Uncle Tom's winterized cottage on Lake Simcoe where he had retired. Brain-addled by years of alcohol abuse, Aunt Molly was wheeled out of Homewood—the same private sanitarium where my father had spent time—for the occasion, even though she and Tom had divorced back in 1976. My parents had been separated just as long, but neither chose to push through a divorce, as if the invisible beast of neurotic guilt forbade it. My father had always hated Christmases and birthdays and the ritual exchange of gifts, for such occasions implied the giving and receiving of tender feelings; yet, against my strong protests, the family insisted on rousting him out of his Scrooge-like cave, even though everyone tacitly knew he could barely tolerate the presence of another human being—especially family.

Tom's cottage was not big enough to accommodate both sides of the family, so we five checked into a nearby hotel before dinner; Mike and I shared a room with Jack while Shelagh stayed next door with Janet. I survived the enforced intimacy of the Christmas dinner table relatively unscathed, but predictably my father did not: When we returned to the hotel before midnight, he was wildly stirred up. All night long, he scurried back and forth between his bed and the bathroom, wolfing down pill after pill, finally falling into a psychotic state, clawing at the curtains of the window in terror. Nothing I said had any effect. At dawn on Boxing Day, my sleepless brother and I drove him back to Toronto and the hospital emergency room; over the frigid hour-long trip—the car's heater was symbolically broken—no one breathed a word.

A week later, on New Year's Day 1984, I fainted on the bathroom floor of my downtown apartment, a stream of dark crimson blood gushing out of my guts. Coming to, I found the strength to crawl from the toilet to the phone and dial 911; an ambulance carried me to the nearby Wellesley Hospital where I was revived by transfusion. The gastroenterologist was baffled as to why I bled so violently, so profusely, as the cyclops eye of his sigmoidoscope did not reveal even a benign polyp in my bowels. When I tried to link my condition to my galvanizing sessions with Peter, combined with the trauma of the family Christmas, he looked at me like I was crazy. His impassive face said it all: What, pray tell, did human emotions have to do with medical science?

But I clung fast to my own truth: My body was, at last, bellowing the bloody scream I had suppressed from birth.

SIX

Let Sleeping Dogs Lie

> For an Irishman, history is nothing if not the performance of myth, and this is perhaps what separates the Irish from the rest of the world, for in our search for a transcendent reality, it is very difficult for us to understand that we're not living in a novel, a legend, or a myth.
>
> WILLIAM IRWIN THOMPSON
> *The Celtic Consciousness*

By the summer of 1991, I was feeling brave enough, at the ripening age of forty, to stare down my Irish roots. With over seven years of healing talk under my belt, and more to come, I was warming to the idea of drilling down into the blood and guts of my ancestral past, then scaling my way back up to the sunlit present.

Sixteen years had passed since my first trip to Ireland. Back in 1975, as I cycled aimlessly through the countryside, I might as well have worn a blindfold; this time, I took a map—a frayed, handwritten genealogical chart that my mother had saved, cobbled together in the 1930s by one of my grandfather's two younger brothers. My great-uncle Sidney, an unmarried lawyer with an office on Vaughan Road, had died in 1942 in his fifties. As I studied his photos, I saw an odd-looking soul with overprominent, Ichabod Crane–like ears and nose; I sensed from my father's blanketing silence that Sid, too, had been chased down by his own headless horseman and come to a dark end.

In my youth, the hobby of genealogy had struck me as a tedious and pretentious exercise dominated by nerdy, anal-retentive, status-seeking social climbers. Yet as I pored over the dead branches of my own family tree, I realized they held great psychological riches: Here lay embedded the fragments of self-knowledge, the secrets of the childhood of one's race. I felt as if an invisible hand had cut our Irish bloodlines with a knife and now I was a surgeon stitching up the wound and transfusing life back into our dying family.

The jottings of my dead great-uncle guided me to the north-eastern Irish province of Ulster—abutting, aptly enough, the desolate vistas of Connaught, the name of my grandfather's laboratory. Renting a car in Belfast, I drove south, passing nervously through a border checkpoint guarded by graffiti-splattered concrete bunkers, rusted barbed wire, and machine gun–slinging British soldiers. Winding my way over placid green hills that seem to camouflage ancient, subterranean currents of bad blood, I arrived in Clones, County Monaghan, a modest market town of 2,500 just one mile south of the border. With the severing of northern and southern Ireland in 1922, Monaghan was one of three largely Protestant counties in Ulster gerrymandered by the newly formed, Catholic-dominated Irish Republic; Clones was now eighty percent Catholic. Since the renewal of "The Troubles" in the late 1960s, the town had been harbouring masked IRA men who launched hit-and-run guerrilla raids into the North. My ancestral village, it turned out, was perched on the strategic edge of "bandit country."

Checking into a hotel overlooking the town square, I sat in the musty bar and chatted up the locals. When I told a condensed version of my grandfather's story to a cynical, black-haired man sipping a pint of Smithwicks, he cautioned me in a harsh northern dialect: "Ah, Seamus, what ya doin' pokin' about the bone-yard? Let sleepin' dogs lie." Another genial soul proved more hospitable. Learning that I was born a Protestant, he pointed to

St. Tighernach's Anglican Church looming across the cobblestones, its spire stabbing the infinite sky like a dagger. Wandering among the churchyard's cracked headstones, I felt warmed by a pleasing melancholy. It was then I found what I was looking for: generations of FitzGeralds, dating back to the 1700s. As I moved contemplatively from stone to stone, jotting down names and dates, I stopped dead in my tracks: A mouldy grave bore my own name—James FitzGerald.

No one in my family had ever mentioned this deep green Petri dish of FitzGerald DNA. I had read that the male Y chromosome is passed on, virtually unchanged, from father to son, generation to generation; that it is boys, therefore, who are most likely to inherit their ancestral traits. Curiosity seemed to be an enduring part of my genetic inheritance, yet it felt as if insidious, invisible forces had been preventing me from discovering my own secret identity. It felt forbidden to know my unknown grandfather Gerry and his achievements, as if I was being forbidden to know myself. And, like it or not, part of myself—my selves—was my repressed Irishness.

I discovered that the surname FitzGerald is Norman French, derived from the first name of the male parent; the prefix "fitz" meant "son of," from the French word *fils*. The name "Gerald" meant "bastard son"; I allowed myself a wry smile when I realized that what seemed like a classy handle, in fact, translated as "son of a bastard son." Another meaning of the name was "spear ruler"—a phallic, piercing image that would resonate in my ongoing exhumations of my grandfather's story.

All FitzGeralds in Ireland were believed to be descended from Maurice FitzGerald of Wales, who accompanied the warlord Strongbow on the first Norman invasion of Ireland in August 1169. A seasoned race of disciplined warriors, the Normans conquered half of the original Gaelic kingdoms. As a reward for his military service, Maurice FitzGerald was made governor of all Ireland, a gigantic fiefdom chiefly composed of the counties of Kildare, Wexford, Limerick, and Kerry. The FitzGeralds built a network of glowering stone castles, monasteries, and fortified towns and gradually intermarried with the wild, green-eyed, red-haired daughters of barefoot Gaelic tribal chieftains, becoming famously "more Irish than the Irish." An early legend told of how the FitzGerald children in Maynooth Castle outside Dublin were abandoned by their parents when the castle was engulfed by fire; when a pet monkey ran inside and rescued the children, its image was impressed on the family coat of arms. It is a fable I cannot fail to identify with.

The Normans were the first to impose order on chaos "beyond the pale" of Dublin, giving the unruly island a centralized administration. As dictated by their high-minded code of chivalry, they deemed cowardice the one great unpardonable sin. In 1261, an arm of the Desmond FitzGeralds founded the windswept, southwest coastal town of Tralee and spawned the Knights of Kerry, who exist to this day. Designed to separate out claims of inheritance by the sons borne illegitimately by the alluring wives and daughters of Gaelic chieftains, the hereditary knighthood is only one of three ever recognized in Ireland or Great Britain. While the English kings waged endless wars in Europe, the Catholic FitzGeralds—the "Geraldines"—were free to rule Ireland. But when the Tudors broke with the Catholic Church in Rome in the mid-sixteenth century and tried to impose their sovereignty on the untamed island to the west, the Desmond FitzGeralds rebelled, severing their traditional Norman allegiance to England with the family *cri de cœur*, "Crom-a-boo"—Dissolve Forever.

In the ensuing Desmond wars, the lush Irish countryside was laid waste. In August 1580, Sir James FitzGerald, a brother of the Earl of Desmond, was captured by the English; on the command of Captain Walter Raleigh, my ancestor was hung, drawn and quartered, and his body parts spread above the gate of the city of Cork to rot for weeks as a moral lesson to all who passed under. The headless corpse of another brother, John FitzGerald, was hung upside down for three years over the River Lee until it decomposed into a skeleton; Raleigh sent the severed head to the Virgin Queen as a "goodly gift to Her Highnesse." The fifteenth Earl of Desmond himself, Gerald FitzGerald, took refuge from his Elizabethan pursuers in Desmond's Cave in the hills outside Cork, only to be hunted down and killed in 1583. In the end, the wild, irrepressible Desmond FitzGeralds were crushed as a great political power and half a million acres of their land in Kerry, Cork, and Waterford was awarded to Queen Elizabeth's Protestant officers.

My own father's middle name was Desmond—my great-aunt Hazel liked to call him "Dinny," an Irish nickname—yet he never spoke of our connection to this gruesomely fascinating strand of Irish history. Then, on my twenty-first birthday—not long after his last, near-fatal suicide attempt—he emerged from his own twentieth-century Desmond's Cave (the TV room) long enough to give me a gold signet ring imprinted with the coat of arms of the Knights of Kerry, depicting a sword-wielding knight astride a rearing horse. He bequeathed this totemic object without a word of explanation; or if he did, it didn't sink in.

The more history I mined, the more I was puzzled by my father's intransigence—and the more determined I became to dig deeper. I learned that another ancestor, John FitzGerald, held an officer's commission in the French and Irish Catholic army of King James II, who attempted to restore Catholicism to England. When James was defeated by the Protestant champion, Prince William of Orange, at the Battle of the Boyne on July 12, 1690, John FitzGerald fled from persecution. Because of the victors' imposition of the severe Penal Laws designed to disenfranchise Irish Catholics—among other things, they were forbidden to own land, carry a sword, vote, hold a profession or public office—John FitzGerald became a Protestant.

His son, James FitzGerald, settled in Clones, County Monaghan, where in the 1730s he built Clonavilla, a three-storey stone house encircled by two hundred acres of fertile farmland. Composed of six bedrooms, sandstone flagstones, a grey slate roof and a trio of chimneys, the gentrified Georgian residence, encased in heavy, off-white limestone plaster, would be occupied by FitzGeralds for nearly two and a half centuries. James struck up a friendship with Jonathan Swift, dean of the Protestant St. Patrick's Cathedral in Dublin and author of *Gulliver's Travels*. When Swift died in 1745, he bequeathed the bulk of his estate to found St. Patrick's, the first asylum in Ireland—a humane, progressive idea in a time when lunatics were publicly displayed and ridiculed like circus freaks. Ever the black humourist, Swift appended a poem to his will:

"He gave the little wealth he had
To build a house for fools and mad;
And showed by one satiric touch,
No nation wanted it so much."

As I made inquiries in the Clones hotel bar, I was disappointed to learn Clonavilla House no longer existed, nor any of the last occupants. Kindly escorting me a half-mile outside the town square to a sprawling property, a local man explained that the original house had been owned by ten successive generations of eldest FitzGerald sons, as dictated by laws of primogeniture. The FitzGeralds had once prospered as a strand of the Anglo-Irish Protestant Ascendancy, their class privileges immunizing them against the Great Famine of 1845–50—the "Irish Holocaust"—when a deadly wind-and insect-borne fungus poisoned the staple potato crop and killed one million Irish peasants by starvation or disease. Within a decade, nearly two million panic-stricken Irish risked hazardous escape to Australia and the Americas; together, starvation and emigration nearly halved

Clonavilla House, the eighteenth-century FitzGerald ancestral home in Clones, County Monaghan, Ireland, destroyed by fire in 1974

the island's population from eight to five million. The average life expectancy of a pre–Civil War black southern slave was thirty-six years; an Irish peasant, nineteen years. Irish graves were ruefully dubbed "the Protestant lease."

For two hundred years, the privileged minority of Protestant Anglo-Irish led a divided existence, proudly claiming themselves Irish yet yearning to be treated as equals by the imperious English. Buffeted by the upheavals of modernity, they gradually withered into oblivion, supplanted by a pragmatic, hard-working middle class; Clonavilla House stood as a microcosm of the Protestant Ascendancy in descent. Neglected by its alcoholic patriarch, William FitzGerald, the house slid into debt and decay in the late nineteenth century. Over the ensuing generation, Clonavilla was occupied by three aging sons of William, all sworn bachelors, who upheld the pretensions of the gentry, addressing one another as "master," even as they lived in collective lassitude and pigheadedness, indulging in persecutory delusions of grandeur, daydreaming of the mythic medieval past when the FitzGeralds ruled as uncrowned kings of Ireland. All the same, the family had remained friendly with the Catholic owners of smaller neighbouring farms, untroubled by the political mayhem of the 1916–1924 period. Unlike countless Protestant houses, Clonavilla was spared the rebel torch.

The FitzGerald brothers let the once-elegant Clonavilla House slowly rot, far exceeding the acceptable limits of shabby gentility; their personal hygiene, nutrition and morals, it was hinted, declined in kind. Echoing my grandfather's brothers, the Irish bachelors died off one by one, childless and heirless. The last and youngest, Harford FitzGerald, an eighth generation descendant of the founder of Clonavilla, stubbornly refused to pass the house on to his nephew, the only existing heir. Instead, in the early 1970s, he sold the place for a pittance to Hugh Tunney, a nouveau riche cattle baron who pugnaciously bragged of his kinship with the American heavyweight boxing champion, Gene Tunney. One day in the fall of 1974, Tunney set the house ablaze, laughing heartlessly as old Harford watched from the roadside, dumbstruck. Covered in the *Sunday Independent*, the cruel event grew into a local legend. Harford died a sad and embittered man in 1980 at age eighty-seven, the last of my Irish Mohicans, my final, tenuous link to an antedeluvian past that lingered only in my imagination.

The night before my return to Canada, I dreamt that my brother, Michael, and I were standing over a sewer grate under a stone bridge at nightfall. My gold signet ring bearing the coat of arms of the Knights of Kerry slipped from my finger and fell through the grate. I heard the ring splash into a pool of raw, stinking sewage. We stood there, paralyzed, in the darkness. Was I willing to abase myself—fall to my knees, heave up the heavy iron grate, then thrust my arm deep into the toxic, germ-infested muck—to retrieve the noble ring? Was my brother? Then I woke up.

The next morning, I took one last walk out to where Clonavilla once stood. I thought about the dream, wondering if my unconscious was making a pun on the word "grate." Like my overreaching father and grandfather before me, was I prepared to risk my own soul, Faust-like, on a chance to grasp a fleeting whiff of "grate-ness"? Was I willing to flounder in the dangerous filth of my own unconscious in search of the ghost of my grandfather? As I brooded, I kicked idly at Clonavilla's foundation stones buried in the mud, as if a cloud of crazy spirits might magically fly out, releasing fragments of acerbic Swiftian conversation. Walking back to my car, I turned to take one last look, and muttered: "It seems I have arrived a generation too late."

On the razed site of our lost, 240-year-old ancestral home, Hugh Tunney was planning to build a slaughterhouse.

SEVEN

I'm My Own Grandpa

I am passing out. O bitter ending! I'll slip away before they're up. They'll never see. Nor know. Nor miss me. And it's old and old it's sad and old it's sad and weary I go back to you my cold father, my cold mad father, my cold mad feary father . . .

JAMES JOYCE
Finnegans Wake

As I gazed out the window of the Air Canada jet heading home, my mind flitted over eight hundred years of Irish history, both national and familial, mad scenes of savage violence rising and falling like jagged spikes on a manic-depressive medical chart. The fiery destruction of the three-storey Clonavilla House, it dawned on me, had eerily coincided with my father's inexorable spiral into madness, a continent away across the Atlantic. I had put away my memory of his terrifying plunges into suicidal psychosis—and the subsequent loss of his house, career, and marriage—as I might shed an ill-fitting suit of clothes. But now I had to admit that my father's fall, and the fathers before him, had marked me more than I cared to know. In the slow passage of my adult awakening, I realized that in my grandfather's and my father's houses there were many mansions, casting insinuating shadows of dim racial memory. I was the third generation, the third story, as yet untold, and even as I chased spectral figures across the Irish Sea, I felt them chasing me.

═

Back in Toronto, I lost no time pursuing the next leg of my journey—retracing the torturous emigration of our branch of the FitzGeralds to the inhospitable British colony of Upper Canada. Again relying on my great-uncle Sidney's chart, I drove ninety miles northeast of Toronto to a picturesque rural town, where I picked up the buried storyline of my Irish forefathers. As I explored the leafy, undulating hills of Millbrook, Cavan Township, a thriving farming community not much larger than Clones, déjà vu enveloped me. The tiny Ontario hamlets surrounding Millbrook—Cavan, Monaghan, Bailieboro—had been named after towns and counties in Ulster by the newly arriving Irish Protestant immigrants because of their striking resemblance to home. The settlers simply picked up where they left off, transplanting their industrious, Old World rural culture—and their sectarian violence—to the densely forested northern shores of Lake Ontario.

A chain of inquiries led me to an unknown second cousin who serendipitously possessed a tattered, leather diary that the FitzGeralds carried on their emigration ship from Ireland. Turning its delicate pages, I learned that John FitzGerald—the forty-five-year old grandson of the builder of Clonavilla House who had befriended Jonathan Swift—emigrated to Canada with his wife, Elizabeth, and eight children on April 24, 1824. When the ship sailed from Dublin, the fourth of the eight FitzGerald children, John, had just turned thirteen. That boy—squinting skyward at the unfurling, windblown canvas sails, his body brimming with God knows what kinds of primal hopes and terrors—was my grandfather's grandfather.

For thirty days and nights, the crowded vessel braved the open seas, then tacked down the vast mouth of the St. Lawrence River, a pulsating, ice-cold artery plunging like a knife deep into the heart of the continent. Thirty miles before Quebec City, they passed Grosse-Île, a barren, windswept island of rock where for decades waves of Irish immigrants were quarantined against murderous epidemics of cholera and typhus. Over fifteen thousand victims of the disease-infested "coffin ships" would perish there.

In the livid heat of the North American midsummer, the FitzGeralds drew into a frontier town of two thousand on the northwestern rim of Lake Ontario. Appropriately, its name was Port Hope. Whatever thoughts assaulted the brain of my thirteen-year-old great-great-grandfather, his descendants can only dream. Trudging inland on foot, the Irish families followed a primitive concession road that gradually petered out into the unforgiving wilderness. Guided by the blazes freshly chopped into pine trees by an advance survey crew, they were compelled to break new trails to reach their destination thirteen miles north. Throwing together crude log cabins to withstand the onslaught of the savage Canadian winter, the settlers' life of exile proved unimaginably brutal. But most were stubborn tenant farmers who were prepared to make the superhuman sacrifice needed to hack out of the forest a civilized, egalitarian future for their Canadian-born offspring.

Young John toiled on the family's hundred-acre farm, then became a storekeeper and fathered seven children. Frontier optimism was undermined not only by killer diseases but intense anti-Irish prejudice. The mad and poor were seen as authors of their own misery; businesses posted signs declaring: "No Irish Need Apply." In Toronto, ninety miles away, locked out of jobs by the militant Loyal Orange Lodge, the outnumbered Catholic Irish had it worse than their Protestant counterparts. Poor diet, contaminated water, filthy habits of personal hygiene, and reckless waste disposal—the city was dotted with stinking public privies—fuelled the spread of disease and death in Irish ghettos. Transiency, chronic drunkenness, and lack of privacy led to rampant promiscuity and interbreeding. Genetic anomalies such as dwarfism, often called "leprechaunism," bolstered the perception of the Irish as a subhuman species. Many such children were quietly killed and buried in the cabbage patches of Cabbagetown, one of Toronto's largest Irish enclaves. Irishness itself became synonymous with disease.

In 1841, as the Irish swamped the financial resources of Toronto's few social-welfare agencies, the city's overflowing county jail was converted into a catch-all temporary lunatic asylum to accommodate the swelling tide of disenfranchised. As recorded on one frigid January day in 1842, over two-thirds of its inmates were Irish natives—a staggering disproportion compared to other ethnic groups. And if some weren't fully given over to lunacy when they first entered the asylum, they would be soon enough.

The mad Irish were typically abandoned by their families; countless neglected inmates died alone in the fetid jails and asylums of cholera, typhus, tuberculosis, pneumonia, and other deadly diseases. Others, off-loaded from rows of famine ships, never set foot in the city, perishing in groaning fever sheds at the foot of Bathurst Street. In the summer of 1847 alone, at the height of the famine, 38,000 sick and starving Irish landed in Toronto, overwhelming a town of only 20,000; by the fall, over 1,100 Irish had perished, their corpses heaved into a mass grave beside St. Paul's Church on Queen Street—built by my stolid Scottish great-great-great-maternal-grandfather, John Ewart—and covered with shovelfuls of lime.

I was dismayed to discover that the FitzGeralds embraced the local Loyal Orange Lodge, a virulently anti-Catholic organization that I regarded, in my own postmodern mind, as a kissing cousin of the Ku Klux Klan; and yet it was part of my legacy. As much a social organization as a religious one, the lodge was originally designed to offer fraternity and continuity in a frigid land of dislocation and isolation, dispensing relief for "the fatherless, the widow, the orphan, and the distressed." Yet the village of my ancestors went a step further, spawning a radical arm of the Orange Lodge that achieved local legend. A semi-secret, hierarchical society given to esoteric passwords, signs, and rituals, the Cavan Blazers were a belligerent gang of young toughs acting as unofficial Protestant militia and political muscle for Tory election campaigns. Over the generations, Orange Protestants tightened their economic and political control over Upper Canadian society. Tolerant in principle, many white Anglo-Saxon Protestants retreated into a collective sensibility of tight-lipped parochialism. In the ceaseless struggle to carve out a happy haven in a heartless world, to conquer the natural world even as they conquered their own natural feelings, an intangible price was paid.

Perhaps all this explains why, when I finally found the gravestones of my great-great-grandparents in an Anglican churchyard outside Millbrook, I experienced a strange elation, an unexpected poignancy. It was as if I had at last discovered the very thing that my father, in his brooding silence, his self-neglect, his family-forgetting, implicitly forbade me from ever knowing. Here, on a grassy hill affording a panoramic view of the countryside, I ran my fingers over the fading words carved on the flat, cracked headstones. My great-great-grandfather John had died on August 30, 1856, in his forty-fifth year, a long life by pioneer standards. A biblical inscription revealed the sole clue to his character—"Mark the perfect man and behold the upright, for the end of that man is peace." As I read, the words seemed to resonate with subtle Freudian sexual imagery. For it was here, bent over the family plot, that I was left with the realization that although I suffered from a paucity of hard facts, I always had poetry to fall back on. Here stood the womb matrix of my own inherited rigidities and fluidities, my paranoid Orangeman locked in a war of nerves with my stoned jazzman, each vying for the high ground inside me.

The headstone bequeathed me one more clue: John's seventh and last child, William FitzGerald, my great-grandfather, was born in Cavan on August 3, 1856—precisely twenty-seven days before his father's death. John had less than a month to cradle his new son in his arms and gaze into his eyes. That meant the infant William grew up fatherless—no small handicap to bear in a cold, tough land.

In my mental free associations, I then realized that William had been born in the same year, 1856, as a Jewish Viennese boy named Sigmund Freud, whose subversive ideas about fathers and sons would eventually roll westward across the Atlantic not unlike, in his own ironic words, the plague. Later, the father of "the talking cure" would darkly declare that the Irish were one race of people for whom psychoanalysis was of no use whatsoever.

═

On the main street of Millbrook, my great-grandfather William—presumably named after the victor at the Battle of the Boyne—worked as an teenaged apprentice in an apothecary shop. His nickname, Will, was apt, for his trade demanded dogged willpower. In the 1870s, the primitive, laissez-faire pharmacy industry was struggling to professionalize itself across North America. Any layman could set himself up as a druggist; armies of charlatans, quacks, and snake oil salesmen brazenly hawked dubious panaceas to a gullible public. In 1879, when the twenty-two-year-old Will took his licensing exams with the newly formed Ontario College of Pharmacy, he finished fourth in a class of sixty with a ninety-six percent average; as I leafed through the sole personal possession surviving from that generation of the family—a crumbling 1869 copy of *Gray's Lessons In Botany and Vegetable Physiology*, inscribed with Will's flowery script—I sensed that facts, books, and education filled, for a time, the vacuum left by his dead father.

My great-grandfather's profession demanded not only manual dexterity—the ceaseless pounding of the crude drugs with mortar and pestle, making large quantities of stock materials for the creation of powders, pills, cachets, and suppositories—but mental discipline. As Will dusted the long rows of bottles lining the wooden shelves, he memorized the Latin terms on the labels and the origin, use, properties, and doses of each of the dozens of drugs. He learned the metric system of weights and measures. He used ingenious machines for cutting pills, then rounding and coating them; for softening corks and pushing corks into bottles; and for measuring powders and for folding powder papers. He made pills by slightly moistening the raw material and rolling it into a long, thin pipe. He cut the pipe into pieces of uniform size and rolled the pieces into balls, which could then be coated with a variety of substances to make them more palatable.

All in all, it was fastidious, exacting work. In a land where illness flourished and only the fittest survived, he carved out a living making a pill for every ill; like the young country itself, a full quarter of it peopled with Irish-born, Will FitzGerald's future was cut out for him.

═

Following my 1991 pilgrimages to Clones and Millbrook, I visited my aging, reclusive father who was still living in the sparsely furnished, one-bedroom rented apartment in a mid-Toronto high-rise where he had retreated in 1975. Twenty years had passed since he had given me the ancestral ring of the Knights of Kerry of the Irish Middle Ages, but only now, in my own Canadian middle age, had the import of it caught up with me.

Since his near-fatal suicide attempts, the loss of his once-thriving medical career, and separation from my mother, my seventy-four-year-old father had regressed into a cracked, brown leather La-Z-Boy armchair—identical to the one I occupied in my weekly meetings with Peter—and "transferred" his slack-jawed, rheumy-eyed lithium-gaze to the TV screen. Blind in one eye and deaf in one ear, his health stripped by years of drug abuse, Jack had long refused to receive old friends; he could tolerate my presence for an hour at most, ironically the standard length of a therapeutic session. I had more than enough cause to write him off, but I hadn't; perhaps I sensed I was destined to write *about* him.

I had become adept at tactfully circling his tough yet delicate psyche, trying to pry loose the tiniest breadcrumb of family lore. But my father had remained immune to my overtures, the psychic cordon sanitaire holding fast. It was if he had withdrawn inside the dungeon of a crumbling Norman castle, pulled up the drawbridge, and condemned himself to the netherworld of solitary confinement.

As I reported my exciting discoveries in Ireland and rural Ontario, I was not surprised to meet a shrug of indifference. Only when I showed him the photographs of my genealogical trip to the Clones boneyard was I able to briefly pry open the vault of his memory. He revealed that in the summer of 1935, when he was eighteen, he had vacationed in Ireland with his father, mother, and sister. The family met their Irish cousins at Clonavilla House, including Harford and his bachelor brothers, the last in the long line.

Jack the jazz aficionado

"I remember our car broke down on a country road," he said in a flat tone, "and a Catholic priest on a bicycle came to our rescue." The image stuck like an oil painting in my mind, but he did not elaborate. Slowly I felt a burning rising in my throat, for my thoughts were flinging me back to 1975 when, as a young man, I had cycled alone and aimless through the rainy Irish countryside. Like a satellite orbiting a dead planet, I had been oblivious of any specific familial tie to the island, or to the ancient stone house torched, only months before, by the cruel cattle baron. It galled me that my father was now revealing that for all these years, he had known about our ancestral home—indeed experienced the place first-hand—yet he had chosen not to tell me. Why, I wondered, had he baptized me with the name of James? Surely because he knew that the family had passed down both our Christian names—John and James—for centuries? For most of my life, my father's embrace of victimhood had deftly robbed me of my natural anger, yet if I finally showed it here and now, I might as well shoot an unarmed man.

Keeping my cool, I gently unfurled another question: "Did you know that your grandfather Will was born in 1856, just a month before his father died? And did you know that Will died in 1917, just months after you were born?"

He simply shook his head slightly, as if to say: "No—and what the hell is your point?"

Then the obvious hit me: I was born fatherless, too.

Now in my early forties, I had long since emotionally surpassed my classically "absent father" who—both before and after his breakdown—avoided all contact with his three children. Now here I was, acutely aware of myself as a compensating presence in his shabby apartment, fathering him, doctoring him, trying to spoon-feed him with small, incremental doses of healing, historical truth. In a weird way, I had become *his* father. If that was true, that meant *I had become my own grandfather*.

As I prepared to leave, I mentioned that I was planning a trip to his father's birthplace, a village northwest of Toronto. Had he ever been there? Again, another ambiguous shrug, no clear yes or no. Then, predictable as nightfall, he lapsed into that familiar deadpan expression that had marked me so indelibly. As if I did not exist.

═

Years earlier, my mother had led me to believe that my grandfather had died in Toronto General Hospital following a Nembutol overdose. Yet when I had recently checked the records, I smelled a rat: Gerry was admitted to TGH on June 16, 1940, yet he died on June 20. I was not surprised that the death certificate indicated he succumbed to a duodenal ulcer, as suicide was still a criminal offence; but it strained credulity that a hospital patient would perish of a drug overdose four days after the fact. Sensing that vital—and likely disturbing—details had been willfully suppressed, I took the subway north to Yonge and Eglinton and made yet another foray into my father's apartment. This time, I was determined to extract the "truth," even if it meant turning a thumbscrew.

Settling into a sofa beside him, I primed the pump with benign small talk—*How about those Blue Jays?*—as was my habit. Sure enough, as I edged towards graver matters, I felt our dialogue drying up fast. Taking a deep breath, I took the leap and asked point-blank: Do you know how your father killed himself?

His face twitched and he looked towards the window. He mumbled that he was a medical student at the time and living at his parents' house at 18 Prince Arthur Avenue—the family was temporarily renting the place to be nearer the university, while renting out Balmoral. He had come home that June day to find in a bedroom on the third floor an empty bottle of barbiturates in his father's dressing-gown pocket. I was on the verge of pressing for more details—Was it he or Edna who found him? Did he call the ambulance? Did it happen in his own bedroom?—but he cut me off. Or maybe I cut myself off. My father's health was declining steadily; due to a mild stroke a few years earlier, he was wobbly on his feet and vulnerable to falls. Perhaps that's why I let him—and myself—off the hook. All the same, when I left, I believed that I had gotten what I came for.

On a subsequent visit, he sarcastically dubbed me "the soothsayer." I took it as a compliment. For a long time, my father had been stubbornly resisting commitment to an institution. During these last, difficult months, he had spent over forty thousand dollars on round-the-clock private nursing care; he was burning up our inheritance—at the present pace, he would be broke within a year—but I had long passed the point of caring. Finally, family pressure prevailed and one day in late May, I wheeled him into the war veterans' wing of Sunnybrook Hospital. I knew he wouldn't last more than a week, and I was right.

By coincidence, that very week I signed a contract for my first book, *Old Boys: The Powerful Legacy of Upper Canada College*, an oral history of our all-boys private school. As with the war, as with his father, my father had rarely spoken of his eight years of boyhood exile, his boarding-house prison of impossible parental expectation, and I had long ago come to understand his silence. I had suspended my growing obsession with the enigma of my grandfather and I was now busy interviewing dozens of UCC alumni ranging in age from eighteen to ninety-eight. Unable to extract any "male truths" from my father, I was diverted into roiling side channels; if my father wouldn't talk to me, others would. Month by month, I was patiently listening to a steady flow of explicit and implicit father-son dramas unfold in all their candour and deception, bravery and intelligence, glamour and cynicism, arrogance and pain. My interview subjects included luminaries of the English Canadian elite—Robertson Davies, Conrad Black, Michael Ignatieff, Ted Rogers, David Thomson, John Eaton—together with a lesser-known cast of characters who were generally less scrupulous about protecting what the Victorians once called a "good name."

In ways I had not anticipated, this demanding task was bringing me bang up against the emotional realities of my own mixed, and still largely unexcavated, paternal bloodlines; I was learning how to become a good listener, something my father had never been. The irony was not lost on me when, as I told my wheelchair-bound father how excited I was at becoming a published author, he barely lifted his eyes from the sports section. My father would live forever as my worst interview.

In his dying days, he passed on his last words to each of his three children separately, as terse and semi-oracular as ever. He told my brother: "I guess I blew it as a father." To my sister, he urged, "Make money!" As I approached his hospital bed one evening, he reserved for me a more sarcastic parting shot: "Here comes Mr. Know-It-All." The remark stung, as I felt I knew next to nothing.

A few days later, he lost his power of speech and on the evening of May 30, 1992, my brother and I visited for the last time. Jack had turned seventy-five the previous day; his mother had died at the same age and I wondered if he was hanging on long enough to join her. We stood over his hospital bed, wordless, watching his unconscious body, curled up in the fetal position, heave with short, panting breaths. As we left, Mike said softly, "We're thinking of you, Dad." Even in a coma, he might have heard him. I said nothing.

When he died at seven the next morning—the time of day he most dreaded—he was alone. The doctor said he was finished off by a rare degenerative neurological condition called progressive supranuclear palsy, which, among other things, affected the nerves of the throat and tongue and thus was also known as "Mona Lisa Disease." The link to La Gioconda's enigmatic smile seemed right: Whatever thin-lipped secrets Jack knew, or ceased to know, he took with him.

Freud said it all for me when he claimed that the most important event in a man's life is the death of his father. My feelings—anguish, rage, grief, longing—thrashed around inside me for weeks, finally giving way to a kind of relief. Soon I was able to pluck my blessings from the wreckage: Jack had been honest by sparing us mutually false gestures of sentimentality. I had held up my end of the relationship, such as it was, and his death only strengthened my resolve to crack the psychological DNA code he had guarded as if his very life was at stake.

When my brother, sister, and I cleaned out his apartment, we divvied up his meagre possessions. I rolled up his mounds of loose change, reminiscent of the quarters I'd purloined as a child. On his mirrored highboy dresser, I found a small pair of black clothes brushes embossed with the initials "JGF," the only items belonging to his father that he had kept. I could only smile at the symbolism—the father habitually "brushing off" the son. My father left without telling me who I was; in doing so, he made me an archaeologist of silence.

═

On an overcast October afternoon over two years later, the family buried my father's ashes without benefit of clergy. The delay was due to the fact that there was no room beside his parents in the mausoleum in Mount Pleasant Cemetery. "Typical," I told Shelagh. "He wouldn't even tell us where he wanted to be planted." As we researched alternatives, I wondered: where were the remains of our great-grandparents? Weeks of sleuthing revealed we owned a family plot in Prospect Cemetery on St. Clair Avenue West, and sure enough, there was Will and Alice, together with their son Sidney, one of my two mysterious great-uncles. Prospect seemed a fit place not only for Jack but Molly as well, for she had died in 1990 after a decade languishing in Homewood and her ashes were also as yet uninterred. The same for their Aunt Hazel who tended them both during their Balmoral childhoods. Although her ashes had long since settled at the bottom of Lake Ontario, we decided to memorialize her name on the same stone as her niece and nephew. I even allowed myself the thought that perhaps, one day, Prospect might serve as the last rest for James, the restless prospector.

With dried, dull-orange leaves crumbling underfoot, I gave a short graveside eulogy before my mother, sister, brother, and cousins, Anna and Patrick. My two siblings and my two cousins had achieved impressive professional success in their lives and were decent human beings, to boot; never would I characterize any of us as victims. Still, I felt the need to speak openly of our legacy of melancholia and madness, suicide and silence, coexisting with our many countervailing strengths. Pushing through a trembling voice, I was surprised when tears spontaneously flowed; even more so when my brother broke the spell of our ancient family inhibitions and placed his hand on my shoulder, a simple, moving gesture I will never forget. By coincidence, I was holding in my hand the first copy of *Old Boys*, fresh off the press that very day. As I dropped my father's urn into the black dirt of the family plot, my blurred eyes scanned the line of a nineteenth-century Irish poem we had chosen for the stone: "Earth has no sorrow that heaven cannot heal."

Then I thought of the trenchant Irishman I had met in the Clones pub three years earlier, the one who had admonished me, the nosey Seamus, to "let sleeping dogs lie." In truth, I was hell bent on waking them up. For I felt a deadness inside me, weighing me down, as if the rusty chain-links of FitzGerald men, stretching back to the Dark Ages, were dumping buckets of flame retardant on the natural passions of my life. With my father gone, I now truly felt like a man alone, or at best a man-in-progress. I dimly sensed the only way out: I must bring my buried grandfather to life, make him as real as if he were walking beside me, close enough to feel his breath on my face. As if reviving a corpse with mouth-to-mouth resuscitation, I felt compelled to reimagine his life, his voice, his actions, creating on the blank page his story as vignette, as dream.

Then, in the climactic scene, I must flush his ghost from the darkness into the light. Then, and only then, would I be truly born into the world.

EIGHT

The Unpardonable Sin

He wasn't clever at all: he merely told
the unhappy Present to recite the Past
like a poetry lesson till sooner
or later it faltered at the time where
long ago the accusations had begun,
and suddenly knew by whom it had been judged . . .

W.H. AUDEN,
"In Memory of Sigmund Freud"

Mere days after my father's death in 1992, Duke Ellington's theme song, "Take The 'A' Train," burst from the speakers of an Irish pub where I was sharing a drink with my brother. The familiar, charging rhythms threw me into the mythological past, before my birth, back when my father was a happier man. A philosopher once said we are all artists in our dreams, and sure enough, that night I dreamt I was performing on stage in a rock concert as a member of Joe Cocker's Mad Dogs and Englishmen tour. I am pounding on the piano keys, playing "Cry Me a River"; Cocker is twitching spasmodically, his face contorted in wild passion, as if jolted by a bolt of lightning. But I can't hear the voices or the instruments—in the midst of communal ecstasy, nothing but silence engulfs me. I gaze into the audience, but the rows of faces are as still as a church. I strain to connect with my fellow musicians—including my brother and sister—but it's as if some invisible hand has cut the circuits. The feeling is familiar, one that occasionally intruded into my sexual life. Sometimes I felt no woman alive could handle my intensity; sometimes, as I struggled for penetration and release, my pent-up lust hit a fever pitch, then switched off with a knifelike finality.

I was now working full throttle on my interviews with UCC old boys. As I did, the imp of coincidence continued to swim through my family bloodlines, the ultimate meanings forever elusive. Shortly after my father's death, I interviewed Richard Howard, a retired headmaster, since deceased, of the UCC prep school. It proved a torturous experience, but I was impressed when he confessed details of a troubled childhood and a depressive collapse in his latter years; he certainly presented a picture of a man hounded by demons, his face as creased and conflicted as my own father's. As our conversation edged towards the darker recesses of the school's history—such as caning and bullying—Howard abruptly invited me to leave, citing his weak heart. Perhaps, I thought, he anticipated that I might broach the disturbing rumours I had been hearing in the course of my interviews.

After its publication in 1994, the book helped catalyze, in a way I never expected, revelations of endemic abuse, sexual and otherwise, of dozens of boys over many years. During the ensuing decade, three former teachers were convicted of sexual offences and the school was compelled to sell some of its art collection and real estate assets to settle a multi-million-dollar class action suit. My father had not lived to witness the controversy that the voices in *Old Boys* aroused. If he had, I'm sure he would have closed his ears, as the school itself initially tried to do. Only in retrospect did I realize I had been unwittingly acting as a self-appointed pathologist trolling with a divining rod for malignant tumours buried in the body politic of my *alma pater*. The darkest undercurrents of my school and family nudged the edges of the madhouse; I did not fully grasp it then, but I was cutting loose one set of baggage in order to free up the strength to shoulder another. Among other things, the experience deepened my embrace of the transformative power of words on the page and the uncanny forces of the unconscious mind.

During the investigations into the scandals, a *Globe and Mail* reporter exposed a disturbing fact: Richard Howard had willfully withheld his knowledge of a master's sexual assault when it was directly reported to him by the first of the victims in the late 1970s. Because Howard did not act, the felonies rolled on for years. As I thought back to our decade-old interview, I realized he had been living for years with the punishing consciousness of his own moral failing; that helped explain, in part, his subsequent breakdown, not to mention his urge to whisk me out the door. Then I remembered something else that gave me goosebumps: As I was leaving his apartment that day, Howard recalled that when he was a stuttering, lonely boy in the 1930s, his father had rented the FitzGerald home at 186 Balmoral from my globe-trotting grandparents. His family occupied the third floor.

Among everything else we were meant to know, we were meant to know this—a young Richard Howard had once slept and dreamt within the same emotional furnace room of our own family secrets, where for generations the voice of the distressed child must remain unheard.

═

On the heels of my father's death and the birth of *Old Boys*, I resolved to smash the family code of silence with all the energy I could muster; if it took me the rest of my life, so be it, for this was no quick fix. I was becoming a living embodiment of one of my cherished quotations: "The Irish don't know what they want in life—and they are prepared to fight to the death to get it."

My genealogical field trips to my ancestral villages had merely laid a foundation; I now started to systematically burrow through assorted medical archives, museums, letters, and photo albums, drilling through the stone wall of my father's denial. Combining my research with my summer holidays, I vowed to retrace my nomadic grandfather's footprints through Europe and North

America, visiting the cities—Buffalo, Baltimore, Boston, Berkeley, New York, London, Brussels, Freiburg, Geneva—where he had trained and worked. I drew up a list of interviews with medical historians and aging former colleagues of my father and grandfather; I planned to visit my grandfather's birthplace—and revisit my own.

Gradually, I was able to digest the full grandeur, complexity, and impact of my grandfather's achievement. I was amazed to learn he was a bona fide Canadian hero, a medical pathfinder of extraordinary drive and vision, his story inextricably bound up in an epic national drama. During the 1920s and 1930s, Gerry rose to the top of his profession as the founder of the Connaught Laboratories and the director of its academic arm, the University of Toronto School of Hygiene, institutions responsible for the saving of countless lives both nationally and globally. The lab and the school rose to become models to the world, yet few Canadians today seemed to know or remember. This struck me as typical of our cautious national character—outside praise failing to register at home.

I learned that the epochal discovery of insulin by Banting and Best in 1921–1922, and its subsequent mass production by Connaught, was only one feather in the lab's cap; over a single generation, my grandfather had boldly conceived and helped build the modern institutional infrastructure of Canada's public health system, making and distributing free vaccines, serums, and antitoxins to all Canadians—a radical challenge to entrenched, profit-driven medicine that shut out the poor. His biological products, together with sophisticated public educational programs, led to the effective control or eradication of a litany of killer diseases, including syphilis, diphtheria, rabies, tetanus, meningitis, typhoid, scarlet fever, tuberculosis, smallpox, influenza, polio, and diabetes. The lines of his CV revealed a career trajectory of such intense, single-minded, self-sacrificing focus, it seemed he knew exactly what he was doing each step of the way. Everything about him shouted: "Go big or go home." I was particularly intrigued that he had started his career as a neuropathologist—like the young Freud—cutting open the brains of pauper Irish lunatics in a fruitless search for the germ of madness.

Precious few of my grandfather's possessions had survived him—high school poetry books, a primitive student microscope, a monogrammed top hat, his World War I officer's sword—and I found myself fondling them with near-fetishistic curiosity. Studying his photographs, I saw a man of archetypal, ascetic Irish features: fair complexion, reddish-brown hair, thin lips, aquiline nose, prominent ears and chin, and soft, hazel green eyes tinged with melancholy. His concave profile, tall forehead, and wire-rimmed spectacles made him resemble his exact contemporaries, Irish Republican statesman Eamon de Valera and the writer James Joyce. In a folder of yellowing newspaper clippings, I found obituaries by colleagues that glowed with obvious sincerity. "Canada has lost one of its eminent citizens," wrote one of many admirers. "We will keenly miss his energetic enthusiasm; his vision; his encyclopedic erudition; his organizing and administrative ability; his incisive and decisive qualities of mind; his shrewd judgment and foresight; his facility of stimulating and leading those with whom he came in contact; his faculty of inspiring the best in work and loyalty among those who served him; his practice and support of the proper principles of academic freedom; his gracious thoughtfulness; and the charm of his personality. . . . We realize that John Gerald FitzGerald can never be replaced, but are comforted in knowing that his good works will be carried on, and will ever endure as a cherished memorial of the unique achievements of a great man."

John Gerald "Gerry" FitzGerald, 1912

All the more strange that our father had never spoken of this "great man" who had not only shaped the consciousness of our family, but the entire country. Doubly strange that when I scanned the groaning bookshelves of medical libraries, I found fat biographies of pioneering Canadian doctors—Osler, Banting, Bethune, Penfield—yet no full account of my grandfather's life. It felt as if his memory, like the ravages of diphtheria or polio, had been virtually erased. Why? What was my father—and medical historians—protecting us from? Why were they so determined to stand between the generations like a pane of stained glass? I was sensing something worse than suicide; something else, something even more unspeakable.

═

I was sensing, too, that like my father, my hard-driving grandfather was not a man given to introspection; few personal letters had survived and I despaired of ever gaining deeper insights into his character. But fortune favours the bold and I kept digging, dreaming of a King Tut–like discovery.

In early 1995 I ventured into a psychiatric archive on the site of the former nineteenth-century lunatic asylum at 999 Queen Street West. I knew my grandfather had worked here in 1907–8 as a young neuropathologist, succeeded by the Freudian champion, Ernest Jones, so it seemed a good place to start prying open the lid of Pandora's box. And so it was, for here, miraculously, came the breakthrough I had been banking on.

As the archivist handed me a folder marked "FitzGerald," I held my breath. Inside, I found sixty intense, confessional letters written by my grandfather in 1939–40, the last year of his life. He was languishing in a private sanitarium in Hartford, Connecticut, being treated for depression in the wake of a failed suicide attempt. The letters were addressed to his close friend, Dr. Clarence B. Farrar, the thin, cerebral director of the Toronto Psychiatric Hospital, a forerunner of the Clarke Institute where, ironically, a generation later, my father was drugged and shocked into submission. Farrar had died in 1970 at the age of ninety-five; the letters had been donated to the archives that very week by his second wife, a woman over forty years his junior. To describe my discovery as "coincidental" seemed a desperately inadequate explanation.

Deciphering Gerry's semi-legible scrawl, I strained to digest the feast all in one sitting. Even as a vicelike pain encircled my skull, in the corridors outside the archives, the strangled cries of medicated patients echoed like the soundtrack of a gothic horror film.

The letters had been written a decade before my birth, yet the tone and idiom seemed so familiar, real, immediate. The distant, unknown stranger who had occupied the haunted house that was my body was suddenly made flesh before my eyes, the past bleeding freely into the present, his soul laid bare before me with a near-unbearable pathos and vulnerability. The lamentations of my grandfather, calling from the end of the 1930s, uncannily mirrored the voice of my father at the end of the 1960s—the same slow slide into mid-life crisis, the same anguished outpourings over the reversals of fortune, the loss of income and status, the upsurgings of panic and self-loathing, the helplessness, the hopelessness, the terror of being "second rate," the paralyzing indecision about whether or not to return to work, all chased by unrelenting thoughts of self-annihilation. The pair of drowning voices thrashed inside my head, fusing and confusing, making no distinction between the separate identities of Gerry and Jack, father and son. In fact, as the voices drilled and swirled and lashed, they awakened a third voice—the sound of my own primal dread. My grandfather's words were pulling me back to the austere Victorian nursery on Balmoral Avenue, the desert of hardwood floors where small children, generation upon generation, are abandoned under the wheels of the repressed. I felt part of some weird trinity of father, son, and unholy ghost; it was falling to me, the third-generation eldest son, to stand and fight.

As I read on, the mystery deepened. In one letter, Gerry wrote, "*I have committed the unpardonable sin—and the penalty is death.*" In subsequent letters, he kept repeating the same phrase obsessively. I felt I had made an electrifying discovery, but what did it mean? Despite his obvious passion for social justice and public service, I had found no evidence that Gerry was a religious man, so the confession of an unspecified sin seemed all the more intriguing. Were these eleven painful words the key to the puzzle? Why did my grandfather, a fifty-seven-year-old man at the summit of a glorious scientific career, martyr himself? What terrible, irresistible forces of fate possessed him? Was it the same force that now drove me to chase him down?

One of dozens of letters written by Gerry from the Hartford Retreat to his Toronto-based friend, the psychiatrist C.B. Farrar, 1939–40

A week later, I visited C.B. Farrar's widow, Joan, the donor of the letters. I felt as if I had stepped into a time machine; for the past twenty-five years she had been living in her dead husband's rambling, three-storey manse at 20 Oriole Road, a five-minute stroll from Balmoral. Joan and "C.B." had "lived in sin" in the house for several years, then married in 1964 after C.B.'s first wife, whom he had never divorced, died; he was ninety and Joan was forty-seven. She remembered the six years of her "educational" marriage as the best of her life, as C.B. read her passages of Shakespeare, Proust, and the Greek classics in bed; took her to lectures and the opera; and snapped photos of her posing nude on the smooth, lakeside rocks of Muskoka. As we chatted in her living room, stuffed with Farrar's massive collection of books, Joan secured my attention in dramatic fashion when she placed on the coffee table a century-old skull of a criminal that once housed a brain her husband had dissected. I could not resist saying: "I thought I was here to pick *your* brains."

Farrar and "Fitz" were the closest of friends and confidants, she revealed, and during those heady, early decades of the century of great medical advances, they had enjoyed "the ride of their lives." She recounted how, one day, as she studied a photograph of her husband in his glory days working with European leaders in his field, she remarked, "Why, C.B., you look so arrogant!" Farrar replied evenly, "You'd be arrogant too if you knew all the people I have known." In response to her own anecdote, Joan murmured to me under her breath, "Oh, the Irish—they're always one step ahead of you."

In the end, she could shed no light on Gerry's "unpardonable sin." Except for one intriguing detail: C.B. had told her that during a stressful time, my grandfather "had an affair" late in his life with a woman he met on one of his many transatlantic crossings. That's all she knew, nothing more. Falling into the arms of another woman seemed wildly out of character for my scrupulously ethical grandfather, but even if he had, it seemed unlikely he would kill himself over it. I felt the source of his guilt and self-hatred ran far deeper, and I knew my quest was only beginning.

When I shared copies of my grandfather's letters with my family, my mother had nothing to say other than she thought them "repetitive." I said, "Maybe he was repeating himself because no one was listening." She looked at me blankly; I suspected Gerry's hamster wheel cyclings had painfully evoked her years with Jack, and so I let the conversation drop. She proved far more helpful when she suggested I approach Dr. John Hamilton, a former University of Toronto dean of medicine now retired in his eighties in a nursing home in British Columbia. As a young pathologist, Hamilton had known my grandfather and later instructed my father at Queen's; he had always been *simpatico* with the FitzGeralds. My mother showed me a letter that Hamilton had written her in 1992 after the death of my father: "I have often thought about Jack—his brilliance in his early medical career and the pleasure of associating with him. I am glad that he did not survive any longer—a torment to himself and to you." My mother had reason to suspect Hamilton was privy to the nature of my grandfather's suicide in 1940, and she was right.

When we first met in Vancouver in 1996, the stooped, grey-haired doctor was loathe to talk to me—a hauntingly familiar experience—deflecting my polite, indirect questions with a patronizing air. Only during a second visit, when I grew passionate about my need to cut through the layers of myth, if only for my own well-being, did he relent. His wrinkled face screwed up in pain as the words fell from his lips; he was a confidant of my grandfather's attending physician, Ray Farquharson, and therefore, he knew the truth. Had I not reached Hamilton in time, the secret would have died with him.

With the shocking revelation came a wave of melting relief, an uplifting validation of my own intuitions and dreams. Yes, I had been right to doubt, to wonder, to question: The Nembutol overdose was indeed a cover story. Together with the discovery of the "unpardonable sin" letters, Hamilton's disturbing confession had thrown me a lifeline. I now believed that my father had not, in fact, taken in the full horror of his father's demise, or if he did, he shut it out; he went to his grave believing, or wanting to believe, the Nembutol myth. For a while, I considered stopping my search right there, but Hamilton had now set loose a new train of questions.

═

As it turned out, at my grandfather's instigation, Dr. Farrar had treated both of the younger FitzGerald brothers, Bill and Sidney, for depression, as well as their cousin, Dr. Herbert Armstrong; as next of kin I was entitled to see their records and so back to the archives I went. All three men fit a similar profile—bright but isolated bachelors seemingly terrified of the opposite sex. Armstrong's story, in particular, bordered on the lurid. A double-gold medalist at medical school, he worked as a GP in the Ontario town of Fergus. His mother, the sister of Gerry's father, was a strict, emotionally rigid disciplinarian who had grown up in the harsh, Irish immigrant outback of Millbrook. Herbert's first breakdown came in 1911 after a thwarted romance and he was admitted for a time to Campbell Meyers's private sanitarium on Heath Street. When a widow pursued him romantically, he developed a pelvic inflammation. In 1927, having degenerated to a shabby state, Herbert was seized with an urge to rush out into the streets of Fergus and scream out that he had been charging patients too much, giving poor service, and generally leading a false, double life. He spent months in Homewood Sanitarium in Guelph where he fantasized that the attendants were feeding him bootleg whisky and conspiring to take him to the slums to marry a dissolute negress who would sue him for support and disgrace his family. Overwhelmed by disturbing sex dreams, he could not bear the sight of women; he screamed obscenities at his nurses.

As I scanned the pages of Farrar's clinical prose, the full tapestry of my paternal legacy unravelled before me. A grandfather, two great-uncles, a cousin, an aunt, a father—all crack-ups. And now from the files came evidence that Will, my great-grandfather, too, was paralyzed by melancholia in his middle age. What was I to make of all this—these unfathomable, unpardonable sins, these sad, unpardonable sons?

I was grateful for Farrar's fastidious notes, even though it seemed the doctor could do little to help his patients. The book I had despaired of writing for lack of hard information now lay within my grasp. I had a self-murder mystery on my hands and I felt as if I had no choice but to assign myself as the poker-faced detective on the case.

But even as I took up my pen, the spectral figure of my grandfather loomed ever larger in the darkened theatre of my dreams, as if he understood—and resisted—the aggression of my truth seeking. More than ever, I knew he lived inside me, just as I knew that if I were to flush his shadow into the light of day, I must put words to page; I must bring out the dead. I was pressing myself into a career as a ghostwriter, sentencing myself to the making of sentences, the slow digging of the tunnel out of my prison cell. Cicero put it best: "Not to know what happened before one was born is always to be a child."

I knew I could not face the task alone, and so each week I mustered the courage to lower myself into Peter's leather chair and joust with the dramatis personae crowding my unconscious. In my sleep, I heard fragments of muffled voices rising to my ears; slowly, tantalizing puzzle pieces began to coalesce bit by bit, rising from my unconscious to my conscious life. Dozens of books accumulated in piles by my bedside and I found quiet kinship with writers tilling similar soil. I came to realize that we are "the spawn of our infinitely regressive family histories" as one author put it; every father is a son, every son a father. Across the generations, we are haunted not by the dead, but the holes left within us by the buried secrets of the father. He bequeaths his unspeakable secrets silently to the son, digging an unmarked tomb of sealed-off knowledge deep in the dirt of his unconscious. Father and son remain partners in crime, until the day the grandson awakes, and speaks.

As I reflected back on that bleak autumn day in 1983, slouched on the park bench with my ex-girlfriend Ann, I realized that the spirit of my grandfather, a full century after his birth, was wedged like an anvil in the heavy space between myself and my most intimate relationships, an enemy of my life's desires. I had much to learn before I felt I had earned the privilege to occupy the rooms inside my grandfather's head, to dare to imagine his last thoughts on earth, to feel and confront his presence inside me. "All death," the acerbic Freud once said, "is suicide in disguise"; and I was daring to tug at the mask. In my dreams, a bony finger pointed to the eternal, paternal sea that conspired to drown me; yet out into the riptide I swam, pulled by promises of rebirth.

My future lay in the past.

PART II

NINE

Science Has No Fatherland

> He who is willing to work gives birth to his own father.
>
> SOREN KIERKEGAARD

On an autumn Sunday afternoon, I drive through dusty country backroads, one hundred miles northwest of the city, to the birthplace of my grandfather. Embraced by gently rolling hills, the population of the farming village of Drayton has remained steady at eight hundred since the days of his childhood. In my briefcase, I carry a tasselled album of family photographs given to me by my mother; turning its pages, I study the black-and-white images of my suspects, keeping silent company with the FitzGeralds, as if I am the ghost and they are the living.

As I tread the sidewalks, I imagine the calendar flipping backwards to 1882. In the bustling Victorian community that shaped Gerry's earliest years, I count six hotels, five churches, and two banks, its simple main thoroughfare lined with factories and shopfronts illuminated by stately wrought-iron gaslights; in his newly published book, *The Gay Science*, Friedrich Nietzsche has shocked the world with the unprecedented declaration: "God is dead." At the crossroads of Main and Wellington Streets, I find myself standing outside a two-storey funeral parlour that once housed the FitzGerald Drug Store. Contemplating the sole surviving photo of my great-grandfather Will, I conjure him up, a thin, unsmiling man with a bushy, handlebar moustache, bent over a counter, puffing a pipe and dispensing a prescription to a customer like a sacrament. What little I know of his character stems from the psychiatric file of his son William, Jr., who spoke of his father as "a good provider and a strict disciplinarian." His sad eyes both defy me and beckon to me, daring me to make something of the phrase "strict disciplinarian."

During his schooling in Toronto, Will had fallen for a bright, vivacious young Englishwoman, Alice Ann Woollatt, who as a teenager had emigrated from Hertfordshire with her parents and five siblings. Her prosperous father, also named William, lived and worked as a gentleman farmer south of Bloor Street. On March 1, 1882, Will and Alice were married in the fashionable downtown Toronto Church of the Ascension at York and Richmond Streets whose congregation embraced many citizens of prominence. The Anglican church was well named; when the rural parvenu Will crossed its threshold, he was taking a step up in the world even as his bride was stepping down. When Will chose Drayton as a place to live and work—Wellington County was poorly serviced by both doctors and pharmacists—Alice agreed to devote herself to a country life with an enterprising young druggist.

The sole surviving photograph of my great-grandmother shows a youthful woman in a tightly corseted dark dress, all buttons and bustle, her hands clasped behind her back. Large, soulful eyes dominate a round, contemplative face. A surviving written fragment, also taken from her son Bill's psychiatric file, testifies to her nature: "She was a lively, buoyant personality who was the really responsible head of the family." On the edge of the village, the FitzGeralds live in a small rented house near the railway tracks. Each Sunday morning, Alice reads vivid biblical parables to a gaggle of village children in the cramped, forty-seat Anglican Christ Church made of white clapboard, newly converted from a union hall after her husband led a local fundraising drive. But soon the swelling of her stomach, deemed vaguely sinful by proper Victorians, induces her "confinement" behind closed doors and drawn curtains. My grandfather is growing inside her . . . and inside me.

Gerry, born in Drayton, Ontario, 1882

It is Friday, December 9, 1882, nine months and one week after the marriage of my great-grandparents. In the modest frame house just outside of town, the smell of snow in the air, Alice lies in a simple brass bed, moaning in pain of biblical intensity, the country doctor leaning over her. As the squalling infant squirms onto the blood red sheets, his wrinkled face is lit by flickering gaslight; then the grim-faced doctor, unaware that the newborn boy in his hands will enter his own profession, reaches for a knife and cuts the umbilical cord. It is Alice's twenty-sixth birthday.

After a second FitzGerald boy, Sidney, is born on August 2, 1884, the family of four rents a larger house closer to town. A cardboard photograph shows two-year-old Sidney sitting at the feet of his four-year-old brother Gerry, emanating the preternatural gaze of an "old soul." Both boys are clad in girlish, Victorian frill—perhaps a concession to the Irish folk legend that told of malicious fairies who stole first-born male children from their cribs. What better way to divert the thieves than dress the boys up as girls?

Alice gives birth to a third son, Reginald William, in May 1886; a daughter, Hazel Kathleen, follows in July 1888. Like countless prefeminist Victorian women of energy and intelligence, Alice's place is inescapably in the home, stoically pining for the high culture of her Hertfordshire youth. The birth of four children in five-and-a-half years is exacting an insidious toll on her health.

═══

In the spring of 1891, the FitzGeralds move twelve miles northwest to the thriving town of Harriston, twice the size of Drayton. They rent a modest two-storey house on John Street, a leafy, tree-lined dirt road near the Maitland River and the neatly manicured greens of the local lawn bowling club. Gerry is now eight years old; Sidney, six; Bill, four; and Hazel, two. A stiffly posed studio photograph survives of the four unsmiling children together with what looks like a large, black Newfoundland dog, bred for the rescue of the drowning; here, I suspect, was born my grandfather's lifelong love of animals.

My great-grandfather sets up his store at 23 Elora Street, a two-storey, red-brick building with tall, arched windows and green awning, only yards from the post office and the busy Coronation Hotel. To supplement his income from patent and proprietary medicines, Will stocks his shelves with an array of books, stationery, bonbons, cosmetics, chinaware, cutlery, postcards, figurines, and British Union Jacks. After three more years of hard work, he is able to buy his first home on a spacious lot just two houses over from his former rental property.

As I pull up the driveway of 88 John Street, I am overcome by a rush of déjà vu. I knock on the door and a smiling, middle-aged man answers. Explaining that I am exploring my grandfather's early life, I am delighted to learn that the present owner is, fittingly, a microbiologist. He kindly shows me through the elegant two-storey, buff-brick house that has changed little since it was built in 1871. I pass through French doors into the living room with polished hardwood floors and twelve-foot-high ceilings. Lingering by the fireplace—a gilded portrait of Queen Victoria hangs above it—I run my fingers over the smooth white Carrara marble, the same brand of death-cool stone that encases Gerry's ashes in the mausoleum in Toronto's Mount Pleasant Cemetery. Exploring the three upstairs bedrooms where the family slept, I am drenched in a feeling of the uncanny; peering at a child's faded initials carved into a windowsill, I take them to be my grandfather's.

As I descend the curving staircase, I pause at a tall sash window affording a sweeping view of St. George's Anglican Church, a block distant, where the four FitzGerald children were each confirmed in succession. Looking out, I conjure up the long gone stable and the horse and cutter where I envision young Gerry brushing down a nervous, sweating mare. Mere yards away, railway tracks stretch south to the destinations of my grandfather's steam-powered dreams. From the silence, I long to resurrect the thicket of the words, the pregnant glances, the movements of bodies long since reduced to the motes of dust on the carpet under my feet.

The FitzGerald children, Bill, Hazel, Gerry, and Sidney, Harriston, circa *1891*

As I ponder whether I will unfurl the darker details of my family history to my hosts, the owner's wife imparts a story of her own. At the foot of the creeking staircase, she calmly informs me that she had recently seen the ghost of a sickly woman, clad in a nightgown, staring out a bedroom window into the winter night. I feel in no mood to doubt her.

In September 1895, three months shy of his thirteenth birthday, my grandfather enters Harriston High School, a squat, white-brick, two-storey, four-classroom building standing a few blocks north of his father's drugstore. The school delivers a sound classical education, many of its graduates moving on to the large cities where they will distinguish themselves in law, medicine, and academia. Gerry is naturally drawn to the chemistry lab, where he squints down the barrel of a microscope and experiments with the same kinds of compounds and substances that litter his father's pharmaceutical counter. After school, in his father's busy shop, he watches the townsfolk routinely spend their hard-earned wages on handmade pills and potions to ease their various afflictions. Stern and formal, Will FitzGerald works long, punishing hours, day and night, six days a week, and expects the same of his apprentices. Gerry takes to the business readily, passing every spare hour after school assisting his father. In the making of pills and medicines, he learns that precision is critical. A single slip of the knife blade, tipping in a tad too much toxic material, courts catastrophe. To visit sickness—or even death—on a fee-paying customer is to commit an unthinkable, unpardonable sin. *Primum non nocere*—first do no harm.

Potent narcotics are legally and widely available. Physicians routinely prescribe opium and morphine as panaceas for everything from fever to colic, measles, smallpox, restlessness, pain, diarrhea, coughing, even teething formulas for babies. Neurasthenic and neuralgic women, given to puritanical introspection and self-reproach, carry vials of laudanum in stylish, self-medicating kits; others discreetly slip a hypodermic syringe—newly invented in the 1870s—into their garter belts, tucked under their long dresses. Languid cults of illness—with phallic overtones of injected, forbidden pleasure—arise in counterpoint to the manic, masculine pursuit of aggressive exercise and moral training. Addiction is acceptable in the upper classes, not the lower; Queen Victoria herself smokes marijuana to relieve her painful menstrual cramps. Not until 1908, as more and more lives succumb to addictive substances, will Canadian law limit the narcotic content of medicines.

Gerry FitzGerald grows into a lean, six-foot-tall, red-headed youth with an omnivorous mind hungry for technical detail. When he isn't grinding raw medicines with mortar and pestle, or delivering them by horse and cutter to the sick in outlying farms at all hours, he buries his face in the textbooks lining the shelves of his father's store. Here is born an enthrallment with the mysterious power of infectious disease and the galvanizing "eureka" moments dotting the pages of medical history. In his first term of high school,

Gerry learns of the death of seventy-two-year-old Louis Pasteur, the French chemist of genius who will profoundly influence his own future path. European lab men are making a string of historic breakthroughs in breathtakingly rapid succession. The world is experiencing the genesis of powerful and sophisticated new sciences—bacteriology, immunology, and preventive medicine—that promise to leave his father's primitive drugstore in the dust.

═

"Every invalid is a physician" claims an Irish proverb. Or perhaps every invalid's son. Every December 9, Alice FitzGerald and her favourite, first-born child, Gerry, share the same birthday. In my imagination, I picture them together, after dinner and the mutual opening of presents, sharing the settee in the parlour, as Alice's eyes whisper intimations of her deepest secrets. Now in her early forties, her hair streaked with grey, she is receding into a dark pool of ennui, rarely rising from her bed, a melancholic shadow of her once buoyant self. For a woman, all human happiness must begin and end at home, a domestic tyranny Alice has borne stoically, and at a price. Year after year, Gerry sees his mother swallow pills from the bottles crowding her bedside table, an array of antidotes and anodynes created by his father's own hand. But as much as Gerry needs to believe in Will's alchemical powers, the truth is painfully obvious: He is failing to save her.

Gerry's mother, Alice Woollatt FitzGerald; a chronic invalid in her later years, she died at 51 in 1907

As restless as a racehorse, Gerry will habitually run roughshod over his reflective, sensitive side over the course of his life. He despises raucous singing or sudden, loud noises. When the crash of summer thunderstorms makes him flinch, he turns his hate on his own nervous vulnerability. Although highly attuned to natural and cultural beauty, he is not the kind of lad given to lazy strolls along the banks of the Maitland River, turning the pages of Tennyson, and contemplating the deep silence of the snow-laden fields. He doesn't have the time. His dreams take another form: He will make himself *useful*.

In the grip of adolescence, Gerry's head swims with impossible, grandiose ideas about the perfectibility of man, tempered by a down-to-earth, tough-minded pragmatism. Under the demeanour of a conscientious, well-mannered young gentleman burns a competitive fire, a hatred of losing, a passionate desire to win at any cost. Prevention is the key. Perhaps extraordinary, worldly deeds will somehow redeem his mother, stranded and stifled in a remote rural town. Perhaps extraordinary, worldly deeds will placate his poker-faced father, the good provider and strict disciplinarian who demands nothing but precision and perfection in the execution of one's work and conduct. Perhaps such actions might win their unalloyed approval, their pride, their gratitude, their unconditional love. Whatever Gerry FitzGerald does in life, it will have to be extraordinary—a cut above.

In September 1899, John Gerald FitzGerald, a gawky, precocious sixteen-year-old country boy wearing a bow tie, stiff white collar, and wire-rimmed spectacles, strides to the train station directly behind his yellow-brick home. Riding on the cusp of a hopeful new century, he sets off on the hundred-mile journey to the University of Toronto where he has been accepted as the youngest member in the history of the medical school.

Leaving the obscure Ontario town behind, the blood red steam engine chugs southward, its piercing whistle cutting the late summer air. In concert with the incessant clicking of the wheels on the steel rails, the credo of my grandfather's hero, Louis Pasteur, blazes his path: "Science has no fatherland."

TEN

Crucible of Ambition

> Athene gave Aesclepius two phials of the Gorgon Medusa's blood; with what had been drawn from the veins of her left side, he could raise the dead; with what had been drawn from her right side, he could destroy instantly.
>
> ROBERT GRAVES
> *The Greek Myths*

It is the fall of 1999 and I have buried myself in the third-level stacks of the old stone-walled science library on the edge of King's College Circle, nudging the University of Toronto medical school. Exactly a century earlier, my teenaged grandfather first descended into this same scholarly underworld of tomblike silence and once again, as I sit alone with the reversing flow of my thoughts, the portals of time past open like a book.

Outside these walls, the gritty Protestant city of Toronto, over 200,000 strong, is known as "Belfast of the North" for its concentration of Irish blood. It is also derided as Hogtown, the self-serving, pork-selling maw of urban commerce, and Toronto the Good for its reputation for puritanical rectitude. Set in the heart of the city, the fast-growing University of Toronto serves as a microcosm of national aspirations, determined to test its mettle against its more glamorous European and American cousins. Here, as my grandfather cuts through the wind-blown autumn leaves of 1899, he is enveloped by the symbols and icons of the British Empire, now reaching its global apogee. As strange, mechanical contraptions called horseless carriages honk and sputter through the campus, an ocean away Dr. Sigmund Freud is publishing his revolutionary book, *The Interpretation of Dreams*, in which he characterizes the dartings of our nocturnal mental theatre as "the royal road to a knowledge of the unconscious activities of the mind." The idea that the dark, primordial horse of the instinctual unconscious dictates the path of the human rider—and not vice versa—proves so outrageous that prim, enlightened Victorians are loathe to embrace it. Surely the force between our ears was, is, *must be* stronger than the force between our legs?

University of Toronto pathology lab, Sackville and Gerrard Streets, Class of 1903

The medical class of 1903 is infused with young blood from small Ontario towns, the future saints indistinguishable from the future sadists. So while Gerry feels at home among dozens of rural brethren, in another way he labours at a disadvantage: the average age of his classmates is over twenty, and he is a stripling of sixteen. Of course, the class is entirely male; women will not be admitted until 1906. The medical school is still hindered by a lack of a teaching hospital affiliated with the university, where students could be trained at a patient's bedside. But with the growing influence of the progressive Johns Hopkins in Baltimore, the idea is picking up steam. On a streetcar that cuts through the grim Irish slum of Cabbagetown, riven with abysmal sanitation and shocking infant-mortality rates, Gerry travels regularly to the 425-bed Toronto General Hospital at Gerrard and Sackville Streets, a run-down, red-brick, four-storey French colonial edifice with five castlelike towers and a dank basement mortuary. As portrayed by a fellow student, "The ancient pile looked dingy, smelled horribly, and was ventilated like a tomb."

In the pathological histology lab, rows of serious young men in jackets, ties and stiff white collars press elbow to elbow at long wooden tables. Placing a stained glass slide under the lens, Gerry peers into his microscope and scribbles notes on the distinctive tissue patterns and textures swimming before his eyes. He reads the epic stories of medical breakthroughs, one tumbling after another—how the Englishman Edward Jenner invented a crude smallpox vaccination in 1798; how, in 1876, the German doctor Robert Koch unlocked the secrets of the anthrax disease cycle, proving for the first time in history that a specific microbe caused a specific disease; how, in 1881, Louis Pasteur attracted world attention by successfully vaccinating a herd of sheep against anthrax; how, in 1882, Koch discovered the bacillus for tuberculosis and in 1883, for cholera; how, the following year, his colleagues Edwin Klebs and Johannes Loffler isolated the diphtheria bacillus. Then, the following summer, Pasteur risked administering an unproven, rudimentary rabies vaccine to nine-year-old Joseph Meister, bitten fourteen times by a mad dog. In dramatically saving the boy's life, Pasteur—in tandem with Koch et al.—laid the foundations of modern preventive medicine.

Fuelled by the death of Queen Victoria and the incessant eruptions of Republican nationalism across the Irish Sea, disillusionment with the burdens of empire ripples through the English body politic. Waves of immigrants—many thought to carry the germs of syphilis and madness—threaten to infect, Dracula-like, the pure veins of hardy Anglo-Saxon bloodlines, and the populace falls into a state of barely concealed panic. Foreign influences are sapping national efficiency; physical, mental, and moral degeneration menaces the survival of the imperial race. The catchphrase "race suicide" peppers popular discourse.

Rallying to meet the perceived crisis, medical officers of health are fiercely lobbying for the nationalization of health care, built around the existing public health structures. Initially, the public health movements in nineteenth-century England and America had been led by social reformers including, but not dominated by, medical doctors. With the increasing professionalization of public health and the epochal discoveries in bacteriology—by the turn of the century, scientists had discovered the microbes responsible for fourteen different diseases—the scene is shifting. Physicians are now elbowing their way to the front, determined to seize the brass ring of public health as their own specialty.

Even as Gerry absorbs dense and compelling new bodies of knowledge—pathology and bacteriology most rivet his attention—the city's dirt roads and natural water table are mired in filth. Toronto's two main river arteries, the Don and the Humber, feed a system of smaller streams and creeks, radiating like blood vessels through the patchwork of wooded ravines and residential neighbourhoods. Taddle Creek, for one, had once snaked its way from Wychwood Park on the high ground of St. Clair Avenue, running downhill through the heart of the university campus, past the biology building, and drained into the harbour. But the creek became so polluted that it had to be covered over and driven underground. The ravages of rapid urban industrialization have enveloped Hogtown in a miasma of overcrowded tenements, scattered refuse, stinking public privies, smoggy air, clogged sewage systems—a perfect breeding ground for killer diseases. And a perfect breeding ground for a new species of warrior class—the formidable public health reformer—wielding a terrible, swift sword.

═

Under the tutelage of Professor John Amyot, Gerry leans over human cadavers laid out on cold marble slabs and cuts open the mysteries of the body with a razor-sharp scalpel. Over his head hangs a sign in Latin—*Hic locus est ubi mors gaudet succurrere vitae*—"This is the place where death delights to help the living." Many of the cadavers fed to Toronto's juvenile medical students come from the Provincial Lunatic Asylum at 999 Queen Street West, two miles west of the campus. The place exerts a special pull for Gerry, whose melancholic, invalid mother, Alice, languishing in Harriston, rarely strays from his thoughts. It is only natural, then, that he would be much taken with the extramural course taught by Daniel Clark, the august, Scots-born professor of mental diseases and the asylum superintendent for the past twenty-eight years.

When it opened half a century earlier, the six-hundred-foot-long lunatic asylum was optimistically proclaimed a paragon of progressive architecture that embodied enlightened Victorian principles for the humane treatment of the insane. The reformers' mandate was clear: liberate the persecuted, demonized mad, chained to the walls of jails, from brutality and starvation, and treat them with sustained kindness and compassion. But in the wake of the catastrophic potato famine of the late 1840s, Toronto had been flooded with poor, starving, bog-Irish immigrants, and many fled straight into the new lunatic asylum. Within a generation, the swelling tide of the mad—Ireland led the world with the highest incarceration rates—has been pumped by the unremitting spread of industrialization, the trauma of immigration, and the misguided policies of government bureaucrats. Plagued by chronic overcrowding and understaffing, the institution has gradually succumbed to the horrors of the unregenerate snakepit.

Gerry, the keen young neuropathologist

When Gerry first steps inside the place, the near-palpable clouds of despair make it hard to distinguish the staff from the inmates. An all-pervasive stink, the legacy of a faulty plumbing system, clings to the halls and walls like some foul fungus. Yet Clark is humane in his attitude to his asylum charges. He abhors the use of physical restraints and advocates modest amounts of alcohol as a therapeutic measure. In opposition to the emerging trend in scientific medicine, Clark does not regard localized pathological conditions in the brain as playing a vital part in the psychoses.

Although the medical students use Clark's own textbook, their education is chiefly clinical, moving through the corridors of the mad, soberly observing and recording their strange gestures, their chilling cries, their frozen silences. They hear the famous story of how the daughter of the firebrand political rebel, William Lyon Mackenzie, the scourge of Toronto's patrician class, was committed here in the 1850s and how Mackenzie clashed violently with the irascible Irish superintendent, Dr. Joseph Workman, over her treatment. And as Gerry follows Dr. Clark through the gloomy wards of the Canadian bedlam, separated according to sex, he does not fail to notice that many of the inmates are immigrants of Irish blood.

The few of my grandfather's possessions that reach my hands decades later include a dusty cache of medical and non-medical books. His personally inscribed 1903 edition of *The English Dance of Death*, a collection of poems by William Combe, hints at Gerry's taste in literature as a newly graduating medical man hypnotized by the mysteries of madness and mortality. I am excited enough to find the book, but as I open it, the very first poem I see takes my breath away. It is called "The Suicide." A coloured plate shows Death personified as a grinning skeleton lying on a rock with a spear at his side, as a distressed damsel hurls herself into the sea. The inscription reads: "Death smiles and seems his dart to hide / When he beholds the suicide."

On the afternoon of June 12, 1903, my grandfather takes his place shoulder to shoulder among the phalanx of students outfitted in robes and mortarboards, crammed into the university gymnasium, to receive his degree. The University of Toronto is now poised to realize its ambition of becoming a national university of world stature, daring to dream of standing on an equal footing with the likes of Cambridge, Harvard, and McGill. Within months, a party of medical luminaries, including two of the fabled "Four Horsemen," William Osler and William Welch of the elite Johns Hopkins Hospital—the accomplished pathologist Welch was yet another mentor figure who would loom large in Gerry's future—will travel to Toronto to preside over the opening of a new three-storey medical building, just south of the library, where my grandfather will eventually return. Distinguished by its white-brick, Renaissance style and twin, domed towers, the new medical school contains the most up-to-date laboratories and a new pathology museum; its enrolment of over six hundred students is one of the largest in North America. Osler challenges the province's two smaller medical schools in London and Kingston to "commit suicide" and affiliate with Toronto in order to create a school of the first rank; Welch urges the university to acquire and control its own hospital and fuse teaching, research, and clinical practice—in short, create a Canadian version of Johns Hopkins. Toronto medicine is being built on a British model built on an American model built on a German model—one that would, over the alchemy of time, evolve into a distinctive Canadian hybrid.

University of Toronto Medical School, 1903

With no promising opportunities for postgraduate training in Canada, in June 1903 Gerry takes the five-hundred-mile train trip southeast to New York City and strides up the gangplank of the SS *Philadelphia*, a stately, three-masted passenger ship of the American Line, where he has signed on for the summer as acting surgeon. Although the world's competing steamship lines cater to the rich, it is the endless flood of "emigrant fever" that keeps the liners financially afloat—and troubles the sleep of the eugenic-minded doctors of my grandfather's generation.

After stops in Southampton and Cherbourg, the ship glides into Queenstown, Ireland, where Gerry first glimpses the verdant landscape of his ancestors. Then he steams back to New York, completing the first of seven round trips across the North Atlantic that summer. One of his prime responsibilities is to ensure that the steerage passengers—one thousand in all—clear health inspection upon arrival in New York. If they fail—and one in fifty will—the "undesirable aliens" face a bleak voyage back to Europe at the expense of his employers, the American Line. The daily images of the benighted shipboard immigrants—mirroring the face of his own grandfather eighty years earlier—burn in his mind, merging with thoughts of the mad Irish funnelled into the 999 Queen Street asylum. As he dispatches the unlucky sick and insane travellers back to the prison house of their beginnings, my beardless, twenty-year-old grandfather experiences for the first time the godlike power entrusted in a freshly anointed member of a rising professional class. He contemplates a noble future poised on the edge of a new century, a man-on-the-move who identifies with the liberating speed and influence of infant technologies—the telephone, the automobile, the radio, the cinema, the airplane. By the fall of 1903, the career choice comes, as decisive as the falling of a knife: neurology and psychiatry. The new will conquer the old; the future will kill the past.

ELEVEN

A Kind of Superman

> Father! Father! Where are you going?
> O do not walk so fast.
> Speak, father, speak to your little boy,
> Or else I shall be lost.
>
> WILLIAM BLAKE
> "The Little Boy Lost"

Boarding a blood red electric streetcar heading due north, my grandfather passes over the crest of the Avenue Road hill and stops at St. Clair Avenue, the upper limit of the city of Toronto. Striding along a sequestered, residential street lined with majestic oak, elm, and maple trees, within seconds he finds himself enveloped by the genteel neighbourhood of Deer Park. To the north looms the landmark clock tower of Upper Canada College, commanding a sweeping view of acres of open farmland and the thick forest beyond; perhaps even then he senses that this compelling magnet for the aspiring bourgeoisie is fated to absorb his unborn son and grandsons.

Gerry's brisk step delivers him to a dignified three-storey, red-brick mansion discreetly set back from the street on a three-acre lot. The facade of 72 Heath Street West appears serene, a pair of elegant chimneys lording over symmetrical rows of shuttered and gabled windows, a red, white, and blue Union Jack fluttering from the roof of a thrusting turret. Yet behind the drawn lace curtains of the country's first and only private neurological hospital, troubled human figures shuffle and tremble and weep.

═

In the space of seconds, five decades pass and a memory pulls me back inside the skin of my eight-year-old self. I am walking home from prep school with a classmate. Approaching Heath Street, my friend points and whispers in the direction of a big house, concealed behind tall trees, where he says crazy people hide, as if marked by some unspeakable shame. I know nothing of my grandfather, yet I do intuit something subtle—as if that abandoned patch of heath is telegraphing waves of ghostly vibrations for me to receive and decipher. My classmate dares me to explore the grounds of the old sanitarium, but I am afraid. I am more interested in the daily five o'clock reruns of Superman, *my favourite television show. I am transfixed by Clark Kent, the bespectacled, mild-mannered reporter shedding his secret identity and transforming himself into the invincible Man of Steel, housed in his Fortress of Solitude, vulnerable only to green chunks of kryptonite, his Achilles heel. Passing the haunted houses of Deer Park, I unknowingly inhale the same winter air as my dead grandfather; running home to bask in the emanations of the cathode ray tube, I digest the cultural lessons of what it means to be a hero.*

═

A stout, handsome man of forty distinguished by a broad forehead and neatly clipped moustache, Dr. Campbell Meyers trained in Paris under the charismatic Jean-Martin Charcot, "The Napoleon of the Neuroses," whose pioneering work with hysterical patients at the Salpêtrière Hospital, using hypnotic trance and the mysterious power of suggestion, deeply impressed the young neurologist Sigmund Freud.

Equidistant between Yonge Street and Avenue Road, the canopied seclusion of Meyers's Heath Street hospital proves a perfect haven for his well-heeled, neurotic clientele. Rejecting cases of insanity and drug addiction, he promises his twenty status-conscious patients that they will never rub shoulders with

Toronto's first private sanitarium for nervous illness, founded in 1894 by Campbell Myers, 72 Heath Street West

unruly psychotics, a stigma, which, he noted, "the laity so much dread." Like dozens of its kind spread across the continent, the well-appointed private sanitarium offers the nervous middle class an attractive escape from the horrors of the public madhouse.

Around the time of my grandfather's birth, the neurologist—the anatomist of the nervous system—had emerged as a distinct medical specialist, seeing all disorders of the mind, from neurotic to psychotic, as diseases of the brain, each with their own unique, identifiable pathologies buried in cerebral tissues. But Meyers swam against the tide in Toronto's medical circles, believing a continuum existed between the mild neuroses and severe psychoses; prompt, early treatment, he said, could prevent the explosion of full-blown madness. So began the fractious turf wars, boundary disputes, and hard questions that endure to this day. How do we classify, diagnose, and treat the full spectrum of mental diseases, running the gamut from garden-variety neuroses to severe psychoses? Do we separate the acute from the chronic, the violent from the gentle, the suicidal from the homicidal, the syphilitic from the addict? Should asylums become hospitals or should hospitals become asylums? Is madness essentially organic or psychological? Or both? What kinds of treatments—electricity, exercise, diet, drugs, surgery, water, or words—can free tormented souls from their private rings of hell? And who is best qualified to minister such relief?

It is a period of obsessive classification of disease states, physical and mental. In 1896, the Munich alienist Emil Kraepelin, a man of cold and aloof temperament, published an exhaustive compendium that laid out diagnostic categories, such as manic depression, that psychiatrists use to this day. Yet despite the upsurge in cutting-edge scientific techniques and laboratory investigations, researchers will fail to link the vast array of mental disorders to specific pathologies in the brain or nervous system. As effective treatments fail to keep pace with the advances in diagnosis, a veil of pessimism settles over the profession. Under the sombre influence of Darwin's evolutionary theories, medicine falls back on the fatalistic concept of cumulative, hereditary degeneration—"tainted blood"—as the likely root cause of insanity.

But by the 1890s, Sigmund Freud, the subversive Jewish outsider, is challenging such pitiless dogmas. Betraying his neurological training, he abandons his ambitious attempt to map out the physical basis of all nervous disease and carves out a bold new *psychological* path, radical in its implications, an epic voyage into the pathogenic underworld of sexual conflict that would form a cornerstone of his controversial new method of psychoanalysis. His suffering patients become his lab. He listens; he learns. He thinks about the sexuality of infants, the repression of emotions, the uncanny grip of memory, the timeless verities of dreams, and the oceanic depths of the unconscious mind. He cultivates the creative technique of free association, asking his patients to give voice to all random thoughts, no matter how senseless or trivial or bizarre or shameful, as if reporting the scenery witnessed from a moving train; by such an unorthodox method, painful truths tumble out and symptoms are alleviated. His message is as disturbing as it is revolutionary: Human motivation, coiled in the deeply recessed webbing of the psyche, acts independently of our conscious knowledge. We are not, he suggests, the captains of our own destinies or the masters of our own houses, guided by the light of reason.

Appalled by such shocking claims, the captains of the medical establishment stand united in their vilification of Freud. An influential American neurologist freely betrays his ignorance of psychological nuance when he admits: "While I can understand the patient falling in love with the nurse, I do not as easily comprehend the nurse falling in love with the sick patient." He scoffs at dream interpretation and the concept of transference, whereby neurotic, sexually stifled patients project their histories onto their listeners, acting out their erotic longings or sulphuric hostility towards their doctor father figures. Or the concept of countertransference—that he, too, the trained expert, might harbour tangled, subterranean feelings for the vulnerable people under his charge.

And so the nascent field of psychology, like religion, fast fragments into warring sects. In Europe, Emil Kraepelin and Sigmund Freud stand as the two pillars of the great schism. In Toronto, the conflict is being played out by the neurologist Campbell Meyers and C.K. Clarke, the bullheaded alienist of the Toronto Asylum for the Insane. C.K. insists mental disease must be treated in the public wards of 999 Queen Street West—or a new, standalone, scientifically based psychiatric hospital—while Meyers is fighting for the establishment of a special Nervous Ward at the existing Toronto General Hospital, bridging the chasm between the listless asylums and general medicine. Believing that neurosis represents the early stage of psychosis—and that the mad incarcerated in asylums are, alas, too far gone to reach—Meyers is convinced that mental derangement can be nipped in the bud by prompt and disciplined medical intervention.

The bulk of Meyers's patients are young and middle-aged women of prominent Toronto families, cushioned by the pillows of affluence yet cursed by high-strung temperaments as taut as piano wire. In a concentrated effort to restore his patients' spent nerves, Meyers imposes an American invention, "the rest cure," a prolonged, costly regimen of enforced seclusion, bed rest, a heavy diet of milk fats, mild exercise, electrotherapy, hydrotherapy, massage, and sedation. He denies them all forms of distraction—letters, newspapers, books—for weeks on end; the composing of poetry, he believes, leads directly to insanity. Family visits are particularly discouraged as they might spoil the patient with too much love. Childlike obedience to the authority of the doctor's will is deemed an integral part of the cure—a tactic that seems to work better with women than men. Dr. Meyers roundly chastises his soul-searching female patients for "discussing their symptoms with non-medical men in a distinctly thoughtless manner." The medical doctor bears a weighty responsibility, to the point of advising patients on marriage and child rearing: "He must be a kind of Superman," writes an eminent American neurologist, "one with higher ideals, more potent inhibitions and wiser in life and wider in outlook than those he is trying to guide." Freud, in contrast, regards stupidity—particularly in doctors—as the one, great unpardonable sin.

═

Stepping into the hydrotherapy room of white-tiled walls and enamel bathtubs, my grandfather tightly wraps a young woman, a chronic sufferer of neurasthenia, in a white sheet soaked in freezing-cold water. Like a mummy, she lies flat and still in the pale cocoon. Gerry advises her not to struggle, as the effort, combined with the body's natural response to the cold, may raise a dangerous fever. But this day, she does struggle, and violently, so Dr. FitzGerald quickly bids a muscular attendant to place a bag of ice on her head.

Next he moves to the electrotherapy room where a portable galvanic-faradic battery, housed in a wooden cabinet, sits on a table. Gerry grasps the electrodes, attached to wooden handles, and moistens their sponge tips. Slowly, he raises his patient's nightdress and places the electrodes on the smooth, chalk-white surface of her upper thigh, the site of her nervous pain. There comes a low buzzing sound, then a tiny shower of sparks sprays from the coils inside the box. As a stream of low-voltage electricity pulsates up and down her leg, contracting her muscles, the ghost of a seraphic smile rises to her lips. Sometimes patients do not know what's good for them; and sometimes they do.

In the creased faces of the genteel women my grandfather attends, images of his invalid mother, Alice, suddenly flare up like ribbons of gaslight, then die away. As he feels for a blue vein in the soft, pale wrist of a patient, as he counts the beats of her pulse, does he glimpse traces of himself reflected back in the quiet pleading of her eyes?

No, such an idea is unthinkable: *He* is the healthy, expert doctor; *they* the sick and weakened patients. They need his help. That is why he feels so strong, confident, responsible. Each morning, as he pulls a straight razor over the contours of his lathered throat, as he knots his checkered bow tie, as he affixes the studs in his celluloid collar, he feels as sane as mother's milk, his thoughts and motives above reproach.

═

Gerry knows that if he is going to improve his curriculum vitae, he must forge important and useful connections in the US, Europe, and beyond. Only one hundred miles southwest, just across the American border, stands the Buffalo State Hospital. Unlike Dr. Meyers's hothouse sanitarium—small, private, and discreet—Buffalo's asylum typifies the sprawling, government-run public institution overflowing with floridly psychotic patients largely drawn from the rough and tumble working class. It is a gargantuan warehouse of the mad—and it has its own pathology laboratory. Here is a chance for Gerry to deepen his quest for the holy grail—the physical roots of madness.

A fifteen-minute streetcar ride propels my grandfather down the Avenue Road hill to Union Station. It is September 1904. As he climbs into the second-class carriage of the steam train, the gleaming gold pocket watch attached to his waistcoat swings rhythmically in sync with the forward movement of the locomotive.

Travelling through the membranes of time, his grandson sinks into a semi-hypnotic state. I look out the train window and envision a giant, quivering arrow arching southward through space, fast as a magic bullet. I imagine its razor-sharp point piercing a precise patch of earth many miles distant, a place I feel compelled to find, as if my sanity—or life—depended upon it.

═

Exhausted after a day researching the history of Campbell Meyers's sanitarium, I walk home and in a late afternoon nap I fall into a dream. An open-decked ship is carrying me down the Avenue Road hill, the streets below flooded by a prehistoric glacial lake. I am accompanied by my friend Bill, a psychiatric nurse whom I befriended in the archives and who shares the name of my troubled great-grandfather and great-uncle. Looking overboard, I see several men floundering in the water—the male line of my family?—and suddenly one is swept under the hull and I fear he is being chewed up by the propeller. I am afraid more will follow. As the boat is sucked into a swirling vortex at the base of the hill, I hold tightly onto Bill's leg, as if binding an open wound.

I wake up and reach for my dream diary. Tomorrow, another day of archival digging beckons.

TWELVE

Gospel Certainty of Relapse

MACBETH: Canst thou not minister to a mind diseased,
Pluck from the memory a rooted sorrow,
Raze out the written troubles of the brain,
And with some sweet oblivious antidote
Cleanse the stuffed bosom of that perilous stuff
Which weighs upon the heart?

DOCTOR: Therein the patient
Must minister to himself.

Macbeth, ACT V, SCENE III

On a chill, rainy October day in 2002, a rental car propels me over the Canadian border into the rusted-out, industrially depressed city of Buffalo. Less than a mile inland from the foaming Niagara River, I turn up Forest Avenue—the original forest is long gone—as the windshield wipers hypnotically sweep away sheets of drizzle. The bleakness of the day mirrors the sight that looms ominously into view: a mammoth, derelict, stone madhouse stretching nearly half a mile long.

I have come on this particular day because a wing of the asylum, closed year round, has been opened to the public for a few hours. The city fathers haven't had the heart to condemn the crumbling old place, and for my own reasons I am grateful, even as a blanket of gloom settles over me. An idiosyncratic revival of medieval architecture known as Richardsonian Romanesque, the Buffalo State Asylum for the Insane is a fortresslike pile of rough sandstone, five feet thick, stained a reddish brown shade of dried blood. At once sinister and enticing, a pair of identical 185-foot towers are crowned by four corner turrets and precipitous copper roofs. Hard to fathom that this grim heap of loss, failure, and decay once embodied the highest of civic aspirations. What in heaven, or hell, compelled my grandfather to toil in this catch-all machine of social correction?

Popping open my umbrella, I meander through the empty warren of ivied walls and barred windows. Tiered like a rotting wedding cake, ten interconnected pavilions recede back from the twin towers in the shape of a giant V. Stepping out of the rain, I slip through the doors of the asylum. Edging down a curving, silent corridor, I cross the time barrier of two past generations, my body and soul suddenly engulfed by the cacophony of the hopelessly deranged. Here I find my grandfather making his ward rounds, moving among the wild-eyed lumps of traumatized immigrants—Irish, Germans, Poles, Italians—nearly 1,700 crammed into a space designed for 660. I watch Gerry, dignified and self-possessed, methodically brace himself against the dark wave of human flotsam, scribbling on carefully ruled columns and charts; like pegs in holes, he is fitting the day's new admissions—blacksmiths, shopkeepers, farmers, nurses, students, governesses, factory workers—into the major diagnostic categories, one by one, each according to the nature and degree of his or her illness. The advanced syphilitics, blind, disfigured, their brains turning to mush, their flesh a grotesque tapestry of chancres and pustules, most compel his attention, oozing the horrifying stigmata of unpardonable sexual sin, the stark judgment of the "leprosy of lust." In a journal article, Gerry notes the fierce resistance to the idea that a single microorganism could be the direct causal agent; many scientists cling to the conviction that alcohol abuse or dissolute, immoral living invites "the ravages of syphilization." The most common, and generally ineffective, chemical treatment inspires the saying: "A night in the arms of Venus leads to a lifetime of Mercury."

Moving from the noisy wards to the womb of the autopsy room, Gerry fatefully encounters the formidable figure of Dr. Adolf Meyer, a thirty-eight-year-old Swiss-German émigré psychiatrist whose thick Teutonic accent one could cut with a knife. It is a meeting ripe with consequence for my grandfather. A rigid, ultracautious, controlling personality possessed of a withering, razor-sharp tongue and basilisk stare, Meyer studied under the legendary Emil Kraepelin in Munich, composing a thesis on the reptilian forebrain; it was such icy qualities that rewarded him with the directorship of the Pathological Institute of New York State Hospitals, founded nine years earlier to study brain sections collected from the state asylums. As the freshly appointed professor of psychiatry at Cornell Medical College in New York City, he regularly travels back and forth across the state, lecturing and training doctors and medical students at the University of Buffalo and elsewhere, and advising lines of investigation in the asylum labs.

My grandfather stands respectfully at Dr. Meyer's elbow and learns how to extract telltale signs of disease from the multiple samples of blood, urine, tissue, and sputum. Goateed and fine featured, the fastidiously dressed professor, addicted to hierarchy and fine detail, cuts a sharp figure; his students either abhor or revere him, and my grandfather numbers among the latter. From the great man himself, Gerry is absorbing the precise techniques of brain dissection, cutting, staining, and imbedding tissues on histology slides with dabs of sticky, sweet-smelling Canadian balsam.

But he soon discovers that the Swiss psychiatrist is trying to be more than a specialized brain doctor. Tiring of a steady diet of autopsies, Meyer is now studying the mad from multiple perspectives. Therapeutically, he attempts to actively guide and re-educate his patients, a process he calls "habit training." He champions the concept of the individual case history, bringing together in a single file a comprehensive tabulation of the patient's life events, from changes in height and weight to political leanings, down to the pattern of wallpaper in his bedroom. Nothing is too trivial to record as such observations might lead to clues as to the patient's state of mind.

Within his adopted country, Dr. Meyer will forge an impregnable beachhead for scientific psychiatry and earn himself the title Dean of American Psychiatry. Significantly, he will replace the literary, old-world term "melancholia" with "depression"—a sleeker, more modern, more precise descriptor. In time, his pluralistic, all-embracing brand of thinking will come to be called "mental hygiene"—a philosophy that will fatefully inform the career path of Gerry FitzGerald.

A handful of critics, however, grumble that Meyer is acting out a messiah complex. A prosaic, plodding writer, his rigid, obsessive fact-collecting is not animated by any overarching philosophy; a man who stands for nothing risks falling for anything. Driven by grandiose rescue fantasies, he is puritanical about matters of sex, barring Freud's ideas of the unconscious mind from his own system, which he claims to be flexible and intellectually inclusive. For, if he were so inclined, Adolf Meyer might have intuited the subterranean forces that drove his own actions, the dark secret that he kept even from his closest intimates: the searing guilt he felt when, leaving Switzerland to seek his fortune in America, he heartlessly abandoned his mother to the swamp of a depression once called melancholy.

═

The spring of 1905 brings news of a momentous discovery. The German zoologists Fritz Schaudinn and Erich Hoffman have uncovered the corkscrew-shaped spirochete, *Treponema pallidum*, the micro-organism that causes syphilis. When the spirochete is found in the inguinal glands of patients suffering tertiary syphilis—the softening of the brain known as general paresis of the insane—it accounts for ten percent of all asylum populations. From the buzz encircling my grandfather's staff meeting, an exciting theoretical question arises: Do *all* forms of mental disturbance derive from some nasty, mercurial, infectious bug? Is there a bug for mania? For melancholia?

The discovery acts like a shot into the bloodstream of Gerry's restless ambition. Shunning the baleful cynicism afflicting so many alienists, he is wafted southward by the summer winds of 1905 to Johns Hopkins, the elite teaching hospital in Baltimore, Maryland. And so, inexorably, the autumn winds of 2002 waft his grandson, briefcase in hand, up the same quaint cobblestone Baltimore streets. On North Broadway, I pause to absorb the gilded grandeur of the most famous hospital in North America, rooted on a breezy hill overlooking Chesapeake Bay. Ornately embellished with Queen Anne trimmings, the four storeys of the red-brick and sandstone administration building rise up into a magnificent windowed dome, crowned by an iconic, needlelike tower erected in 1889; heavy green shutters, changed little by time, hang drowsily from the tops of the windows of the adjoining pavilions, as pigeons flutter among the brace of chimney tops.

I step inside and wander through the corridors. For me, this place holds two ghosts, their paths criss-crossing and colliding on the busy marble floors: a son chasing a father, a grandson chasing both, the years 1905, 1942, and 2002 melding as one. Here in 1942, my father spent seven months as an intern, living in the apartments above my head, struggling to live up to his father's lofty reputation. Here, punished by loneliness, he seduces one of his instructors, a married physician, even as he is engaged to his first wife, Caroline—an act of low betrayal that, had his principled father been alive, would have cut him to the quick.

In the hospital archives, I spread books and papers before me, reimagining the forgotten year of 1905. The hand-picked senior faculty is led by the quartet of extraordinary men known as "The Four Horsemen"; like Gerry's father, three of them are named William. The pioneering surgeon-in-chief, William Halsted, a sarcastic bully addicted to morphine, considers his operations works of art; the gynecologist-in-chief, Howard Kelly, a zealous religious fundamentalist, kneels in prayer before anaesthetizing his patients. William Osler, the physician-in-chief, is a native of Bond Head, Ontario, a village thirty miles north of Toronto; by 1905, the saintly fifty-six-year-old icon, a humane, optimistic man of immense personal charm, astonishing memory, lively wit, and intuitive gift for healing, has long occupied the high ground of living legend.

Finally there is William "Popsy" Welch, the pathologist-in-chief, a gentle, blue-eyed, cigar-smoking bachelor of Irish blood. Educated at the finest German laboratories, Welch is revered as an original investigator in bacteriology, inspiring teacher, and medical statesman; a stocky, goateed figure whom my grandfather will claim as a role model and lifelong influence. His biographers describe Welch as standing alone in his uncanny ability to spot raw scientific talent, guiding the young with fatherly precision to the ripe fields he himself could not harvest. His lab sets the tone for the whole of Johns Hopkins, the nucleus around which the rest revolves. Emotionally reserved, he never raises his voice above a tranquil, even tenor; possessed of extraordinary powers of concentration, he loses himself in his work for hours on end, cigar ash accumulating in the crevices of his waistcoat. Pathology is the pure science of medicine, and Welch the pathologist is himself a man pure of heart, living the life of spartan simplicity demanded by his exacting ideals. By middle age, he harbours only a single regret—that no woman ever "set his desk to rights." His protegé and collaborator, Simon Flexner, the inaugural director of the Rockefeller Institute of Medical Research, will later write of Welch: "Surrounded always with people, popular, adept at swaying men, the bachelor scientist moved on a high plane of loneliness that may have held some of the secret of his power."

As Gerry reaches to shake the hand of the great man, his pulse races and his eyes widen like a cat's. For a fleeting moment, my grandfather catches an unspoken message in the older man's New England accent and sea blue eyes: Both of them carry a similar maternal legacy, although Welch's is the weightier. His mother Emeline, a chronic invalid, pined for the loving attention of her busy husband, a Connecticut doctor; withdrawing to her bed, she suffered bouts of punishing physical and mental anguish that drove her into the arms of religion. Following the birth of her son William, Emeline declined rapidly, even as a team of Welch doctors—there were five in two generations of the family—ringed her bedside. She died six months later, abandoning her infant son to his lifelong hunger for medical excellence.

═

One day, as my grandfather doles out pills and medicines from his post in the Neurology Dispensary of Johns Hopkins, a thin, cerebral, thirty-year-old psychiatrist strides through the door and introduces himself. As Gerry shakes the hand of Clarence B. Farrar, he cannot yet know that fate has delivered a fifth horseman into his life, the one with whom he will travel the furthest. At five foot nine and 130 pounds, the pale, soft-spoken American presents a cautious, meticulous figure, a receding, clean-shaven chin, and an affected mid-Atlantic accent enhancing an air of cosmopolitan sophistication. At first blush, he seems a bit of a cold fish, a patina of arrogance masking a smooth, immature face. Or is it just shyness? In the figure of "C.B.," my grandfather divines another kindred Irish spirit, a father as much as an older brother.

Over the ensuing months, exchanging stories and ambitions, the two men will find much in common. In his new colleague's demeanour, Gerry discerns a familiar blend: the culture and sensitivity of his mother, the aloof asceticism of his father. Eight years Gerry's senior, C.B. was born in 1874 in Cattaraugus, New York, an upstate village in sight of the Adirondack Mountains that matched the size of Drayton, Gerry's Ontario birthplace. Farrar's father, Thomas Jefferson Farrar, a successful businessman of Irish Protestant ancestry, owned a three-storey frame house fronted by an open verandah—a picture of upper-middle-class gentility. Young Clarence took an early interest in psychiatric textbooks, avidly turning the pages hour by solitary hour; one Christmas, he asked to be given the works of Dante as a present.

In his village of nine hundred, the boy precociously estimated that one in three families were emotionally disturbed; his own childhood, however, he regarded as happy.

Gerry, top right, and C.B. Farrar, bottom right, Sheppard Pratt Hospital, Baltimore, 1906

Finishing the final two years of his arts B.A. at Harvard University, he fell under the lasting spell of two legendary figures: the philosopher George Santayana and the country's first professor of psychology, William James, a brilliant depressive who, like his father before him, suffered from bouts of neurasthenia and sporadic thoughts of suicide. Farrar spent one year at Harvard Medical School, then transferred to the recently opened Johns Hopkins where he was inspired by Osler. Graduating with an MD in 1900, Farrar moved widely across Europe, swimming freely in the Petri dish of intense organicism; it was a time of exploding interest in neurohistology, the microscopic study of brain tissue. He worked closely with Alois Alzheimer, the man who discovered the form of senile dementia that would take his name, and the Freiburg psychiatrist Alfred Hoche, a promoter of a doctrine of immutable syndromes whose ideas about eugenics and euthanasia will one day compel the attention of rabid Nazi ideologues.

But C.B. spent most of his time at the famous Heidelberg psychiatric "klinic" of Emil Kraepelin. Acidly caricatured by a rival psychiatrist as "a North German village schoolmaster writ large," the austere Kraepelin was not concerned with the causes of insanity but rather its rigorous observation and classification, and brooked no intuitive speculations on the interior of the human mind. "Every attempt to understand the mental life of another person in its inner workings," he declared, "is fraught with manifold sources of error"—a statement that Farrar embraced as a lifelong credo.

Farrar will, like William James, acquire five languages; in his voluminous writings, he will draw upon his refined appreciation of theatre, music, poetry, and literature. An obsessive bibliophile, his enormous personal library would eventually rival that of Sir William Osler. Farrar holds several cross appointments at Johns Hopkins and the Sheppard and Enoch Pratt Hospital, an elegant private sanitarium standing on a wooded hill seven miles north of the city. Recently appointed the head of the brain lab at Sheppard Pratt, he knows the place is in perpetual need of clinical assistants, for the turnover is rapid. When C.B. suggests FitzGerald apply, his friend is quickly accepted.

═

On the surface, Sheppard Pratt could pass for the stateliest of exclusive health spas. The central administrative building is flanked by two grand orange-brick edifices, each nearly four hundred feet long, separated by a swathe of green lawn, each housing seventy-five patients. The institution is a member of the Ivy League of private mental hospitals on the eastern seaboard—a genteel southern cousin to the McLean Asylum of Boston, founded in 1818, and the Hartford Retreat, founded in 1824, each catering to the New England aristocracy. Stained-glass oriel windows squint down from quaint, wedge-shaped towers encircled by latticed balconies; below, patients recline in the Adirondack chairs that dot the rolling lawns as staff serve gourmet meals on silver trays.

On the top floor of the hospital, perspiring in the oppressive Maryland heat, Farrar and FitzGerald move back and forth from each other's apartments and converse deep into the early morning hours. Gerry has already given C.B. a nickname—Geraldine—a whimsical, homoerotic play on the Norman Irish ancestry of the FitzGeralds as well as the ruby-lipped, golden-curled grand opera star, Geraldine Farrar, who only months earlier had made a spectacular stage debut in *Romeo and Juliet* at New York's Metropolitan Opera House.

On the morning of Good Friday, April 13, 1906, the same week of the death of his father, C.B. invites my grandfather to observe his examination of Miss M., a thirty-four-year-old sufferer of dementia praecox since the age of eighteen. Farrar generally attributes mental illness in women to their periodic cycles. Irritation of the menses, the uterus and ovaries, he believes, affects other organs, including the brain, causing "menstrual psychosis." He orders strict isolation and bed rest for the afflicted souls under his care, denying them letters and visitors.

In a sun-filled parlour, the trio of figures, two sane and one mad, take their seats. Characteristic of his prolific writings over the next six decades, Farrar, the strict Kraepelinian suspicious of depth psychology, will transform a stenographic clinical record into his particular brand of mandarin prose:

"As we see the patient before us, she is utterly degraded in every particular. All sense of modesty and shame has been absent during our period of observation. She has absolutely no regard for her personal appearance, is extremely filthy in her habits and is given over to the most disgusting practices of urodipsia and coprophagia [the ingestion of her own urine and feces]. Nothing that can be said to her makes the slightest impression, so far as her feelings are concerned. She leads a vegetative automatic existence, interrupted by occasional, sometimes violent impulses, and in this condition we shall expect her to remain until carried off by some accidental infection, perhaps in old age.

"Following is recorded a *causerie* of one of the patient's more accessible moods. Her part of the dialogue was uttered fairly reflexively, without the slightest show of feeling and with no change whatever in her set, almost expressionless features.

Dr: I should like to ask you a few questions.

Pt: I don't want to answer turkey buzzard's questions.

Dr: How old are you?

Pt: As old as my tongue and as young as my teeth. How old are you? You are an old hog out of the field. How did you get in the house?

Dr: Do you know me?

Pt: Why, yes. You're either a young hog or a petrified cannibal.

Dr: And this lady? (*Indicating a nurse.*)

Pt: That's a poor insane woman.

Dr: And this gentleman? (*Indicating Dr. FitzGerald.*)

Pt: Why, pig, p-i-g; pig, pig, pig, pig . . . (*Uttered in quick succession after the manner of calling pigs to dinner.*)

Dr: Why are you so insulting?

Pt: I don't want to move off the parlour chair.

Dr: What is your Christian name?

Pt: (*Gives her whole name correctly, adding her address in detail.*) I want to go home in that carriage. Angels will drive up for me.

Dr. You are a poet, aren't you?

Pt: I am a splendid poet in my own way. Now I am a very fine poet.

Dr: Do you want to go home?

Pt: Yes.

Dr: Why?

Pt: I think it would be the worst place for me.

Dr: What do you want to do when you go home?

Pt: Play. I want to be an angel instead of playing here. No, I want to go walk the streets of Baltimore. I want to go out to buy some things for myself and for other people. I would not want to go into a young man's room and ask him if he wanted anything for his children.

Dr: Do you know what day this is?

Pt: The 14th of November.

Dr: Did you know that today is Good Friday?

Pt: No, it is not. I suppose you go wherever you are included, if you came across a Kickapoo Indian. A hog with no morals would not move me off that chair there.

Dr: Do you know the president of the United States?

Pt: No.

Dr: Who do you think it is?

Pt: Why, I am the president of the United States. I was once upon a time.

Dr: What is the name of the president?

Pt: Why, George Washington.

Dr: The president now?

Pt: They don't have any president of the United States. They have a whole lot of fools up in the Senate—stand up and squeal like so many fools.

Dr: What do they have in place of a president?

Pt: A hog. Twelve idiotic statesmen. I don't like to talk to a skeleton. Women are fools. People used to keep quiet at church. There are certain rules.

Dr: Have you ever seen God?

Pt: Yes, I see him whenever I look in the glass.

Dr: Have you ever been married?

Pt: No, sir. I never had any desire to marry a man like you. I never could find anyone worthy of me.

Dr: Who discovered America?

Pt: I discovered it.

Dr: *(Repeats the question, severely.)*

Pt: I didn't do it intentionally.

Dr: What language do you speak?

Pt: I never speak any language of anger unless there is cause. Anger. Languor. L-a-n-g-u-o-r. Languor.

Dr: What is the first book in the Bible?

Pt: Genesis is the first book that I wrote. It is right to eat the fruit that was said to be forbidden fruit by the insane.

Dr: Have you any money?

Pt: No, not since you robbed me of it, you old orang-outang. I don't wish any red dress—a pretty blue dress. One for myself and one for my child.

Dr: Have you a child?

Pt: Yes.

Dr: But you are not married.

Pt: Single women can have children.

Dr: How long have you been here?

Pt: I have stayed here. They tell me to stay. *(Turning to the stenographer.)* What dirty, low Dutch cabbage are you?

Dr: Do you know Faust?

Pt: *(Mute.)*

Dr: Did you ever see Mephistopheles?

Pt: I see him now. *(Looking at Dr. Farrar.)* He has poked himself into the same room as me."

═══

In the fall of 1906, my grandfather receives word of a plum position in Toronto. Dr. C.K. Clarke, the newly appointed superintendent of the 999 Queen Street West asylum, is looking for a well-trained young pathologist to assume the position of clinical director, start up its first lab, and teach psychiatry to medical students. As a native Canadian, Gerry is a natural fit for Clarke: with three solid years of clinical and lab experience under his belt, he has worked at three different kinds of asylums, small and large, private and public, Canadian and American; and like his newfound friend Farrar, his star is rising in the psychiatric firmament—an up-and-comer working among the down-and-outs. As Farrar notes of Gerry: "The young man's restless mind would never be content with any activity savouring of routine."

A wintry dusk falls over December 28, 1906. Fresh from lecturing the hospital staff on Greek and Roman art, Farrar has organized a farewell party for FitzGerald, who is due to leave for Toronto on the morning train. The staff dining room holds a full quorum—doctors, nurses, matrons—buzzing with a brand of giddiness only a madhouse can inspire. On past nights, happy and sane among the sad and mad, Gerry and C.B. sometimes painted their cheeks with pieces of burnt cork and entertained the patients with a rollicking, blackface minstrel show.

For those who know my arriviste grandfather of burning ambition, horse racing is an association that will crop up again and again—the image of the galloping thoroughbred of refined bloodlines, whipped and blinkered, charging into the unknowable future, fixed on its final goal. Accordingly, C.B., the sly fox, has composed a mock menu that plays on Baltimore's Pimlico Race Course and the Preakness Stakes:

"Program of Meeting of the Sheppard and Pratt Jockey Club, 470th day. Starter, Dr. John G. FitzGerald. Soiree Mangeante to commiserate the sudden departure of Dr. FitzGerald from the United States of this country to Canada." The entrees include:

Pimlicolas (60 to 1); Sauce Ta! Ta!; Sauterne; Roast Turtle Dove (Stuffing à la FitzMoonshine); Choux-fleurs don't bother me; Family Squab(ble) padded.

"As I rise to descend to this occasion," Farrar addresses the tipsy staff, "I am I hope thoroughly unconscious of the solemnity and indignity of my position." He gently roasts his new Canadian friend for a few minutes, then wonders aloud "how one so tender in years can be so tough in experience."

"Dr. FitzGerald and I (I must speak kindly of him now . . . it's the last time perhaps I ever shall) have often tried to turn the tide of conversation into channels of literature, art, the beauties of nature and Baltimore, and biblical history, but in vain we have been beaten back from ditch to ditch, catching cold feet from trembling in the ditch water, back to table talk avidly anatomic and pathologic. He and I have at times been compelled to leave the table in haste, leaning on each other for support, our tears full of eyes, our throats full of arti-chokes, and our ears full of cotton. I fancy for the general uneasiness that there is a tendency on such an occasion to conceal sad and sober thoughts under pathologic mirth.

"But soon there is to be a vacant chair and our last and only comical assistant will have flown. We shall miss his optimism and his weather prognostications. We shall miss the sunshining of his countenance and the moonshining of his actions. We shall miss the stripes on his stockings. We shall miss his English jocularity. We shall miss the lithesome footstep, the gentle prattle of his voice and the merry ring of his napkin.

"FitzGerald—Plain case. Recurrent. Good prognosis for present. But gospel certainty of relapse."

In my Baltimore hotel room, a memory flings me a decade back to May 1992. Just weeks before the death of my father, I had awoken from a vivid dream, the kind that demanded my close attention. I scribbled it down in my diary and labelled it, "The Racehorse and the Doctor." I brooded about it for two days, then explored it in my weekly session with Peter.

A thoroughbred racehorse lies flat across an operating table, unconscious, ready for surgery. I stare at its front legs slit open by the scalpel, its bones, sinews, and veins exposed. With calm authority, the doctor intones, "The stride of the racehorse must be shortened." Horrified, I imagine this to be a bad thing; eventually I realize that, in fact, it is not.

The dream seemed to be telling me that both my grandfather and my father perpetually overreached themselves. Their intellectual selves far outdistanced the emotional and spiritual—and I was in danger of doing the same. Their talent and achievements were real, but they habitually trampled on the subtle wisdom of their inner voices. As a boy, I sensed this, but as I grew up, I imposed impossibly inflated expectations on myself, lost in clouds of Apollonian grandiosity. I expected far too much of myself, then far too little.

I linked the dream to my short-lived glory days as a skinny thirteen-year-old track star, setting school records. I thought of my father's sudden, unexpected attention to me, his cracking of the whip that led to the operations on the Achilles tendons in my heels. More than I knew, as a boy I had dug in my heels against my father's misguided desires. Stored inside me, the dream emerged to bring winged Mercury to heel. The dream restores, the dream forgives, the dream heals . . .

THIRTEEN

I Am a Mass of Flesh, and You Are Another

Where the insane are concerned, the public are not only indifferent, but terror-stricken, and very often heartless.

CHARLES KIRK CLARKE

On the day after New Year's 1907, my grandfather tramps along the sidewalks of Toronto's Queen Street West, clouds of factory soot darkening the snow underfoot and the air overhead. On the north side of the heavily travelled artery, Gerry passes rows of drugstore windows displaying every conceivable brand of quack remedy, the shadow profession of his upright father, still toiling in his village apothecary back in Harriston. As he strides up the curved driveway of the looming asylum, the puffs of Gerry's hot breath crystallize in the frozen air; suddenly, a mix of noxious odours assaults his senses—the fumes spitting from the chimney flues, the malfunctioning water closets, the crowds of filthy patients, disinfectants of carbolic and lye, the rotting aromas emanating from the smoky basement kitchens. If madness possesses a smell, it is the vile, frigid stink clinging to the walls of my grandfather's latest place of work. Even the brass address plate—999—suggests an inverted satanic numerology, the tattoolike stigmata of the hopelessly deranged.

Inside, he threads his way to the office of Dr. Charles Kirk Clarke. Two years earlier, "C.K." had been anointed the new superintendent of the asylum and he is now on the verge of assuming the first professorship in the University of Toronto's new department of psychiatry. Looking for all the world like the German philosopher Friedrich Nietzsche—broad forehead, walrus moustache, intense gaze—Clarke, nearly fifty, stands as game and vigorous as a pearl-white stallion; in no time, Gerry will esteem him as a father figure. No small irony is attached, for he cannot know that this man of flesh and blood embodies the seed of a future institution—"the Clarke"—where, six decades on, the soul of his own unborn son will die a death of a thousand cuts.

═

Clarke has hired Gerry as the asylum's first pathologist and clinical director, and both men know why: the nineteenth-century era of custodial "moral treatment" has failed dismally. On the day of my grandfather's arrival, he encounters the once high-minded, white-brick palace reduced to a roiling cauldron of despair—overcrowded, understaffed, violent, and unregenerate.

The indomitable Clarke is bent on a messianic mission of Protestant hue: Madness must be treated like any other disease, the earlier and more aggressively, the better; and the work needs to be set in an academic milieu. He is determined not only to recover the spirit of high optimism that drove the early years of moral treatment, but carve out scientific respectability for psychiatrists—and with it, professional autonomy.

Gerry's early mentor, C.K. Clarke (1857–1924)

My grandfather numbered among the new breed of trained scientists of the mind and his first duty will be to create a lab, modelled on C.B. Farrar's back at Sheppard Pratt. "The morbid conditions must have definite basis in the diseased brain," Clarke insists, "and until we know what that is, there must be no end of groping." Changing the name of the institution from "Asylum for the Insane" to "Hospital for the Insane," he has put his faith in the modern generation of early twentieth-century scientists moving swiftly up the ranks. In young Gerry FitzGerald, he has found a ready disciple.

Born in the village of Elora, just six miles east of Gerry FitzGerald's birthplace, C.K. Clarke was, like my grandfather, a hurricane of disciplined energy who took up manly pursuits to mask a sensitive temperament. A watershed moment occurred when the young Clarke was serving under his brother-in-law, William Metcalf, the medical superintendent of the Rockwood Asylum in Kingston. Their progressive policies included an "open door" policy that would tragically backfire. On August 13, 1885, a patient, Patrick Maloney, slipped into a new cottage where Clarke and Metcalf were working. With a purloined dinner knife, the mad Irishman stabbed Dr. Metcalf in the stomach. Risking his own life, Clarke disarmed the assailant and carried his mortally wounded brother-in-law in his arms to the main building, a distance of 150 yards, screaming for help. But it was too late.

Only the day before, the twenty-eight-year-old Clarke had submitted his resignation to enter private general practice. Instead, upon Metcalf's death, Clarke fatefully accepted the position of superintendent "in order to protect several hundred defenceless creatures from a political hireling who might be pitchforked into the position." On the dull edge of a single dinner knife, the historical course of Canadian psychiatry was irrevocably diverted.

Throughout his career, the phenomenon of mental deficiency—"feeble-mindedness"—fascinated Clarke. A confirmed eugenicist, he feared that Canada would be overwhelmed by an "enormous current of alien degenerates"; until 1902, there were no medical inspections of the tide of immigrants entering Canada and Clarke was now lobbying hard for trained psychiatrists to act as medical inspectors. Believing that Great Britain was deliberately using Canada as a dumping ground for defectives, Clarke wrote that "an analysis of their characteristics is an interesting study in degeneracy of a type we rarely see in Canada. Sexual perverts of the most revolting kind, insane criminals, the criminal insane, slum degenerates, general paretics, in fact weaklings of all objectionable types are represented." Clarke did not disguise his personal antipathy for Jews, Gypsies, Blacks, and Chinese; knowing politicians were anxious to cultivate the vote of the new immigrant, he would temper the blade of his caustic rhetoric and wait until the political climate was more favourable to his ideas. The great German philosopher Friedrich Nietzsche, who died of syphilis in an asylum in 1900, had said it all: The advancement of civilization depended on the exercise of cruelty.

Scientifically, Clarke knows exactly where he stands: Psychiatrists need to solve a medical, biochemical problem, not a psychological one, as Freud and Jung, who will first meet in this very year of 1907, are proposing. Like most of his peers, Clarke will come to dismiss psychoanalysis as "sex problems ad nauseam." Clarke hungers for a medical breakthrough, a gleaming arsenal of magic bullets on which to depend; nothing feels worse than feeling useless. He longs to build a new clinic, based on Kraepelin's in Munich, where trained doctors can study patients scientifically, free of government-imposed administrative burdens. Close to the university and the general hospital, dedicated to education, research, treatment, and record keeping, the hospital would hold seventy-five patients, mainly voluntary and easily supervised; they would avoid the stigma of asylum incarceration and could be reached before they became untreatable, chronic cases.

Initially favouring the idea, the provincial government withdraws its support when the alienists at the other provincial asylums protest that the new clinic would undermine their own professional status. Clarke is castigated as a selfish empire builder, trying to monopolize the most treatable patients and divert the rest to the public asylums. In the end, professional dissension and the indifference of the general public kills Clarke's clinic—for the time being, at least. Clarke's passionate and long-standing hatred of his own father, a prominent provincial politician, might explain his blanket disdain for *all* politicians; but to acknowledge such a "transference" of feelings risked passing ammunition to the thinkers of the equally detested Freudian patriarchy.

═

So it is that in the numbing winter of 1907, Gerry FitzGerald embraces the aberrant world of 999 Queen Street West, its pills, patients, and politics, eager to leave his mark. In his second week, he writes his friend C.B. Farrar back in Baltimore:

999 Queen Street West Asylum, Toronto

> "My dear Farrar, I hereby deliver the first epistle of John to the Ephesian, to mein fuhrer, to whom I am so much indebted. Two weeks ago this afternoon, we started in and we have been busy sowing and a lot of seed has fallen on good ground. My own duties are, until we get our laboratory material (all of which has been ordered) clinical . . . We have shared our conferences, three mornings a week so far. I have held the floor and have had three beautiful type cases, one of psycho-motor depression, one of maniacal excitement and one of pure paranoia. We record the opinion and file it in the case on a special conference sheet; in this way it must be done . . . All the men are enthusiastic and the future looks promising."

In February 1907, C.K. Clarke publishes the inaugural issue of the *Bulletin of the Ontario Hospitals for the Insane*, the first psychiatric journal in Canada; young Dr. FitzGerald is enlisted as a co-editor. Over the next twenty-one months, Gerry will publish nine scientific articles in other Canadian and American journals. In "Modern Methods of Treating the Insane," he deplores "the spectacle of non-medical, minor public officials dictating the policy of institutions, guiding the medical officers, in fact, gentlemen, doing everything but making morning rounds." Week after week, he travels eastward to the university campus to lecture undergraduates as a demonstrator in psychiatry; back at the asylum, he is delighted by the accumulating reams of scientific data, a cavernous factory of Meyer-esque fact brimming with the sexual and religious angst permeating puritanical Edwardian Toronto.

In one voluminous forty-five-page case history—one of dozens I was able, for a steep fee, to pry loose from the Ontario Archives—Gerry assiduously collects the clinical details of "an extremely interesting character . . . an alien, one of the undesirables in the immigrant class." He is thirty-six-year-old James Peerless, a reclusive Englishman suffering florid bouts of manic-depressive psychoses. As I study his photograph clipped in his file, I note that his doleful face uncannily resembles that of Gerry's father, Will.

Gerry performs an exhaustive medical examination and then asks Peerless, a literate man with a poetic bent, to write down the story of his life in his own words as part of his therapy. Drawn to Gerry, the young man churns out reams of scribble, including poetic descriptions of his "wild and scaring nightmares," and a portrait of his intimidating father who was responsible for the certification of lunatics back in his English hometown. As I turn the pages, I realize that Peerless's haunted childhood experiences presage my father's—and my own—on Balmoral Avenue. I smile when Peerless diagnoses Gerry's own manic work pace and urges him to slow down for the sake of his own health:

> "Dr. FitzGerald—Apart from Theology, apart from Science, apart from Friendship, I am a mass of flesh, and you are another. You are healing my flesh well, and I hope in turn

your own is treated well and that you have strength of mind so to shape your present environment animate & inanimate to fit your best interest, especially the ultimate. I rather fancy you have not a good hobby apart from your work to provide your mind sufficient recreation; I should think that although you are not by any means of sporting tendency, that skating or tobogganing even should be moderately indulged in."

The Canadian government has recently passed a law that immigrants in charitable institutions who had arrived in Canada two years earlier are eligible for deportation. And so, after a few months treatment, my grandfather sends his "undesirable" patient back to England with the pronouncement "much improved."

In another case, I find Gerry cutting open the brain of a depressive Irish woman who had killed herself by slashing her throat with a razor. He records that another Irish woman "tells me she is possessed of the devil, that something came out of her mouth and blighted the flowers on her father's grave. She has had very many somatopsychic ideas recently, complaining that she has lost various organs and she has committed the unpardonable sin." Hour after hour, Gerry perches his spectacles on his furrowed forehead and peers at the stained, microscopic sections of brain and rivulets of cerebrospinal fluid. Perhaps here lie hidden the dark physical traces of the memories of the mad; perhaps here the genesis of sin itself can be revealed

Tranquil public relations stills, circa 1907, *mask an overcrowded, understaffed institution in desperate need of reform*

by the secular miracles of the scientific method. Meanwhile, here, as in most asylums across North America, the cadavers of dissected inmates are buried in numbered graves to protect their family names.

In March 1907, my grandfather reads the sensational press reports of a robust, buxom, blue-eyed, thirty-eight-year-old Irish cook arrested in New York City. The previous summer, Mary Mallon had unwittingly infected six members of the household of a wealthy Manhattan banker, vacationing at their Long Island summer home, with typhoid fever from the meals she prepared. An intestinal bacterium, typhoid was widely associated with poverty and filth, transmitted by food or water contaminated with human feces or urine; how could such a plague attack a clean-living, upper-class family?

Police and public health officials had spent months on the trail of "Typhoid Mary," their bumblings worthy of a comic opera. When they finally cornered their fleeing quarry, she resisted arrest with leonine ferocity. At the New York City Health Department, Dr. William Park, chief of the bacteriological lab, will find a pure culture of typhoid in a sample of Mary's stool. In Germany, Robert Koch had recently discovered the phenomenon of the "healthy carrier"—a person who suffered no symptoms yet readily infected others, undetected. Yet Mary stubbornly refuses to believe she is dangerous, raging against the trampling of her free rights and the prospect of being confined against her will. Exploiting the angle of the filthy immigrant menace—a lowly domestic cook, a woman, and even worse, Irish (and therefore mad)—the press runs lurid editorial cartoons showing Typhoid Mary breaking skulls and bones like eggshells into a skillet; her case generates a public health *cause célèbre* and a heightened awareness of the fledgling science of bacteriology. The first monitored healthy carrier of an infectious disease in North America, Mary Mallon will spend the rest of her life incarcerated on a bleak island off Manhattan. To the end, she will rail against the doctor who tested her, famously asking how he would like to be labelled "Typhoid William Park."

Gerry is fascinated by the case; he knows that many infectious diseases—typhoid, syphilis, diphtheria, malaria, smallpox, and pneumonia—can cause sudden, transient attacks of insanity. As he absorbs the strange tale of the tragically willful Irish cook, he does not know that destiny will soon deliver him to the lab door of her nemesis, the celebrated Dr. Park.

═

On Sunday, December 9, 1907, Alice FitzGerald lies gravely ill in her brass bed in the yellow-brick house at 88 John Street in Harriston, the same brass bed where she had conceived, and given birth to, her four children in quick succession. It is her fifty-first birthday, the same day her oldest and favoured child, Gerry, the glamorous neuropathologist, turns twenty-five. He has taken the train up from Toronto for a visit, and as he perches on the edge of his mother's sickbed, as he whispers words of comfort in the place of his own conception, he senses it will be the last time mother and son will ever share their birthdays together.

He is right; only eleven days later, just before Christmas, Alice succumbs to a stroke, or so says her death certificate. Released from her long years of invalidism, her body is transported to Toronto and buried in a shaded plot beside her parents in Prospect Cemetery on St. Clair Avenue West, a few blocks north of the Queen Street asylum.

Over the ensuing months, Will FitzGerald falls into a grief-stricken funk; his drugs had failed to save his wife, and now they do little to soothe his own deepening melancholy. In an uncharacteristically reckless move, he sells the family's empty nest on John Street and his busy drugstore on Elora Street—the solid middle-class life he has worked so hard to achieve—and exiles himself in the town of Bradford, about thirty miles north of Toronto. His second son, twenty-four-year-old Sidney, is burying his head in law books at the University of Toronto; Bill, twenty-two, is shuffling papers behind a teller's cage at the Bank of Toronto; and Hazel, the youngest at twenty, has recently joined her brothers in

Toronto, seeking work. Will, the hard-headed, fatherless disciplinarian, was more dependent on his wife, sick as she was, than he or anyone else realized; over the next decade, mired in melancholia, unmoored from hearth and home, he will seek the sympathy and solace of his four grown children, weighing them down even as they each strive to rise in the world.

═

Ever restless, my grandfather yearns for more extensive training beyond the bleak, foul-smelling back wards of 999 Queen Street West. Stoked by Clarke's and Farrar's vivid stories of Germany, he sets about studying Hugo Lang's *German-English Dictionary of Medical Terms.*

Simultaneously, he is learning the language of love. On a summer visit to the medical school at the University of Western Ontario in London, 120 miles west of Toronto, Gerry finds himself crossing South Street and touring the wards of the Victoria Hospital. Family legend has it that it was here he first meets a refined, chestnut-haired young lady his own age, reclining in bed, recovering from some long-forgotten malady. Edna Leonard is a graduate of the private girls' boarding school, Havergal College, and heiress to a foundry fortune. Gerry takes Edna's pulse, then holds her hand; having lost his invalid mother only months before, perhaps he knows, even in that moment, that she will one day become his wife.

That same summer of 1908, Gerry makes an impetuous move: Without first securing a position elsewhere, he informs Dr. Clarke that he is resigning his duties effective October 1, although Clarke himself had given impetus to the decision when he confided, "There is no future in psychiatry for a man of your talents." His blood fired by the confessional agonies of his patients, the death of his mother, the collapse of his father, and the promising, pale-skinned figure of Edna, my grandfather now begins an intense period of international travel and training, perhaps, paradoxically, escaping his troubled, widowed father in order to find a way to save him.

C.K. Clarke resigns himself to the loss of a promising protegé, but wastes no time landing a stellar replacement. On September 25, 1908, a short, pale, immaculately attired twenty-nine-year-old Welsh neurologist arrives in Toronto from England, installing himself in temporary digs at 85 Bloor Street East. His impressive credentials sullied by unproven accusations of sexual scandal with his patients, Ernest Jones is making a fresh start in the New World. He clicks instantly with Clarke and FitzGerald, the Canadians intrigued by the newcomer's intense, searching eyes, bursting energy, precise speech, and brisk, no-nonsense, military bearing.

Born of humble origins in a Welsh village, Jones burns with an ambition not unlike my grandfather's. As a child, he had suffered nightmares; after a religious crisis as a teenager, he became an atheist. Self-confident and intellectually precocious, he entered university at age sixteen, where he took a medical degree and keenly absorbed the works of Darwin and Huxley. He was drawn to neurology and the mysterious diseases of the nervous system that resisted existing treatments; using hypnosis and the latest psychotherapeutic methods, he becomes deeply intrigued by "a man in Vienna who actually listened with attention to every word his patients said to him."

While Jones was working in the Munich clinic of Freud's adversary, Emil Kraepelin—he diagnosed the German as a coarse fellow with a "gruff manner who showed no sensitiveness or sympathetic insight whatever into his patients"—he got wind of C.K. Clarke's plans for a new psychiatric clinic in Toronto. Clarke arranged for Jones to inherit my grandfather's battery of positions at Queen Street, and then some: teach pathology and psychiatry at the university and Toronto General Hospital; conduct pathological exams on asylum patients; and co-edit the *Bulletin*. From Toronto, he will launch a voluminous correspondence with Freud, the two men exchanging seven hundred letters over the next thirty years, each struggling to improve his knowledge of the other's language.

Standing under the neoclassical dome of the Toronto asylum on a fall day in 1908, my grandfather bids adieu to Ernest Jones, his bristly Celtic kinsman, to whom he has quickly transferred his affections, as he does to all dedicated men of science. They agree to keep in touch, but for now, Gerry is rushing off to Boston, exploring fresh professional pastures. As Jones studies the face of a potential convert, the thought crosses his mind: Will this lean and hungry Canadian, flirting on the fringes of brave new worlds, one day fall into bed with Sigmund Freud?

FOURTEEN

Salvation Lies in the Most Concrete Records

> The search of an investigator for the Unpardonable Sin;—he at last finds it in his own heart and practice.
>
> NATHANIEL HAWTHORNE

September 2003: I am ensconced in the archives of Harvard University, bent over aging onion-skin documents that hold traces of Gerry's youthful, darting movements. On nearby Longwood Avenue loom the neoclassical marble columns of the Great White Quadrangle through which, on a brisk October morning in 1908, my twenty-five-year-old grandfather bounded up the steps of the new pathology building. If he found the austere grandeur of this place inspiring, I find it, nearly a century on, a little too great, a little too white.

Gerry has come to Boston, that most Irish of American cities, to pick yet another fevered brain—this time, Dr. E.E. Southard, a polymath of astonishing intellect, a prolific writer, philosopher, psychologist, neuropathologist, bacteriologist, etymologist, philologist, and intercollegiate chess champion. Over six feet tall and two hundred pounds, Southard dazzles his peers with his profound powers of concentration, his superhuman memory, his brilliant and ceaseless turns of conversation. He possesses enough ironic self-awareness that he has nominated himself a charter member of Harvard's "Hypomania Club."

"His extraordinary acuteness of mind, his play of humour, his powers of observation, and his gentleness, all these had a most profound effect on me," my grandfather will later write of the mentor who will meet a premature death at age forty-three, when he is infected with a deadly bacterium. "His friendship was one of the pearls of great price which in my brief sojourn in Boston I acquired."

Three times each week, a steam train carries the two men thirty miles north to the Danvers State Insane Asylum; within months, its name will be changed to the more palatable Danvers State Hospital. The red-brick institution lies a short distance from the village of Salem, the site of infamous witch trials three centuries past; and it is here I have now come, together with my partner, Katy, a kind-hearted psychotherapist of German immigrant blood who entered my life two years earlier. Her own respect for the power of unconscious forces—not to mention her passion for Victoriana—is proving of inestimable value in my searchings. Under a radiant sun, the pair of us circle the decaying, long-abandoned madhouse stretching hundreds of yards across a thickly wooded hilltop, its landmark 130-foot spire and sloping green-slate roofs casting knifelike silhouettes over the weed-choked lawns. The eerie stillness and emptiness mimic the background scenery of a silent movie and as my eyes scan the maze of interconnected buildings, the walls over three feet thick, the image of 1,300 shuffling inmates, mostly Irish, most hopelessly mad, bursts open like a wound. Inside the pathology lab, the ghostly figure of my grandfather hurls himself into fastidious experiments, autopsies, and clinical conferences. Here the enlightened brain doctors of Harvard have replaced the priestly robe with the white lab coat; here, "sin" has been rechristened "neurosis" and "psychosis."

Gerry is inspired and guided by Southard, a furious germinator of ideas and catalyzer of research; a colleague says Southard's lists of projects would keep twelve men busy for one hundred years. Bent on proving that psychiatry has outgrown its reputation as the bastard child of medicine, Southard and FitzGerald inject horse serum into the brains of guinea pigs, inducing violent and fatal reactions. Over their shoulders gleam two thousand glass jars holding carefully dissected brains, organs, and spinal cords immersed in Zenker fluid; wooden filing cabinets bulge with hundreds of autopsy reports. Southard has built up a gigantic index of symptoms drawn from seventeen thousand case histories. "Salvation" he assures my grandfather, "lies in the most concrete records."

On the front steps of the crumbling asylum, I pose for a photograph, my lips forming an incongruous smile. It is then that it hits me: My grandfather was *happy* here. Yet as I linger beside a solitary apple tree, I feel the boarded-up windows, sad towers, and hollow silence conspiring to rebuke my presence, as if I am once more a child in the nursery in Balmoral; as if I am not, and never was, welcome on this earth—a feeling I have carried inside me ever since I began to remember. Then, a blue police car rolls up the hill and a forbidding, pistol-packing officer wearing the darkest of sunglasses curtly orders us off the property; public health officials have declared the place toxic, infested with rotting asbestos. I am among the last to see the old asylum up close, for within months it will be condemned and the site occupied by luxury condominiums.

That night, back in Boston, Katy and I celebrate my fifty-third birthday. We ride the elevator up to the Top of the Hub, the fifty-second floor of the Prudential Tower, the city's tallest building, where we drink in the panoramic scene. In a dream hours later, I am rising up an elevator to the fifty-third floor of an unknown building. Stepping out into a room crowded with people, I feel dizzy, unsteady on my feet; prudently, I decide to turn around and take the elevator back down to the ground to recover.

Inexorably, nature dictates that the son must pass and surpass the father, claiming the high ground of his biological destiny. Yet within my own family, something unnatural has unfolded across the generations. Fifty-three was the age at which my father silently climbed the stairs to the third floor of our home in Forest Hill, lay down on my brother's bed, and thrust a lethal dose of morphine into his arm. Perhaps my unconscious is delivering a coded message: My fiercely competitive father and grandfather lived far too much "in their heads," in their straining, Icarus-like aspirations for elevated power and knowledge and status, they defied the natural forces of gravity and their mortal limits. The subtle wisdom of flesh and blood and body was calling them down from the rarefied air to mother earth. But they did not listen.

═

Prompted by Southard, my grandfather invites Ernest Jones to consider working at the neurology department of Harvard Medical School with a view to joining the staff of the newly conceived Boston Psychopathic Hospital, to be run by Southard when it opens three years hence. It is something Gerry himself is seriously considering. But Jones turns Gerry and Southard down; he is shifting away from lab-based brain science and diving deeper into the unruly waters of psychoanalysis. He is appalled by the "colossal ignorance" of American doctors; he has not met a single one—except the Austrian-born New Yorker A.A. Brill—who has read Freud's epochal *The Interpretation of Dreams*, nearly a decade old. Jones's strategy is to keep up the scientifically respectable neurological work; then, when he introduces psychoanalytic ideas, the medical profession would be more inclined to listen, inoculated with a series of small doses. Yet he is not yet prepared to hand over the reins of psychoanalysis to lay people. "As soon as therapy becomes separated from pathology and from diagnosis," Jones remarks, "it loses its scientific basis and risks falling into charlatanism."

Most traditional psychotherapy is ineffective, he argues; psychoanalysis offers "a deeper insight into the essential nature and origin of the morbid phenomena." He believes that hypnosis or persuasion merely blocks "the outward manifestation of the underlying pathogenic idea"; the techniques of psychoanalysis—free association, dream interpretation, and analysis of the transference of feelings rooted in the past—effectively treat the symptoms by ferreting out and removing the underlying pathogenic idea, making the conflicting wishes of the unconscious mind conscious.

Scientific objectivity is gained through the close, painstaking observation of the transference and countertransference of thoughts and feelings, often sexualized, between therapist and patient—an "inside job" more attuned with the feminine, empathic, intuitive, artistic, and subversive.

On September 10, 1909, Sigmund Freud, Ernest Jones, Carl Jung, Sándor Ferenczi, and Abraham Brill converge on Clark University in Worcester, Massachusetts, an hour's train journey west of Boston. It is Freud's first and only visit to North America and on this famous occasion he is reputed to have uttered the ironic remark: "Gentlemen, I bring you the plague." Despised in Europe, the fifty-three-year-old Freud is delighted to be treated as an equal by the American elite—his conference co-speakers include Adolf Meyer, Gerry's old teacher in the Buffalo asylum—and as Freud steps to the podium to deliver his lectures, it feels, he says, "like the realization of some incredible daydream." Instead of the sex-crazed loose cannon many of the audience had expected, they encounter a conservative, Old World gentleman of precise manners and speech, given to succinct, circumspect observations. "We ought not to exalt ourselves so high," he declares in one sample of his astringent style, "as to completely neglect what was originally animal in our nature." Freud is still suspicious of the depth of Jones's commitment to the movement as he seems all too willing to compromise the radical sexual material in the service of wooing converts. Freud confronts Jones about his loyalty, then in a conciliatory gesture comes to the train station to see him off to Toronto. Jones had earlier invited Freud to stay with him

Sigmund Freud (seated left with colleagues Carl Jung, seated right, and Ernest Jones, standing middle), at seminal psychoanalytical congress at Clark University, Massachusetts, 1909: "Gentlemen, I bring you the plague."

in Toronto, but he declined. Henceforth, Jones will promote Freud with a near-rabid devotion.

So it is that my grandfather stands astride the seminal, conflicting factions raging in psychiatric circles across Europe and North America: the established biomedical model, viewing all mental illness as organic, biochemical, or genetic in origin; and the emerging, revolutionary challenge of psychoanalysis and cathartic talk therapy, forged by Freud and his small band of devoted acolytes.

The battle lines are being drawn. Of the North American psychiatric elite, Farrar and Clarke remain immune to the siren song of the Viennese sorcerer, and Meyer, Prince, James, Southard, et al. keep a tentative, respectful distance. Only a handful of men, led by Jones, Putnam, and Brill, embrace Freud unequivo-cally. Open-minded and freethinking as many are, members of the medical profession look askance at the fledgling phenomenon of psychoanalysis, with its unsavoury pansexualism and murky talk of the unconscious mind. Sustained, unstructured listening is not the kind of skill valued by doctors trained to treat illnesses they regard as organic; they fail to warm to the suggestion that symptoms are clues, full of meaning, rather than something to be medically labelled and eradicated. The neuropathologists, as a rule, dismiss the Freudians as no more scientifically credible than an unruly tribe of witch doctors. For when they peel back the mushy layers of brain tissue, they see, clear as day, no shred of evidence for the existence of the unconscious mind.

Ten days after Freud's return to Europe, my grandfather has reached a crossroads in his career. He will continue to confer with Ernest Jones, but after five years of steady work in five different asylums—public and private, massive and modest—he has become intensely frustrated that he can do little or nothing to help the nervous and the mad. Influenced by his new friends in Boston, he puts his faith in the potential of the microscope and accepts a post as lecturer in bacteriology at his old alma mater, the University of Toronto.

My grandfather is reaching a crossroads in his personal life as well. A week after returning home, he receives a letter from Ernest Southard: "Please convey my very best to the fair Anonyma! And permit me to tell her she is very, very lucky to get a man who is both sane and idealistic!"

"The fair Anonyma" was, of course, my grandmother.

Another Harvard colleague, Frederick Gay, writes Edna:

> "My dear Miss Leonard, I have known about you for about a year now but it was only the other day, by direct appeal, that I was able to extract your name from FitzGerald. Just think of it, my dear young lady, I have been working for a year with a heavily-in-love young man whose status was obvious to even so short-sighted a person as a married man!
>
> "I need scarcely rehearse to you the personal charm of the man you are to marry but it may please you to know that we too have appreciated it. My interest in FitzGerald is far deeper than any mere enjoyment in being with him. He has chosen a splendid field of work and has made a worthy start in it. He has already innate the first essential of a scientist—honesty; he has acquired the second, namely a judicious critical sense; and the rest—power to work and a proper development of original thought it will be your privilege to see increase as the years go on. You know that your life will be happy and I can assure you that it will be interesting and intellectually satisfying."

Ah, yes—*intellectually* satisfying.

==

In the fall of 1909, my grandfather returns to the dreaming spires of the medical school of the University of Toronto. He finds a place to live at 17 Oriole Gardens, a quiet street in Deer Park within sight of the four-faced clock tower of Upper Canada College and only a short stroll from Campbell Meyers's sanitorium on

Heath Street. Only fifteen months earlier, Gerry was guiding young undergraduates through the labyrinthine fields of psychiatry; in the labs overlooking the greensward of King's College Circle, he now faces rows of medical students drawn to the cutting-edge science of bacteriology. While he will never fully surrender his drive to understand and heal insanity, he is now widening his focus to include the full range of infectious *physical* diseases, spurred by the growing cries for the radical reform of the ever-worsening state of Canada's public health.

Between his classes, Gerry sometimes catches the flashing eyes of the indefatigable Ernest Jones as he rushes down the corridors, carrying sheaves of papers, and the two men pause to talk shop. Besides traces of Celtic blood, they share disciplinarian fathers, doting mothers, and a single-minded obsession with high standards and efficiency; as Jones's father liked to say, "There's always room at the top." Jones is an accomplished chess player and figure skater, and unlike the more puritan Canadian, he indulges in noisy singsongs, puffs foul Turkish cigarettes, and attracts women like a magnet. As if compensating for his five-foot-four-inch stature, Jones, "the little Napoleon," is forever trying to best his colleagues, to the point of suffering an omnipotence complex that masks a terror of not knowing, of being seen as wrong. He hates the extremes of the Toronto weather, the blazing summers and frigid winters, but the intellectual climate he finds even worse: "It was not merely that I found myself back in the biblical and Victorian atmosphere of my boyhood—that would have been bad enough to someone bent on emancipation—but it was the dead uniformity that I found so tedious: one knew beforehand one's opinion on every subject, so there was a complete absence of mental stimulation or exchange of thought."

Often tactless and disingenuous, Jones is readily embroiled in fierce departmental struggles while Gerry remains a more relaxed political animal, relying on the gentler arts of charm and persuasion. Jones busily foists the controversial works of Freud and Jung on both medical students and the asylum staff of 999 Queen Street

West. As he will later write to Freud: "Two hundred innocent youths are being severely inoculated with psycho-analytical doctrines under the official auspices of the University of Toronto." Using the city as his base, Jones makes various "raids across the border" to address assorted medical and psychological societies in American cities; every summer, he sails to Europe to read papers before an international congress. Freud is relying on Jones's networking and administrative acumen to build a global movement.

Throughout that year, Jones and FitzGerald work and lecture independently, yet cross over into areas of mutual interest, particularly syphilis and psychotherapy. Jones now occupies a three-storey house at 407 Brunswick Avenue in the wooded Annex neighbourhood at the northern tip of Toronto. His "only extravagance" is a collection of five thousand books that overflow into every room in the house. His self-described "harem" includes his wealthy, charming but troubled Dutch-Jewish mistress, Loe Kann; his sisters, Elizabeth and Sybil; and two female servants. Addicted to twice-daily injections of morphine because of a chronic kidney ailment, Loe is given to cycles of manic depression and is bedridden much of the time, where she nurses a burning hatred for her mother from whom she inherited her substantial wealth. Jones passes Loe off as his wife, but the wagging tongues of Toronto the Good know the foreign doctor is brazenly living in sin.

Here Jones will write *On the Nightmare* and *Hamlet and Oedipus*, his most original and influential works. "All the beliefs about the Nightmare, in whatever guise," he concludes, "proceed from the idea of a sexual assault which is both wished for and dreaded." One outraged Toronto doctor, Herbert Bruce, rails that the *Hamlet* paper, probing the slippery crevices of the incest taboo, can only have been composed by a sexual pervert. When Jones reprints a lubricious case history in 999 Queen Street West's *Bulletin* in April 1910, it arouses a storm of controversy from which he will never recover. He explicitly details the sexual life of a thirty-nine-year-old manic-depressive Toronto woman and diagnoses a perversion in which she "identified the act of fellatorism with the partaking of the holy sacrament," resulting in "a complete and exhausting sexual orgasm" as she sucked the rim of a cup of water. Even though Clarke publicly supports Jones, he is appalled: "Any ordinary reader would gather Freud advocates free love, removal of all restraints, and a relapse into savagery." Quietly, Clarke removes his unrestrained colleague as co-editor of the *Bulletin*. Canada is proving a barren breeding ground for the seeds of the talking cure.

═

On Monday, April 4, 1910, Ernest Jones, still hopeful of pulling my grandfather into the Freudian camp, scribbles a note acknowledging Gerry's impending wedding and overseas honeymoon:

> "I wanted of course to wish you bon voyage for a pleasant time in Europe, and to convey my warmest congratulations and well wishes on the other little matter. I have thought out a rather original wedding present for you, namely a complete set of Freud's works. As marriage is at times an adventurous enterprise, no one contemplating it will be harmed by an addition to his knowledge of the human soul. Unfortunately it took longer to find the books in Germany than I had calculated, so they will not arrive in time. That is not altogether a loss, for I daresay you won't want to be bothered reading much German in Brussels."

Gerry and Edna marry in London, Ontario, April 1910; Ernest Jones presents Gerry with the complete works of Freud as a wedding gift

Five days later, Gerry marries "the fair Anonyma" in her hometown of London, Ontario. A formal photograph shows the wedding party arrayed under the leafless trees of early spring. Because of the time delay of the camera shutter, the rows of faces are forever fixed in unsmiling expressions. Gerry stands next to Edna, whom he has nicknamed "Ted," his hands are not wrapped round his bride's waist but tucked behind his own back. Sidney, articling in Toronto, looms behind their sister, Hazel; their father, Will, and brother Bill, are both strangely absent.

This same month of April, Sigmund Freud's Vienna Psychoanalytic Society holds a symposium on the phenomenon of suicide. On the verge of bitterly parting company with Freud, Alfred Adler theorizes at length about feelings of inferiority, revenge, and anti-social aggression. Wilhelm Stekel relates the act to masturbation and its attendant guilts; then he articulates the famous principle that will dominate the discourse for years to come: "No one kills himself who has never wanted to kill another, or at least wished the death of another."

Freud suggests that the mystery of suicide will not be better understood until more is known about mourning and melancholia. The pathological mourner who has lost a loved one to sickness or death, he is beginning to suspect, incorporates the beloved within his own thoughts; in time, he comes to believe he has killed the object of his desire and therefore, inescapably, must atone for the crime. Like the ghost of Hamlet's father, a ruthless grand inquisitor courses through the dreams of the grieving and the aggrieved, demanding bloody revenge.

FIFTEEN

The Travel Bug

In the night you hide from the madman
You're longing to be;
It all comes out from the inside
Eventually.

STEELY DAN
"Here at the Western World"

The year 1910 proves a bellwether for Canadian public health. With the founding of the Canadian Public Health Association and mounting calls for a national Pasteur Institute in the wake of a serious rabies epidemic in Toronto, the newly married Gerry FitzGerald envisions himself as the spark plug of a national movement.

Speed is of the essence and speed is his personal forte. In an extraordinarily intense three-year period, my grandfather pinwheels back and forth across North America and Europe, a man possessed. As a part of a "working honeymoon" in the summer of 1910, he toils in the iconic Pasteur Institutes in Brussels and Paris, sacrificing his holidays on the altar of a single-minded goal: siphoning off wisdom from the minds of inspired men. In Brussels, he befriends the fastidious, Nobel Prize–winning scientist Jules Bordet, a man of icy logic who is pioneering the healing properties of blood sera. In Paris, Gerry collars the ascetic, hawklike fifty-seven-year-old Emile Roux, a tubercular, devilishly goateed bachelor and bohemian who stands as a stark contrast to his mentor and partner Pasteur, the bourgeois Catholic family man. Roux is famous for his trail-blazing work on diphtheria, a disease of gruesome reach: A thick membrane clogs the throat, slowly strangling its victims, usually children, like a faceless serial killer slipping through the cracks in bedroom windows. The germ cuts down the high- and low-born, the celebrated rich and the nameless poor. In the 1890s, Roux had pioneered the first antitoxin pulled from the blood of sheep, an epic moment in the history of medicine. Making the intuitive leap characteristic of all great scientists, he had discovered that diphtheria germs dripped out a lethal poison of mind-boggling power: he estimated that one ounce of the purified diphtheria toxin could kill 600,000 guinea pigs or 75,000 dogs. His antitoxin was no accidental discovery: a deadly scourge was being arrested—though not yet fully eradicated—by steady, deliberate, painstaking lab work. Not only could the medicine prevent disease, it could stop it in its tracks—the world's first scientific cure. As a field marshal in the Pasteurian army, Roux personified the motto: "One disease, one germ, one vaccine."

Gerry, right, at the Pasteur Institute, Brussels, summer 1910, watching a colleague receive the injection of an experimental rabies vaccine

Here, too, Gerry encounters the wild, semi-unhinged genius, Elie Metchnikoff, the father of cellular immunology. An unkempt, irrepressible atheistic Russian Jew forever racing ahead of himself, Metchnikoff suffers violent migraines and often contemplates the relief promised by suicide. Like a manic character vaulting out of the pages of a Dostoevsky novel, he throws together sloppy animal experiments, driven more by stabs of feminine intuition than classical scientific rigour. While gazing at bowls of starfish and studying their digestion passing through transparent larvae, he spun a fantastic, off-the-cuff theory of how the human body resists the assault of germs. In a flashing, epiphanic moment, he understands that the wandering cells in the body of the starfish larvae not only eat the food, but malignant microbes as well. Without a shred of experimental evidence, he clings fast to a fixed idea: Humans carry a natural immunity to all diseases. Whistling the symphonies of Beethoven, he sets rings of fire around scorpions to show that they do not commit suicide by stinging themselves and he gulps down beakers of virulent cholera bacilli to prove that blood had no relation to immunity to the disease. He insists he experiments best when surrounded by a harem of voluptuous, nubile girls: "The truth is that artistic genius and perhaps all kinds of genius are closely associated with sexual activity."

Here in Europe, Gerry struggles to master the making of rabies vaccine, diphtheria antitoxin, and smallpox vaccine. A slide survives of my grandfather standing tall in his white lab coat, watching a Gallic colleague jab a mustachioed man in the fleshy folds of his stomach with a freshly prepared dose of rabies vaccine. Coincidentally, during Gerry's first week in Brussels, the legendary German Jewish bacteriologist Paul Ehrlich, renowned for smoking twenty-five strong cigars each day while eating next to nothing, makes a momentous announcement from his lab in Frankfurt. After years of numbingly repetitive experiments working with hundreds of different chemical compounds, he discovers Salvarsan 606 for the treatment of syphilis, a "magic bullet" designed to single out and pulverize the toxic organism at which it is aimed. The "cabalistic designation" of 606, Gerry notes, refers to the number of attempts Ehrlich made to find a chemical remedy active against the corkscrew-shaped *Treponema pallidum*. The invention of the world's first effective chemotherapy generates enormous publicity—the kind of breakthrough that my grandfather had dreamt about as a boy in his father's apothecary shop. The drug is clearly a godsend; at his old haunt, the 999

Queen Street West Asylum, a full quarter of all male admissions suffer from tertiary syphilis, acting out appalling, Goya-esque scenes of irredeemable madness. The advance will also spawn another unintended consequence with a direct bearing on Gerry's future: With the production of more sophisticated chemicals, doctors will grow more dependent on drug firms, only caring if a medicine works, not how it works. Salvarsan will dramatically change the economics of medicine: Complex, costly drugs will be prepared by private, specialist producers who will charge what they want; only the rich will be able to afford them.

The following summer of 1911 pulls the young Canadian to the medieval walled city of Freiburg, the cosmopolitan capital of the Black Forest. At the Institute of Pathology, he rubs shoulders with its director, Ludwig Aschoff, destined to become the leading figure in German pathology in the first half of the twentieth century—and eventually a Nazi eugenicist. At the nearby Freiburg Psychiatric Clinic, my grandfather meets Alfred Hoche, who has influenced innumerable psychiatrists in Europe and North America, including Meyer and Farrar. In 1920, Hoche will publish a hugely successful book, *The Permission to Terminate Life Unworthy of Life*, the seed of Nazi euthanasia policies. Doctors were doing the mentally defective a service by relieving them of their useless lives, Hoche believes, for they were already mentally dead and a drag on the economy.

Later that same summer, together with his bride, Gerry crams in a side trip to London and the Lister Institute, his tight travel schedule designed to kill several birds with one stone. Barely breaking stride, he sets his sights on yet another foreign culture, nearly halfway across the world. By boat and train, the State of California lies a daunting seven thousand miles west of the Old World; yet it is here he will go, to his fourth major lab—French, German, English, and finally American—within eighteen months, taking up a two-year position as an associate professor of bacteriology at the Hearst Laboratory at the University of California in Berkeley.

Newlyweds Gerry and Edna, 1911

By the spring of 1912, he streaks back east for another summer-long "working holiday," this time in America's largest metropolis. Manhattan is a tinderbox for a public health catastrophe of unprecedented scale. From Ellis Island, a relentless stream of bewildered immigrants carry the multiple plagues of cholera, smallpox, tuberculosis, typhus, and diphtheria; rows of corpses tied up in burlap bags line the docks. Two-thirds of the city, now the most densely populated in the world, is crammed into dank, unventilated tenements, emerging to work enervating fourteen-hour days. New York lacks public schools and crime flourishes unchecked; homeless, ragged urchins ramble like wild animals among the back alleys, bathed in the stench of overflowing shithouses.

Gerry has come here because the Americans are leaping ahead of the Europeans in the practical management of diphtheria, improving on the methods of Behring and Roux. Temperamentally, Gerry shares much with the forty-eight-year-old director of the New York City Department of Health, William Park. The leading American authority on diphtheria, Park had lost his frail, semi-invalid mother when he was seventeen, instilling a pervasive loneliness that will never leave him. Staid and methodical, patrician and precise, possessed of a naive, boyish honesty, he lives with his half-sister and half-brother; like William "Popsy" Welch of Johns Hopkins, Park remains a confirmed bachelor fervently devoted to a single mistress who answers to the name of science. Gerry is awed by Park's achievement: Not only has the American made antitoxin five hundred times as potent as the Europeans' and slashed its cost by nine-tenths, he has broken down the production line into tasks performable by ordinary, non-scientific workers, transforming his lab into a Ford-like factory of unprecedented reliability and efficiency. He distributes his medicines free to the poor in New York and sells it elsewhere. The model is set: Find the bug, make a cheap and safe vaccine, mass produce and distribute it, monitor its effects, and train others to do the same. The first organization of its kind in the world, Park's lab proves so successful that the private drug companies and the city's private physicians join forces in an attempt to quash it.

Gerry is impressed with Park's work, but he also sees the system's limitations in funding, manufacturing, and distribution, as well as shortcomings in the relationship between universities and public health research and education. Twenty years after the first clinical use of diphtheria antitoxin, its use has only mildly stemmed the overall incidence of the disease; in 1913, New York City alone will suffer over 1,300 deaths. Not only does the antitoxin have to be used early to be effective, it typically inflicts assorted side effects, particularly nasty skin rashes, and families fiercely resist the enforced quarantine of the sick, the bumptious Irish most fiercely of all.

In the fall of 1912, Gerry and Edna return to Berkeley for a second academic year under the benign California sun. During the long journey west, he broods over the whirl of his recent experiences, closely weighing the cost difference between the public model of free distribution by government health departments and the private market. Only the wealthiest of Canadian families can afford to pay for the potent antitoxin imported from Park's New York lab. Could he make the miraculous medicine in Toronto? Could he improve it? Could he make it universally available?

Gerry's extensive travels have driven home a stark fact: Canada is one of the few remaining Western nations that still depends on other countries to supply its preventive medicines. Thousands of sick Canadian children are dying because their parents cannot pay the exorbitant price of imported American and European products. The murderous epidemics afflicting his own country fill him with alarm and disgust, but he knows he does not yet possess the resources to stem their deadly flow. For Gerry FitzGerald, knowledge itself seems like a powerful, spreading virus; and viruslike, Faust-like, he hungers to gather even more. Why, he wonders, has Canada failed to build a self-sufficient arm of the Pasteur Institute?

Then a small miracle of serendipity delivers part of the solution. In November, Edna's sixty-year-old father, Charles Leonard, the iron foundry baron, dies in London, Ontario. Edna stands in line for a substantial inheritance. To my grandfather, money is a means to an end, simply another sharp-edged tool, used for good or ill; in his restless mind, dollars glow like burning coals, stoking the iron furnace of a white-hot ambition.

═

I have assigned myself the task of visiting almost all of the cities in the US and Europe that dot my grandfather's curriculum vitae and I am happy in knowing I will fall short of the killing pace he set for himself. In one of many memorable moments, in the summer of 2004, I am guided by a kindly, silver-haired professor of pathology into the original Freiburg autopsy room. I stand before the semicircular bank of wooden seats where my grandfather once gazed at the same Latin motto—"This is the place where death delights to help the living"—that he had first seen as a medical student in Toronto. As we linger over a shining aluminum slab where countless cadavers have been spread over the decades, my host looks me in the eyes and intones: "This room is exactly as it was in your grandfather's time—except the autopsy table was made of marble."

Such experiences give me a powerful sense of witnessing, of realizing, of colouring in the blanks. If this were all a dream, how would I interpret it? Gerry is telling me something, if I care to look; something in the perfect symmetry of his arrowlike trajectory, something moving, something metaphorical. As I move from plane to car to hotel bed, my own internal, nocturnal glass lantern show clicks over, real and imaginary, filling the white void of a family myth erased by time and forgetting. Most of my Gerry-dreams contain a recurring theme: He stands silent before me in a ghost white lab coat, but as I approach him, he's too busy to stop and talk, and melts away. Often he's like a sped-up Charlie Chaplin figure in a black-and-white silent movie, arms and legs jerking stiffly.

Yet the more facts and images I accumulate and digest, the more I dream. The more I dream, the closer I come, each time straining to catch up to my driven, dreamless quarry, thin and sharp as a scalpel blade, if only to slow him down, to confront the part of himself that is part of myself. Freud believed that a man addicted to world travel was, at root, trying to escape his father. Or maybe his grandson.

═

Shaking off the dirt of scandal, Ernest Jones leaves Canada for good in May 1913. Over his dynamic five-year tenure in Toronto, he has churned out seventy articles and toured the US and Europe like a whirlwind, spreading the gospel of psychoanalysis; he attributes his high productivity to his deep unhappiness in the soul-crushing colonial backwater. In his zeal to help liberate Toronto women from the shackles of Victorian sexual propriety, Jones has felt the sting of such an intense puritanical backlash that it has sped his departure from the city. For decades afterwards, resentful psychiatrists will continue to disdain the tactless, fiery Celt as "Freud's Rottweiler."

That same month, Gerry returns from Berkeley to take up his new appointment at the University of Toronto. His former undergraduate teacher, John Amyot, has invited him back to assist in the production of Canada's first indigenous rabies vaccine in the provincial lab over the summer, then serve as associate professor of hygiene in the fall. In his medical journal writings, Gerry compares the success of the diphtheria antitoxin with his old obsession: the prevention of madness. "Why is it not possible to apply the same principle of prophylaxis in psychiatry? Why not strive to prevent the development of such cases rather than be satisfied with the decision that there is 'insanity in the family'?"

Back home, Gerry catches up with his younger brothers, both of whom need his counsel. Sidney has recently opened a private law practice at 42 Vaughan Road in the city's west end; Bill, the youngest brother, works as a teller at the Dominion Bank at 1004 Queen Street West, sometimes looking out the window across the street at the grim asylum whose threshold, perhaps even then, he senses he will one day cross as a patient.

The two brothers now live with their morose, widowed father, Will, at 315 Avenue Road. The house stands on a depression of land at the bottom of the hill, just south of a massive, gabled, yellow-brick, Gothic revival mansion, Oaklands, for the area is flush with oak trees. Erected by a wealthy merchant, Oaklands's crenellated battlements and ornate fretwork seem to scorn Will, the proud, entrepreneurial white-haired Irishman, and his fall from grace. Now fifty-seven, he spends hours slumped in his chair in a gruff silence, ringed by his sons and daughter, his soul deadened by his own fatherless darkness. They all seem bent on cheering up their grim-faced parent, as if refusing to be pulled down into the gravitational field of a dying planet. Or so goes my General Theory of My Relatives.

Gerry's father, Will, the melancholy village druggist, with Hazel and Bill, 1911: "A good provider and a strict disciplinarian."

None of them dares say it aloud, but the FitzGerald family has pinned its hopes on the eldest son, Gerry, and his racing, lancet-sharp intellect, watching his steep ascent in the world with a mixture of amazement and unease, envy and optimism, imagining a place in the sun at the crown of the Avenue Road hill, the shaded place of my own impending birth.

═

After an evening of writing, I unwind and relax with a beer, watching a bit of Late Show with David Letterman. *I am feeling frustrated that I will never know more of my great-grandfather Will, the cipher of Harriston whose dark influence I intuit yet cannot make real. I am watching the actor Bill Murray, nominated for an Oscar for the film* Lost in Translation, *who is making a rare appearance on a talk show; he is talking about his five sons and his late awakening to his neglectful parenting.*

Falling asleep, I am transported back to the sunroom (the "son" room) of 75 Dunvegan Road in Forest Hill, where my father spent so many hours in the 1960s, locked in symbiotic embrace with the TV talk shows, clicking the remote control in monkish silence. At the far end of the room sits Bill Murray, remote and sphinxlike. I walk over and present him with a copy of my finished book on my grandfather's life, but he's bored and indifferent. Suddenly, a crowd of people sweeps him out of the house; I feel the need to follow him, to reach him—like the paparazzi, I want to hound the slippery, elusive celebrity.

As he slips out the door, Murray lets out the oracular words: "I'm part of a group of disturbed characters."

I awake; my unconscious "talk show" is over. I think of my great-grandfather Will, toiling in the shadows of his Victorian drugstore. I think of his son, my grandfather Gerry, the driven maker of medicines. I think of Gerry's son, my father, Jack, the addict of jazz and narcotics, seeking a mirror in the crystal ball of the television screen; the talk is nothing but show. My mind drifts from dot to dot, son to son. I think of Jonathan Winters, the witty word wizard, the lion in winter, committed to the Hartford Retreat and refusing the invasive treatments that will consume both my father and grandfather; I think of all the flip, manic comedians my father worshipped, as if they—and we—were part of a secret tribe of blood brothers. The interpretation of dreams, the soul of the talking cure, has failed to reach the doctors of my family, but I am plumbing it for all it is worth.

I realize that actors share much with doctors; they are made to feel like superior beings, like gods. Yet, most of the time, inside, they feel the same as the rest of us—less than godlike, less than omniscient, less than perfect. We star-gazing mortals raise our heroes into the clouds, then devour them like a swarm of locusts.

In the end, someone must pay the bill.

PART III

SIXTEEN

The Miracle in a Stable

A fixed idea ends in madness or heroism.

VICTOR HUGO

My grandfather's return from California to Toronto coincides with a turning point in the history of the city—and his own life path—for on June 19, 1913, the magnificent new Toronto General Hospital officially opens at the corner of College Street and University Avenue. Relocated from the slums of Cabbagetown to the southern edge of the university campus, the fourteen-acre, 670 bed institution is the largest in North America, its classical dome, flanked by the orange sandstone solemnity of Sick Children's Hospital and the Queen's Park legislature, forming an instant landmark.

With powerful symbolic force, the spanking new hospital has displaced the northern end of the teeming slum, St. John's Ward, spawned by high rents and low wages and infamous for its concentration of filth, sin, and disease. Stretching 142 acres east to Yonge from University and south from College to Queen, "The Ward"—its very name suggested a roiling asylum back ward—had rolled right up to the steps of city hall as if to mock the impotence of the city fathers. The influx of "dirty foreigners" had terrified the largely homogeneous, white, Anglo-Protestant, middle class; they vilified the slum as a breeding ground for disease and immorality where a destitute, degenerate under-class of over ten thousand poor Italian, Jewish, and Polish immigrants—a population three times denser than London's slums—recklessly indulged its animal sexual appetites. The ghetto was judged intrinsically pathological, rife with contagious disease, graft, sloth, pauperism, harlotry, Sabbath-breaking, gambling, alcoholism, theft, swearing, and sexual abuse of children. This demonic, seething mass of proletarian vice was nothing less than an omnivorous bacillus threatening to infect and subvert the civic body of good, clean WASP neighbourhoods.

And so an "army of occupation" of hard-headed moral reformers—doctors, clergy, and influential lay people—banded together to take decisive action, bent on socializing the "great unwashed" to the prevailing Protestant value system. Platoons of sex hygiene educators urged family members to sleep separately to ward off the magnetic forces of erotic arousal and moral contagion. "The crowded couch of incest," one crusader cried, "infests the warrens of the poor." A man's sanitary surroundings were seen as a direct barometer of his moral, intellectual, and spiritual condition; sociology did not yet exist as a social science, but if it did, it would be called *bacteriology*. And so, logically, the fervent social bacteriologists methodically cleared the way for the new hospital backed with Methodist money, evicting the depraved and deprived eastern European, Italian, and Jewish immigrants from their tenements, and literally ploughing the earth to cleanse it of germs.

As Gerry prepares to resume teaching at the university, this time as an associate professor of the fledgling department of hygiene, the country is struggling to pull itself out of an unprecedented public health disaster, an epic war not only against deadly germs but political inertia. Since the turn of the century, Canada's rudimentary public health apparatus has wallowed in perpetual crisis, overwhelmed by rapid industrialization and immigration. The chief causes of child mortality in Canada remain preventable communicable diseases. Municipal and provincial boards of health, understaffed and undertrained, are reactive and uncoordinated; the formation of a federal health bureaucracy is still years away.

Passing 400,000, the population of Toronto has doubled since the turn of the century, reshaping Hogtown into a formidable industrial hub that sows affluence and misery in equal measure. While "The Ward" has been singled out as a hot spot, the rest of the city is not yet out of the woods: General infant mortality rates have spiked as high as twenty percent and an estimated five to fifteen percent of the population is infected with venereal disease. A tuberculosis outbreak of record proportions cuts down hundreds of lives. Thriving in the sunny months of summer, polio, or infantile paralysis, has become epidemic for the first time, threatening to kill and cripple untold numbers of the young; no treatment or preventative exists to combat the baffling intestinal virus that mysteriously attacks mainly affluent, middle-class families—including the children of the medical profession itself. For decades, the lakefront has been reeking like an open sewer, a perfect breeding ground for all kinds of water-borne infections. Water treatment plants are seen by taxpayers, politicians, and the press as costly and unnecessary, a stubborn stance that infuriates health reformers. As the debate drags on, the incidence of typhoid fever in Toronto triples that of London, England, and citizens die like vermin.

Ordinary Canadians fear crude, often impure preventive medicines more than disease and a bitter battle breaks out over compulsory smallpox vaccination. Protesting the unequal application of vaccination laws between the rich and poor, the Anti-Vaccination League charges that, in addition to infringing civil rights, compulsory immunization is being used as a class weapon by elite public health doctors against working men who sometimes lose many days' wages suffering from severely ulcerated arms, a dire side effect of the vaccine. A doctor who formerly advocated vaccination recanted his position when he saw the hideous effects on children: "Mother after mother bared the arms of their little ones to show me the vicious, ulcerous sores, so large a silver half dollar would drop into them, and deep nearly to the bone." Compulsory vaccination, he charges, is akin to torture and murder.

In Toronto, the growing hub of the Canadian public health reform lobby, my grandfather busily cultivates strategic alliances with key political players, including C.K. Clarke, now installed as the university's dean of medicine. But Gerry's chief ally is a forty-six-year-old devout Roman Catholic, Dr. John Amyot, who decades earlier had taught him the rudiments of the autopsy knife in the chilled basement morgue of the old Toronto General Hospital. Amyot has forged links with two important men: Dr. John McCullough, the provincial medical officer of health, now standing at the height of his career and basking in an international reputation; and his legendary counterpart in Toronto, the iron-willed, Irish-blooded Charles Hastings. Having lost his own young daughter to typhoid from infected milk, Hastings has been leading the attack on the city's epidemic-ridden slums with Presbyterian intensity—imposing quarantines, demolishing thousands of putrid outdoor privies, pushing to chlorinate drinking water and pasteurize milk supplies, and educating the recalcitrant masses on the basic principles of hygiene.

For over a decade, Amyot has laboured under primitive conditions in the provincial lab, cleaning the glassware and feeding the animals himself. Only a few weeks before Gerry's return to Toronto, the lab had moved to a gabled, three-storey Victorian house at 5 Queen's Park Crescent, opposite the newly built Toronto General Hospital. In 1912, Amyot convinced the faculty of medicine to establish a new two-year diploma course in public health. Only a single person—an earnest, curly-haired young graduate student in biochemistry clad in wire-rimmed glasses and homburg fedora—stepped forward to apply. The appearance of twenty-three-year-old Robert Defries, a third-generation Torontonian, will prove a pivotal moment in the history of Canada's nascent public health movement.

Defries's character was shaped by early hardship. When he was four years old, his father, a successful brewer, was hitching his horse to a carriage one morning when the animal suddenly reared, driving his chest into the harness shaft. The family savings were exhausted on medical bills; when he died, his widow and two sons retreated to a shabby boarding house at Yonge and Bloor where wee Bobby and his older brother risked catching diphtheria from toxic gases leaking from trapless sewers. The boys were raised by the strict hand of their widowed mother, Agnes, a pious teetotaller who will wear black the rest of her life. A forceful, no-nonsense character, the stern and iron-willed Aggie, standing five feet two inches tall, relentlessly pushed both her sons to become ascetic, self-sacrificing doctors.

Young Bobby quickly learned to suppress his own needs to serve the needs of others. A self-effacing evangelical Presbyterian, he was caught up in the Christian social gospel movement sweeping North America at the turn of the century. Bobby believed in turning prayer into action, taking to heart the example of Jesus healing the sick; he believed in the regeneration of the corrupt cities and bringing the kingdom of heaven to earth; he believed in the church as custodian of the Anglo-Saxon virtues. A holy warrior for social justice and muscular Christian ideals, he interpreted the Bible literally, swallowing whole the tale of Jonah and the whale; yet he was also a broad-minded man who refrained from proselytizing.

Defries wanted to reconcile the religious teachings of Christ with the scientific ideas of Darwin, yet he saw the purely humanistic approach to life—Gerry FitzGerald's approach—as ultimately limited: "Humanism does not take into account that man is sinful by nature," he declares. "We underrate man's proneness to evil and the necessity for moral law." A committed bachelor, Bobby lives with his mother and takes lunch with her at home every day; a picture of her and his dead father hangs over his bed. Every Sunday, he will escort Aggie to Bloor Street Presbyterian Church until her death in 1942; such was a son's undying gratitude to Agnes Defries for taking care of him as a fatherless boy.

With his arrival in 1913, Gerry becomes the first and only full-time member of the department, as Dr. Amyot still spends much of his time working in the provincial lab. Even though the department lacks a lab of its own, four like-minded men—Amyot, McCullough, FitzGerald, and Defries—are already daring to dream of a national role for the embryonic department of hygiene.

That summer of 1913, as my grandfather pours out his ideas in an agitated stream, Amyot lends a receptive ear, an act that stokes the eloquence of the young doctor. For Amyot, it is an easy decision: He hands Gerry the keys to the provincial lab at 5 Queen's Park Crescent, together with the services of his technician, Mr. William Fenton. After weeks of work in the close August heat, Fenton volunteers to test the safety of an experimental rabies vaccine; taking no small risk, he rolls up his shirt and watches Gerry jab him in the stomach, the first of a series of injections that sting like an angry wasp. As if to remind his loyal assistant of the nobility of their work, Gerry unfurls anecdotes of his days at the Pasteur Institute, working side by side with the European trailblazers; only natural, then, that he proudly dubs the brave Bill Fenton "the Canadian Joseph Meister." The test proves successful and its implications profound—not just for rabies victims. Canadians sick with a host of infectious diseases will, sooner than later, no longer depend on expensive, often substandard foreign medicines to save their lives.

═

Emboldened by his rabies success, my grandfather turns his energy to diphtheria, the leading killer of children under fourteen. Despite the importing of expensive antitoxin from American drug firms, the death rate has remained as high as twelve percent over the first decade of the century, claiming thousands of Canadian lives. Diphtheria remains easily misdiagnosed, confused with other throat inflammations. Inadequate dosages of antitoxin mean the death rates will remain high. The proliferating demand for the medicine stimulates many commercial firms and government departments of health around the world to manufacture mass quantities, but as production levels grow, international concern mounts over the lack of uniform standards of potency.

In Canada, the federal government has levied a fifty percent duty upon imported serums; American antitoxin is almost unprocurable by Canadian doctors because of its prohibitively high price, upwards of eighty dollars per course of treatment, the equivalent of several weeks wages for most families. "The children of any but the wealthy," laments McCullough, "are left to die of diphtheria." Rather than see their patients asphyxiate to death in slow, agonizing degrees, some doctors pay for the antitoxin out of their own pockets. One Toronto physician records the story of a family who can afford only a single dose of antitoxin for their two infected children, presenting the parents with an agonizing choice. Tragically, one child is treated and lives; the other dies.

Infected kids are quarantined in their bedrooms with a sheet soaked in disinfectant hanging in the doorway; for their safety, the family is isolated as well. The invalids' dishes and utensils are sterilized, bedclothes boiled, their books, toys, and teddy bears burned. Yet such precautions often prove fruitless; "the strangler" can kill several children in one household within hours, as parents hover by the door, watching helplessly. Mercifully, not every child dies; in many cases, their blood possesses a natural antitoxin that renders the bacillus harmless.

Gripped by a sense of urgency, Gerry makes his case to the university's board of governors. If they can do it in New York, why not in Toronto? Waving his arms in the air, his words tumble out in a quickening stream, his eyes darting anxiously back and forth, gauging the reactions of the dignified, bewhiskered millionaires arrayed before him. "I'll make diphtheria antitoxin right here!" he exclaims, slapping the oak tabletop with the palm of his hand. "If I make it cheaply enough, the provincial government can buy it and distribute where needed, free of charge."

At first, Gerry's moral passion meets a hesitant response. The idea of an academic institution facilitating the commercial production of high-quality, low-cost biomedical products—for *free distribution* to the entire population—is radical and unprecedented. Yes, the board will consider the proposal, but it will take time.

Knowing he needs many months of work to make an effective antitoxin, and confident the board will eventually come through, Gerry plunges ahead on his own initiative. Without consulting the board, he accepts an offer by Billy Fenton to use the empty lot next to his home at 145 Barton Avenue, just north of the university, near Bloor and Bathurst Streets, to build a makeshift horse stable. Naturally, Edna is eager to help her husband, too. She understands the notion of noblesse oblige; besides, one of her own uncles perished in childhood of diphtheria. Edna agrees to donate money from her dowry. Naturally, he is thrilled by his wife's beneficence—doubly so when she reveals she is now carrying their first child.

After inspecting Fenton's empty Barton Avenue lot, Gerry spends six hundred dollars to build a simple, two-storey stable made of galvanized sheets of corrugated metal; then he stocks it with lab equipment and houses a colony of guinea pigs in the loft. He buys four aging horses—saving them from the glue factory—and leads them into a small injection room in the stable. He names the first decrepit old nag, costing three dollars complete with halter, Crestfallen, for its sad eyes. As the snows of early December fall, Gerry has spent three thousand dollars of his wife's money.

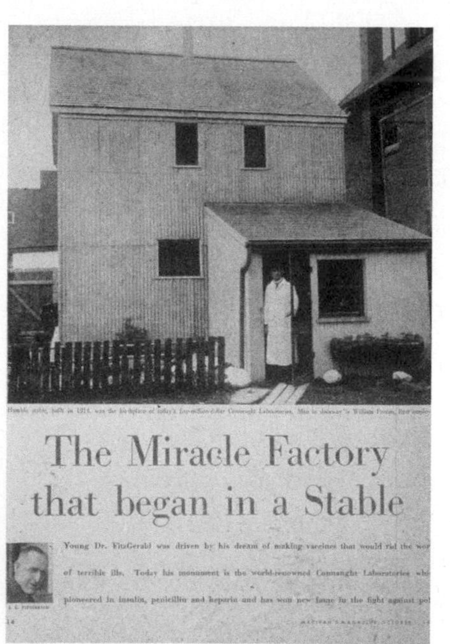
The Miracle Factory that began in a Stable

Young Dr. FitzGerald was driven by his dream of making vaccines that would rid the wor
of terrible ills. Today his monument is the world-renowned Connaught Laboratories whi
pioneered in insulin, penicillin and heparin and has won new fame in the fight against pol

Horse stable erected at 145 Barton Avenue, Toronto, 1913, where Gerry made Canada's first indigenous diphtheria antitoxin

On December 11, 1913, two days after his thirty-first birthday, my grandfather starts work, assisted by two staff members of the Ontario Veterinary College.

After weeks of filtering out quantities of lethal poison from culture broths, Gerry picks up a large hypodermic needle containing a minute but deadly dose of diphtheria germ. Slowly he walks over to Crestfallen, standing oblivious in his stall; as Fenton holds the halter, Gerry sweeps back the horse's mane and feels for the main artery in its massive neck. Then, steadying his hands, he slowly pushes in the sharp point of the needle, knowing a slip exposes him to an agonizing death.

As the poison flows into the horse's bloodstream, Gerry holds his breath. The men watch and wait, knowing the animal's system is invisibly forming antibodies that are attaching to, and then neutralizing, the invading toxin. After four consecutive months of incremental injections—once, he breaks a needle, luckily without mishap—the four horses build up immunity to diphtheria. They can now tolerate without discomfort an injection strong enough to kill off an entire cavalry brigade; the hyperimmunized blood is now "ripe."

Next comes the bloodletting. On March 14, 1914, Gerry picks up a razor-sharp lancet and cuts open an artery in the neck of Crestfallen. With a rubber hose, he sucks out ten litres of blood, then pours it into large sterile bottles capped with rubber stops. After sewing up the horse's neck, he gingerly loads the precious red sap into the trunk of a coal black Model T and drives it the thirteen blocks southeast to the basement lab of the U of T medical building.

The space is cold, dark, and cramped, mould streaking its concrete walls; a coal furnace has been removed to make more space. Here, the two women dearest to him stand by his side as helpers: his twenty-six-year-old sister, Hazel, who has taken a crash course in bacteriology and washes the stacks of glassware, together with the stalwart Edna, as pregnant as the lab itself. After forcing the antitoxin through a porcelain filter to remove all traces of bacteria, Gerry instructs his sister to hold down a squealing guinea pig as he injects the antitoxin to test its potency. The strength of the toxin is estimated in an ingenious way. First, Gerry determines how much toxin kills a guinea pig of a given weight; then, by dosing guinea pigs with just enough toxin to kill, mixed with varying quantities of antitoxin, he finds the exact quantity needed to save its life.

The final stage is the "filling" operation. In the words of a colleague, Gordon Bates, who will pen an article on the lab in *Maclean's*: "An assistant sat before a glass cabinet, visored and armoured in sterile mask, cap and gown, gauntleted in sterile rubber gloves like a veritable knight of the operating room." He fills countless vials and syringes with the magic medicine, "ready to be sent on their errand of mercy across the country."

That same month, the Provincial Board of Health awards a contract to Gerry's fledgling lab. The production cost of the antitoxin is twenty-five cents per thousand units; the government agrees to purchase it for thirty-five cents—representing a ten-cent profit margin for the lab—and distribute it to doctors and local boards of health. On March 31, 1914, Gerry sells his first batch of antitoxin to Ontario's board of health, directed by his friend John McCullough, at one-fifth the going commercial rate. Orders soon begin pouring in from all over Canada; in the months ahead, the production costs will be cut even further. In the words of Gordon Bates, freedom from infectious disease is now potentially "in reach of everyone."

Flushed with his coup, Gerry is ready, once more, to browbeat the leaders of the University of Toronto. To the board, he details his production costs, his earnings from selling antitoxin to boards of health, and how much he anticipates in sales across Canada. He calculates he needs six thousand dollars for the university to assume the work, confidently predicting: "The work will be self-supporting within three to six months, probably less."

His plan is threefold: the lab will prepare a spectrum of high-quality, low-cost preventive medicines and arrange for their sale and distribution across Canada through government departments of health; a research arm will develop new products to meet the constant threat of infectious diseases; finally, all the work will be intimately connected with undergraduate and graduate teaching in the Department of Hygiene, passing on the torch of knowledge to future generations. The wholly unique organization will form a non-commercial, self-sustaining part of the University of Toronto—a paragon of a privately endowed public dispensary of biological products, tied to research and teaching.

No precedent exists anywhere in the world of an academic institution operating a commercial drug firm. Yet, crucially, the revolutionary plan marches in step with the scientific spirit of the university and its traditional role of public service. An ounce of prevention is, indeed, worth a pound of cure. How can the good men of the academy refuse the equivalent of their very own biomedical hydroelectric plant? It is almost too good to be true: Here is a young man so passionately committed to his altruistic medical work that he is prepared to risk his own health and his own wealth—or rather his wife's—to achieve his end. FitzGerald is a rare bird, possessed of both the instincts of an incisive, cost-cutting entrepreneur and the values of a scholar. There is only one hitch: Can the cautious, impecunious university risk supporting such an ambitious and unconventional project?

═

South of the border, synchronistic events are unfolding. In May 1913, John D. Rockefeller, Senior, the founder of Standard Oil, widely despised as the incarnation of tooth-and-claw capitalist rapacity, astonishes the world by drawing up an audacious plan to give away half of his billion-dollar fortune, the largest ever accumulated, most of it to medical science. Within a few short years, the New York–based Rockefeller Foundation, a charitable trust of unprecedented global scale, will become a kind of surrogate government, the largest philanthropic organization in the world, the greatest lay benefactor of medicine in history, and a pivotal influence on my grandfather's destiny.

Not yet forty, John D. Rockefeller, Junior, is elected president of the Rockefeller Foundation in May 1913 and installed in an office at 26 Broadway. The foundation chooses to focus on medical education and public health, spreading its largesse in unprecedented fashion beyond its national boundaries and across the globe. Having recently emerged from a prolonged nervous breakdown, Junior believes his father is innocent of all criminality, a gleaming white knight astride a white horse; he is humbly prepared to shine his shoes and carry his bags, if needed. While his father retires to the golf course and restores his overworked body to health, Junior shoulders the dark legacy of managing the largest concentration of wealth in the world. Wracked by migraine headaches worsened by a punishing conscience, he will try to reconcile the warring values of religion, science, and business; caught in the vortex of dynastic expectation, he will struggle to redeem, through sheer force of will, the unpardonable sins of his father.

═

In April 1914, the University of Toronto votes approval of Gerry FitzGerald's ambitious plan for a self-sustaining laboratory devoted to public service. The Faculty of Medicine agrees to take over financial responsibility for the innovative enterprise and on May 1, 1914, the Anti-Toxin Laboratories in the Department of Hygiene are born. With pocked, concrete walls; flaking overhanging pipes; warped hardwood floors; vault-like, ice-cooled walk-in refrigerators and curtainless windowsills caked in layers of coal dust, the dank basement lab will win no beauty contests. But because it is self-supporting, few funds are needed from the university; the operation will never be a financial burden. *The Canadian Journal of Medicine and Surgery* enthuses that the new lab is "free as air and entirely untrammelled with red tape." Gerry's foresighted plan to use boards of health as a distribution partner makes the government an instant ally. Nearly everyone stands to win—the government, the university, and above all, millions of Canadian citizens, regardless of social class or income. Understandably, general practitioners and druggists—men like Gerry's own father—feel less enthused, for they depend on cash customers for their livelihoods. The commercial drug manufacturers have even more cause for alarm; the high-minded idealism of young Dr. FitzGerald cuts so sharply into their bloodstream of profits that their businesses may not live out the year.

═

On June 5, 1914, my grandmother Edna makes her way down to the maternity ward at the newly opened Toronto General Hospital on University Avenue, where the smell of fresh green paint still lingers in the corridors. Only a few hundred yards along College Street, Gerry is carefully carrying bottles of horse blood down the steps of his basement lab when the phone call comes. Dodging passing streetcars, he races over to the General still wearing his white lab coat. Within hours, he is gazing into the blue eyes of a squalling baby girl; like him, Molly is red-haired and fair complexioned. Hospitals, even the newest, seethe with killer germs; but Molly is healthy as a horse.

Molly was born in June 1914, just weeks after the founding of the Connaught Laboratories

Gerry, the new father, is soon back in the basement lab, grinding out long punishing hours of dangerous work with primitive equipment. The critical refining phase of the antitoxin begins on June 15, 1914, as ten-day-old Molly gurgles in her crib; the tests will continue for over a year.

One by one, Gerry hires doctors, technicians, and secretaries, many of whom will reward him with decades of unwavering loyalty. He envisions his humble lab blossoming into an efficient factory for a wide range of products—typhoid vaccine, tetanus antitoxin, and antimeningitis serum—to meet all conceivable demands. Simultaneously, he envisions a new home for his new family. In the wake of a residential building boom, Deer Park, a thickly wooded suburb of freshly tarred dirt roads, has recently been annexed to the northern edge of the city. Coinciding with the opening of the lab and the birth of Molly, work crews break ground on a long, narrow lot, 58 feet wide and 135 feet deep, abutting the cinder playground of Brown Public School. Over the steamy summer months, a three-storey house of fifteen rooms and seven fireplaces, costing eight thousand dollars of Leonard money, assumes its inevitable shape. Perhaps it all began there and then; perhaps the amiable-enough stacks of bricks and innocuous buckets of mortar, rumbling up the driveway on the backs of flatbed trucks, concealed a toxic intent. The container of my father's childhood awaits his conception, as it does mine.

Balmoral, Balmoral,
A story in search of a moral.
Quiet and cold as Father Death,
I hold my breath and
thrash the cast iron bars
of my throatless dreams.

═

On June 28, 1914, as the timbers of 186 Balmoral Avenue lock into place, an assassin's bullets once more divert the course of history in the city of Sarajevo. Loosening his stiff white collar in his basement lab, my grandfather sweats out the long summer days, anxiously awaiting news. Little Molly is ensconced in the nursery, attended by a nanny; with a pair of scissors, Edna bends over her daughter's head, snips a lock of her soft red hair, and presses it into the pages of an album that one day I will inherit.

On August 2, Germany invades Belgium and occupies Brussels, the home of Jules Bordet's Pasteur Institute of Brabant and the site of Gerry's memorable summer of training four years earlier. The Germans mete out a deliberate policy of terror, murdering thousands of innocent civilians and torching their homes; masses of refugees flee in panic. Canadian Prime Minister Robert Borden calls the war the "suicide of civilization."

Friends and colleagues in peacetime, German, French, and English scientists now face a long, bleak period of alienation. Engulfed in a tornado of life-changing events, Gerry has little time to take it all in. Within the span of a few weeks, he has felt the contortions of four successive births: a laboratory, a baby girl, a house, and a world war. And it is the war that now holds him most in thrall. Gerry wants to join the army and do his duty for king and country, but that means abandoning the lab and suspending production of all his life-saving medicines. For weeks, he agonizes; all that he has fought for—a public health infrastructure of potentially national scope—is suddenly threatened with being strangled in its cradle.

But as the Three Fates would have it, my grandfather is saved by the timely intrusion of yet another benefactor, yet another man of wealth and taste. This time, it is a hale and hearty whisky baron, bearing the blood red symbol of the Canadian Red Cross.

SEVENTEEN

Born with the Dead

We die with the dying:
See, they depart, and we go with them.
We are born with the dead:
See, they return, and bring us with them.

T.S. ELIOT

I conjure the image of my grandmother Edna busily hanging curtains and arranging pieces of polished rosewood furniture, putting the finishing touches on the new house on Balmoral that is fated to haunt my childish dreams. It is September 1914.

"The war will be over by Christmas," comes the oft-repeated wish, but the snows of December drive home the devastating reality: Ninety percent of the original expeditionary force—ninety thousand men—have been killed or wounded. Not even the best medical minds anticipated the deadly impact of wound infections in the trenches of Europe. Lurking in the tilled, fertile soil of Belgium and France, gas gangrene infects even minor shrapnel wounds. "Despite great advances in antiseptic methods," one dismayed doctor complains, "we seem to have gone back to the infections of the Middle Ages."

Tetanus alone is causing five percent of all deaths, its bacillus far more common than rabies and equally as dreadful—two hundred times as toxic as strychnine. Despite the efforts of European labs, medicine is scarce; antitoxin imported from the neutral United States costs an exorbitant $1.35 per dose. To meet the crisis, my grandfather recruits his young colleague, Robert Defries, who hammers out a plan: The university lab will produce enough serum for every British and Canadian soldier for sixty-five cents per dose—half the cost of medicine imported from the New York City Department of Health. The quoted price is so low that the British War Office cables back for confirmation.

Despite the dispensation of a five thousand dollar federal grant, the university's first ever, the institution lacks the funds to shepherd the lab through a world war; and so President Robert Falconer approaches several Toronto multi-millionaires, including Colonel Albert Gooderham, scion of the iconic Gooderham & Worts, the largest distillery in the British Empire, for a private donation. On a winter morning in 1915, the tall, robust fifty-three-year-old philanthropist tramps across the snow-encrusted lawn of King's College Circle, soon to be occupied by trim military tents, rows of biplanes, and full-throated young recruits practising bayonet thrusts on the tennis courts. Cutting a striking figure in his tweed plus-fours, glinting spectacles, and waxed, dagger-point moustache, Gooderham passes through the portico of the twin-towered medical building and grandly inquires: "Where will I find Dr. J.G. FitzGerald?"

"In the basement, as usual," comes the reply.

Down the stairs, Gooderham finds a cramped, damp room; inside, he makes out the lean, white-coated figure standing beside his sister, Hazel, both working madly to fill orders and hire staff to meet the crushing demand overseas. Over the next ninety minutes, my grandfather voices his high hopes with a feverish eloquence. Gerry agrees to prepare the tetanus antitoxin at cost, but he needs more space. Showing Gooderham around, he explains his plans to use the federal grant to convert the old boiler room into a horse stable to accommodate the swelling clamour for medicine. The colonel pauses and curls his lip; a boiler room is not fit for horses. "I suppose for twelve thousand dollars or sixteen thousand dollars," he muses, "it would be possible to purchase a few acres of land and build a stable and some laboratories there."

In early April, Gooderham drives Gerry up to the hamlet of Fisherville at Dufferin and Steeles, along with a real estate agent. Bumping along twisting dirt roads for twelve miles north, the three men come upon a long abandoned fifty-seven-acre farm. Here they find a derelict barn, red-brick farmhouse and decaying grist mill once driven by the west branch of the Don River. "The whole property," my grandfather later records, "conveyed very strongly the impression of long-continued neglect and disuse."

After the men briskly survey the land in the cutting April air, Gooderham removes his pipe from his mouth and turns to Gerry.

"Where would you like to go to now, doctor?"

"I don't know that I'm interested in seeing any other property, Colonel. One acre would be fine."

Gooderham shoots back: "How about the whole property?"

"That would be fine, too," my grandfather replies with a smile.

The whisky baron explodes with laughter. "The man says one acre is fine; then show him fifty-seven and he says it's fine!"

Gooderham immediately cuts a cheque for $75,000 for the purchase of the property and the erection of a lab and a superintendent's residence; later he will add a gift of $25,000. Such largesse is a remarkable act of faith in a young, untried university professor.

Robert Defries has agreed to help Gerry with the tetanus production, but only temporarily as he has his heart set on working overseas as a medical missionary. As the two men stroll through the grassy farmland, Gerry listens impatiently to Defries's plans for the future.

"Bobby," Gerry urges, "you can do all that later. Just now I need you badly."

The contrasting personalities of FitzGerald and Defries strike a fine chemical balance, the reckless Irish plunger stabilized by the prudent Presbyterian plodder. Like Pasteur and Roux, Gerry has found his ballast in the indomitable, curly-headed Bobby Defries, the devout man of science who hates sin as profoundly as any germ.

═

In August 1915, a lengthy article, "Lowering the Cost of Life Saving," published in the general circulation magazine *Maclean's*, rains praise on the new public service lab. Churning out trainloads of medicines stretching from Newfoundland to British Columbia, a man could make a mint of money from such an enterprise, the author, Dr. Gordon Bates, opines, but FitzGerald is a man of vision. Bates reports the lab "commenced selling at what a private drug corporation would call a suicidal rate of 25 cents per 1,000 units and has again cut this rate to a fraction above 20 cents, about one-fifth the actual [commercial] rate." By the outbreak of war, the lab is supplying a smallpox vaccine at two-thirds the price of private firms. The lab expenses are paid and a profit shown, to boot. In time, expenses will be slashed even more dramatically. In Bates's oft-repeated phrase, preventive medicines will soon be "in reach of everyone."

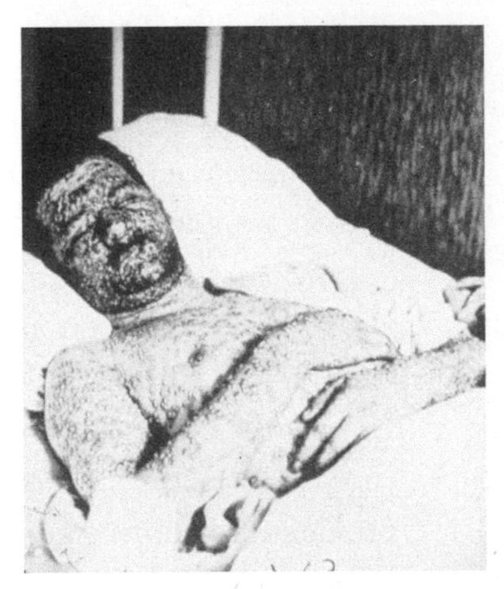

Victim of smallpox, "the speckled monster"

Druggists and private drug companies across Canada and the US jealously watch the effects of my grandfather's radical cost-cutting. Rumours spread that the University of Toronto's medical products are prepared by incompetent students, their stables and labs are ill-equipped, and that their inferior medicines cause more adverse reactions than commercial ones. Druggists sometimes refuse to carry the university's products, hindering distribution to general practitioners. But by early 1916, the University Farm has become the official source of biological products in Ontario and has practically eliminated the competition of commercial firms. Remarkably, the first year of the province's free distribution program of rabies, smallpox, and typhoid vaccine; diphtheria and tetanus antitoxin; and antimeningitis serum will cost only forty thousand dollars—a quarter of the price of the imported commercial supply. A total of 250,000 doses of tetanus antitoxin alone are passed out, cutting tetanus deaths to zero.

═

Concurrent with his lab and teaching work, Gerry commands the bacteriological unit at an army training camp in the Niagara peninsula. Here he injects serum into the spines of soldiers afflicted with meningitis and swabs the suppurating wounds of dozens of amputees suffering wound infections. He trains thousands of soldiers in rigorous hygienic discipline, from purifying water, burning garbage, and ventilating barracks, to boiling dishes and utensils. Even so, he manages to contract a touch of pneumonia during his patient rounds.

Vaccination parades take precedence over all other military rituals. "The effect on morale was excellent," notes Gerry, "if the band played during the time the inoculations were being done." He is most concerned by the soaring levels of venereal disease, a thorny, intractable sexual, social, economic, moral, and medical problem of daunting complexity. By 1915, nearly one third of the Canadian Expeditionary Force in France is infected with syphilis and gonorrhea, clogging the overtaxed military hospitals. When the waves of apple-cheeked Canuck farm boys descend from the troop trains, local prostitutes swarm the boys with maple leaves sewn on their uniforms, for the Canadians are paid five times as much as English soldiers. Other armies record half the Canadian rate of infection, the British only five percent.

Yet the conditions of war are enabling public health officials to break through the long-entrenched wall of silence surrounding VD, and to wield a form of control denied them in peacetime. As the captive audience of the army becomes the first to receive instruction on the use of prophylactics, some doctors worry that the public might think they are endorsing prostitution; the church, for its part, sees "the leprosy of lust" as fit punishment for unbridled sins of the flesh. But before a female audience, Gerry endorses condoms as the only realistic policy: "We put it to the men in this way: We are not providing you with this prophylaxis in order to condone your conduct or to in any way lessen the consequences which you might suffer from. We do it because we do not want to lose you as a fighting unit." Recently promoted to the rank of major, Gerry chairs a military committee that pushes for a barrage of pamphlets, posters, movies, and lectures. Glass lantern shows, packed with lurid images of blindness, madness, and bodily corruption, are specifically designed to arouse the necessary degree of alarm and action. The population must struggle, Gerry says, "so that the more than ninety percent of clean young Canadians who are fighting our battles may realize that everything possible is being done by those of us remaining here." From 1916 to 1918, the rates of VD, the third-highest killing disease in the Canadian army, will drop by two-thirds.

═

On May 29, 1917, as the fate of a fragile civilization hangs in the balance on the Western Front, I walk along the dreamscape of College Street and up the front steps of the Toronto General Hospital. In the maternity ward, I find my chestnut-haired grandmother, Edna, ghost-pale, propped up in bed by a bouquet of pillows. Society allows her the privilege of motherhood, but not, as yet, the right to vote.

In her arms, she cradles her newborn son, John Desmond Leonard; beside her, John Gerald beams with fatherly pride. He has achieved his wordless wish—the seed and blood of the unfolding future, a good, brave son to shoulder the intangible freight of his own Christian name. Yet in the flush of her own joy, Edna feels an intrusive shard of pain, for the primal experience of birth has flushed up the memory of the loss of her own mother when she was but two years of age.

I hear a knock at the door; my grandparents turn their heads to catch the grinning, chalk-haired figure of C.K. Clarke, now the hospital's superintendent, extending a fat cigar. Donning a white cotton mask, Canada's preeminent psychiatrist picks up the war-born baby boy in his arms and tickles his stomach. Though my father's constitution will prove delicate and allergy-prone, Clarke chooses to believe he is holding a hardy, pale-skinned Protestant male, a specimen of superior racial bloodlines, untainted by defective genes.

As my grandmother places Jack in the bassinet beside her bed, I bend down to gaze upon my father's infant bundle of flesh, wrinkled and fretful as an old man's, confined in its pine, casket-shaped box of time. In a split second, the span of his seventy-five mortal years collapse before my eyes; suddenly I feel an urge to cry out, to howl in brilliantly articulate protest, to surrender to the truth of words and passions that will set us all free. But it is only animal cries that I feel, primal and wordless, catching in my throat, as if stuffed back down by mailed, medieval Irish fingers, cruel as the blades of knives.

═

A bleak, rainy day in October 1917 comes alive in the theatre of my thoughts. Variously clad in frock coats, top hats, and military uniforms, a line of dignitaries—a governor general, a provincial premier, an Anglican bishop—ascend a tented stage that overlooks my grandfather's newly erected laboratory. Nearly a year after it was first occupied, Colonel Gooderham is formally presenting the fifty-seven-acre farm as a gift to the University of Toronto; it is capped by a $75,000 research endowment from the province of Ontario. At Gooderham's

Horses grazing at Connaught Laboratories University Farm, Steeles Avenue and Dufferin Street, officially opened October 1917

request, the enterprise is baptized with a new name, The Connaught Anti-Toxin Laboratories, after his friend, the Duke of Connaught, a favoured son of Queen Victoria and the former governor general of Canada.

Following the ceremonies, Gerry ushers the guests through the hundred-foot-long, two-storey main building crowned by a churchlike bell tower and a Union Jack flapping in the autumn rain. Covered with Spanish-style grey stucco and a sloping roof of slate studded with dormer windows, the place echoes the Gothic character of the FitzGerald home on Balmoral Avenue. Inside, the party inspects a fully stocked laboratory, a living quarters, an operating room, cold storage rooms, and stabling for twenty-seven horses. On the concrete floor of the operating room, pools of equine blood stand out like dark crimson wine. In a hayloft above, dozens of rabbits and guinea pigs, destined to be sacrificed in ceaseless experiments, scratch and claw at their wire cages.

That evening, a wiry, cold-blooded pathologist and bacteriologist of German-Jewish extraction takes the stage at the university's domed, two thousand–seat Convocation Hall and hails the contribution of the unique Canadian lab. Gerry had met the fifty-two-year-old Simon Flexner the previous year in New York to pick his brain on his groundbreaking meningitis serum, but on this night the American is his special guest. Flexner is an original researcher in his own right but it is his inspired stewardship of the Rockefeller Institute of Medical Research, a legendary place likened to the atmosphere of a theological seminary where hand-picked, eccentric geniuses of Nobel Prize calibre struggle to unlock the mysteries of life, that history will judge his greatest accomplishment. A colleague once observed that Flexner possessed "a logic far beyond that of most men, final as a knife." After bleeding horses repeatedly to produce serums and antitoxins, it is customary to turn them out to pasture; but Flexner sells them to the slaughterhouse or relentlessly bleeds the animals to death, harvesting every last drop of the precious medicines. "No one can run an institution" he believes, "unless he has the capacity to be cruel."

Yet of all the important personages who attend the ceremony that day, it is Gerry's sickly father, leaning on a cane—perhaps the same one that in my imagination lashed the backs of his three imperfect sons—who most commands Gerry's attention. What is his achievement all about if not the tacit need for a father's love and approval? To assure a national character for the Connaught, Gerry is already racing ahead of himself, planning a body composed of the medical officers of health of all nine provinces, allowing him to keep key decision-makers abreast of his unfolding vision. A public health system, after all, is only as strong as its weakest link. Earning a reputation as an integrative thinker, Gerry is obsessed with lashing together the disparate regions of the vast, unintegrated homeland into a unified, coast-to-coast network.

As father and son pose together for the photographer, both men know the grizzled, aging druggist, born fatherless in 1856 in the unforgiving Upper Canadian outback, is not long for this world; yet he has survived sixty-one years, and a nervous breakdown, to see his eldest son attain something of a miracle. In the snapshot I hold in my hand, Will and Gerry stand stiffly at attention, the son in his officer's uniform radiating the aura of a supremely confident man who knows exactly where he is going, the father by his side grim in the certain knowledge of where *he* is going. I stare at the image and it stares back; in the empty space that divides the generations, unspoken lines of tense, terse dialogue, crackling with suppressed violence, remain as invisible to the eye as a murderous virus.

Within weeks of the ceremony, Will suffers a stroke and is admitted to a public hospital occupying a gabled, three-storey, red-brick Gothic revival mansion at the corner of Wellesley and Sherbourne Streets. On December 11, 1917, two days after his thirty-fifth birthday, Gerry hovers over his father's body, just as he had hovered, six months earlier, over the crib of his newborn son, caught in the twilight world that links the generations with an invisible cord. Only weeks earlier, he had asked the university for permission to volunteer for duty overseas. At last the time has come to do his bit and put to rest the worst internal, father-borne accusations that could befall a fit young man: *slacker, shirker, coward, conchie, deserter.*

Will lies still, withholding and inscrutable in his dying as he had been in his living. As the last breath slips from his father's lips, a jet of feeling stabs Gerry's chest. He has seen countless people die, old and young, men and women, saints and sinners, mad and sane—but a father's dying equals no other. If he could have found a different kind of courage, the unmilitary bravery of words, the son might have risked putting it to the father plainly: *What do you expect of me? What must I do to deserve your love? Why can I not shake the hopeless feeling that no matter what I do in this life, it is never good enough?*

But the questions, and the answers, never come.

Jungians would call it synchronistic. That very same week in Vienna, Sigmund Freud publishes his seminal essay, *Mourning and Melancholia.* When Freud pronounces the death of a father as the single most important event in a man's life, Gerry FitzGerald pays no heed. His grief, his relief, fused with his father's, and his father's father, and so on back to Father Ireland—the dead weight of it all feels overpowering—and so all of it must wait. Other, more pressing voices call him into the future.

═

Braving an unusual winter thunderstorm, my grandparents slide up the snowy driveway of 93 Highland Avenue, a grand stone Rosedale mansion owned by David Dunlap, a gold-mining magnate. The house will, in the years to come, accommodate successive presidents of the University of Toronto; but on the evening of January 26, 1918, an event of singular importance is unfolding.

At the front door, C.K. Clarke greets Edna and Gerry; although the latter is dressed in his officers' uniform, he is still months away from being posted overseas. Stepping into a spacious living room ideally suited to entertain large numbers of people, they begin a round of introductions to over thirty hand-picked members of the academic and medical establishment, a who's who of elite Canadian social reformers determined to impose order on a chaotic postwar society.

Beside Dr. Clarke hovers a tall, excitable thirty-two-year-old doctor, his eyes darting nervously about the room. Working with Clarke at the new psychiatric clinic at TGH, Clarence Hincks has masterminded this special evening, personally inviting this impressive array of doctors and philanthropists to the gilded mansion of Toronto's leading gold baron. Hincks has already formed a bond with Gerry as both men were born into small Ontario villages and entered the University of Toronto medical school as precocious sixteen-year-olds. A thin, frail only child, Hincks was indulged and spoiled by his iron-willed Irish mother; his father was an itinerant, firebrand evangelical Methodist preacher given to bristling attacks on drinking, smoking, gambling, sex, and Sunday streetcars. Growing up, the boy became ever more aware of competing with his father for his mother's attention—"as though mother had two sons." In his first year at medical school, the sensitive Hincks suffered the first in a chain of nervous breakdowns. In a pattern that will ceaselessly repeat itself, he will retreat every summer to a family cottage in the wilds of Lake Rosseau in Muskoka, then return to work refreshed. "Work is the greatest therapy," he believes. "When you stop working, you die."

Only months earlier, Hincks's life was changed forever on a trip to New York. There, he heard an electrifying speech by Clifford Beers, the handsome Yale alumnus and former asylum patient who in 1908 had published an influential autobiography, *A Mind That Found Itself.* In a vivid novelistic style, Beers documented the abuses he suffered in a string of private and public American asylums and the spontaneous recovery he experienced. Exploding like a supernova within the asylum community, the book led to the founding in 1909 of the National Committee for Mental Hygiene, a body of assorted reformist academics, psychiatrists, physicians, and social workers who ignited the first medical-lay reform movement in history.

Hincks devoured Beers's autobiography in a single sitting. Although he had never been incarcerated in an asylum, Hincks imagined himself the Canadian incarnation of Beers and, on the spot, invited Beers to Toronto. Beers, in turn, asked Hincks to join him in creating a worldwide movement, with Canada the first country of occupation.

The two men soon discovered their characters to be similar to an uncanny degree. Extremely sensitive and easily hurt, they both suffered breakdowns in early adulthood; charismatic and highly creative, warm and charming, both are possessed by a sincere sense of mission, superhuman energy, and a dogged tenacity in the face of opposition. They are swallowed by periods of apathy and mute inertia that give way to frenzied, loquacious activity; the downswings typically interrupt their work from anywhere from days to months. They hunger for attention and approval; despite morbid self-consciousness, they have a gift for management and forming loyal friendships and alliances, hiding primal fears under the mask of a vigorous and charming personality. As long as they are not laid low at the same time, the two men make a formidable team.

At first, the Canadian and the American will be magnetically drawn to each other, calling each other "Beersie" and "Hincksie"; but soon they will descend into stormy fights and disillusionment. Both men are prodigiously gifted fundraisers; both are achieving—and will achieve—much in their life's work, despite many deep and painful cycles of depression and mania. Yet, ironically, in the years to come, Clifford Beers will remain completely unaware of the mental travails of his Canadian friend.

As the distinguished guests settle in their seats on this winter evening in Rosedale, Dr. Clarke introduces Beers as the keynote speaker. Standing before the crackling fireplace, Beers mesmerizes his audience with florid descriptions of his bouts of paranoia, suicide attempts, and the similar ordeals of his three brothers. He recounts how, at the Hartford Retreat, the private sanitarium where imaginative literature was considered harmful to disordered minds, he scribbled incessantly, consuming the institution's entire supply of stationery. Resorting to swathes of wrapping paper cut into two-foot-wide strips, he released a flood of words over a distance of fifty feet. Suffering the indignity of force-feedings and straitjackets, drugged against his will, he imagined himself being tried and condemned for a litany of crimes and made several half-hearted attempts to end his life. Craving the respect of men of status, he fired off letters to the governor of Connecticut and President Teddy Roosevelt. His head swam in delusions of grandeur; at the peak of his euphoria, he harboured the grandiose idea of creating a billion-dollar charity, supported by business tycoons like Rockefeller, Morgan, Carnegie, and Vanderbilt, that would wipe out all known diseases. He saw himself as a man on a mission, a selfless tool of God, disinterested in personal gain.

Then he had a "born again" experience, spontaneously snapping out of his mental afflictions as if awakening from a nightmare. Determined to push through his big ideas, he boldly lobbied psychiatrists and medical statesmen of impeccable reputation. The exacting Swiss professor Adolf Meyer closely vetted Beers's manuscript, making many changes that aligned the book with mainstream medical thinking. Beers co-operated at first, for he was thrilled to have gained the attention of men of prominence. In Baltimore, C.B. Farrar reviewed the book with clinical detachment, declaring that it should "not be taken without a liberal grain of salt." He saw Beers as an unrecovered maniac whose credibility was suspect; his mind, in fact, had *not* yet found itself. Most psychiatrists remain loathe to surrender their power to a layman, particularly one of dubious sanity.

Winding up his speech to the assembled Canadian elites, Beers says his achievements had come in the face of the opposition of his family and friends who saw the public airing of his insanity as akin to committing an unpardonable sin. His mission is to awaken the conscience of the world; and it seems to be working like a dream, for the audience is listening in rapt attention. When Beers speaks of the courage he had to muster to speak of such intimacies so openly, the Canadians erupt into spontaneous applause. Overwhelmed, he finds the ovation embarrassing and as he sits down, my grandfather takes the floor. With near-equal fervour, Gerry speaks of the urgent need to replicate Beers's organization in Canada. All heads nod in agreement; as the first committee formed outside the US, it will plant the seed of an international movement. Within minutes, over $24,000 is raised; my grandparents pledge $600 to the cause. A *Globe* reporter quotes an anonymous member of the audience: "I never saw people so enjoy being asked for money."

That spring, the Canadian National Committee for Mental Hygiene is officially formed with Clarke appointed medical director and Hincks his associate; executive offices open at 143 College Street, a short walk from Clarke's psychiatric clinic in the General Hospital. Beers is ecstatic; in Clare Hincks he has found a kindred spirit, a man as adept as he at the delicate art of "button-holing the tycoons." As a colleague of Hincks later writes:

> Unlike other prophets who care not for gold,
> Clare Hincks has a greed that's quite uncontrolled.
> Of the pickpockets' college he ought to be dean,
> But it's all in the interests of mental hygiene.

On June 22, 1918, as a devastating flu epidemic crests in England, my grandfather steps foot on the soil of northern France, attached to the Royal Army Medical Corps. From the shores of the English Channel to the foot of the Swiss Alps, five hundred miles of man-made trenches cut across the belly of western Europe like a livid scar. This grotesque outdoor insane asylum writ large is called the European "theatre of operations." Gerry arrives as General Ludendorff's massive spring offensive is sputtering; the Allies are launching counterattacks and seizing the upper hand. Yet in the unholy year of 1918, the British will suffer more casualties than in any other year of any previous—or future—war.

Gerry takes command of Number 39 Princess Patricia Mobile Laboratory, one of eighteen deployed as part of a general reorganization of the five British armies. No more strenuous and unforgiving work falls to medical officers than to the bacteriologist. Tracking down infectious disease wherever he finds it, my grandfather's specialized knowledge is indispensable, his work never done. He is not ignorant of the mortal dangers of his job—round-the-clock shellfire, skulking clouds of poison gas, and the silent insinuation of killer disease—yet he takes some comfort in the knowledge that the Allies, and his own lab back home, are subduing germs far more efficiently than Germans.

Major FitzGerald (front, centre) commanding a Mobile Pathology Lab, British Fifth Army, Western Front, 1918

To move closer to the firing line that the mobile lab can't reach, Gerry jumps in a motorcycle sidecar and careers past advancing munitions trucks; he risks being captured or killed, for at any moment, the stalemate of trench warfare might break open into rapid movement. In the distance, my grandfather sees men sliding on their bellies like mud-encrusted reptiles; he hears the muffled boom of shells and chokes on the drifting clouds of smoke; he hears the piercing shrill of tin whistles, signalling another suicidal bayonet charge over the top. He sees the bodies of horses, the spine of the supply system, lying in pieces, severed heads hooded in canvas gas masks. He sees the orderlies, clad in their peaked hats, black boots, and brown puttees, red crosses sewn on their arms, thread through the vast crimson carpet of wounded and dying, lugging stretchers. Bearing their satchels of sterilized dressings and boxes of morphine, they kneel down and thrust needles into the flesh of the fallen. The all-pervasive stench defies description: the lime poured on the limp carcasses; the stink of suppurating wounds mingling with the disinfectant splashed in the latrines; the acrid smell of putrefied flesh mixed with piss and shit. On the scattered pieces of carrion, swarms of bluebottles, rats, and lice feast.

Rumbling up to field hospitals, Gerry moves through a tableau of moaning, bloodied men, the dying and the freshly dead, men tortured by boils, psoriasis, jaundice, ringworm, infected eczema, and trench foot, men burying their faces in books and morphine-laced handkerchiefs mailed by wives and lovers. He sees the self-inflicted wounds of deserters; he recognizes the stark face of lunacy from his days in the asylums back home. Fixing his gaze on infected wounds, he scrapes off the bacteria, then hurries back to the mobile lab where he lays out his day's catch. Through the barrel of the microscope, he squints down on the flow of sputum, pus, blood, urine, and stool, fishing for signs of septic organisms. It is delicate, responsible work as the army surgeons, awaiting his results, must decide whether or not to operate. From his bag, he pulls torn bits of tissue and severed body parts, each showing the effects of wounds, gas, and disease; preserved and classified, they are destined to occupy a glass case in the London museum of the Royal College of Surgeons.

Each night, my grandfather's fitful dreams are disturbed by the rumble of munitions trucks, exposed to enemy shellfire; somehow, when he awakes each morning, his nervous system a-jangle, he drives himself onward. The war demands a stoical endurance, a numbed tenacity whose final strength and appeal is to a sense of ethical superiority. Yet Gerry FitzGerald is not alone in sensing something else, something obscenely thrilling, some unspeakable amoral feeling of aliveness glowing like a defiant ember in the guts and slime and blood lust of blind slaughter; the feeling of surviving when others' bodies explode into pink mush; the sweet illusion of invulnerability. The shock of war feels so real, so pervasive, so grotesque, there is no question of not surrendering to his lifelong fascination with the extreme; nothing matters but war.

Nine decades on, struggling to grasp the enormity of my grandfather's war, I turn once more to still photographs.

I am gazing at Gerry posing with seven comrades in front of his mobile lab, a red cross painted on its side. A smiling sergeant stands behind him, holding a small black dog. In polished gaiters and boots, peaked hat, his gloved hands resting on a wooden swagger stick, Gerry stares resolutely at the camera, as if to deflect my recognition of a hidden delicacy.

I imagine the image leaping to life like a resurrected corpse, exploding with noise and smell, bombs and blood and mad, frenzied movement. I salute my grandfather but he cannot see me; he does not have the time to dream me, the writer safe and distant in another century, the one who must dream him, the one who feels compelled to inhabit his body like a parasite, seeing what he sees, eating what he eats, feeling what he feels.

But I am a peacenik child of the '60s, one who has never known war except through the safe screen of books and photos and films. Suddenly a memory spirits me forward from the summer of 1918 to the spring of 1970. In a rare moment of togetherness, I am sitting beside my father in the darkness of Toronto's Hyland Theatre. It is mere weeks after his first suicide attempt; although our mother has protected her children from this fact, I have intuited something amiss all the same. As we watch the taboo-breaking MASH—the first Hollywood film to permit the utterance of that primal, exhilarating word "fuck"—I sense that something about mobile ambulances, war-torn bodies, and the black humour of stressed doctors, strikes a chord with him. As I listen to the movie's lilting theme song—"Suicide is painless; it brings on many changes, and I can take or leave it if I please"—I try to discern a signal in my father's eyes. But he gives nothing away.

In my grandfather's war, doctors on both sides of the trenches are attacking yet another infection swamping entire battalions:

epidemics of cowardice. To the soldiers' minds, shell shock is as real as scarlet fever; worse, in fact, for they can take no vaccines for madness. In 1916, Gerry's old friend C.B. Farrar had emigrated to Canada to take charge of a specially built hospital for shell-shocked veterans in Cobourg, seventy miles east of Toronto. The hospital offers whirlpool baths and leisurely sports, but Farrar also wields the notorious wire brush to jolt soldiers back to reality. Military psychiatrists generally blame the effects of shell shock on overprotective mothers who have smothered the masculinity of the North American male. The solution, Farrar believes, is simply a question of restoring manly willpower; a man who loses his nerve is clearly a shameless malingerer. Hospitalization only encourages such men to escape the responsibilities of work as they have escaped death in the trenches. The cure is simple: Treat a shock with an even greater shock, for cowards clearly possess a constitutional inferiority, some defect of tainted blood.

In military hospitals across the civilized world, psychiatrists strap down soldiers, paralyzed with mute terror or quivering with hysteria, to receive the treatment. Attaching the pad electrodes to the mouth, head, spinal column, or extremities, the doctors smoothly hand crank a battery cell as if making a casual phone call, then shock the men with bursts of voltage. Teeth rattle, hair stands on end, sweating faces contort in agony, arms and legs twitch violently, muscles turn rigid as stone, screams pass unheeded. Sometimes, men are thrown back with such force that they rip the leads out of the battery. Afterwards, the soldiers are returned to the trenches, their shredded memories bleeding into the fall of night.

The British Army uniformly recoils from the "talking cure" that is both German and Jewish, yet not all doctors rush to embrace the barbarity of electric shock. For Ernest Jones, working in obscurity in private practice in London, cut off from Freud by the war, it proves no small breakthrough to have some doctors admit that nervous breakdowns might be psychoneurotic in character, the natural outcome of strong, brave men broken by the slings and arrows of relentless stress. With the coming of peace, the forced recognition of thousands of cases of "war neuroses" will gradually infiltrate the entrenched defences of medical orthodoxy and help put psychoanalysis on the map.

═══

In baking-hot August weather, the British Fifth Army slogs fifty miles through Bethune, Lille, and Tournai; my grandfather's mobile lab rumbles behind, hard on the heels of the advancing lines of infantry. On September 28, during the fight for Cambrai, Captain Fred Banting, an earnest Ontario farm boy Gerry had taught as a medical student, is assigned to clear the wounded from a forward area with horse-drawn ambulances. As enemy shells whistle down, Banting dresses the leg of a wounded corporal; glancing out the door of the barn where German prisoners are being held, he sees a German major standing on the farmhouse steps, smoking a cigarette, suicidally oblivious of the danger. Suddenly a shell hits, ripping off the head of the enemy officer; a steel fragment from the same shell tears through Banting's forearm, severing an artery and nearly splitting his limb in two. A few hours later, groggy with morphine and the loss of blood, he forces himself back to work tending the wounded, his bandaged arm throbbing painfully. A fellow officer screams: "Are you out of your mind? Are you trying to kill yourself?"

Recovering in hospital in England, Banting is plagued by horrifying dreams. The news that he has won the Military Cross is undercut by the fact that, like thousands before him, his wound has become infected. A young surgeon-lieutenant, an officer of inferior rank, matter-of-factly tells Banting he is going to cut off his right arm on the spot.

"No, it will be alright," he stubbornly insists. "I'll treat it myself."

Destined for a place in the history books, the obscure twenty-seven-year-old Canadian doctor does not take kindly to being treated like just another patient.

═

On November 11, 1918, the British Fifth Army occupies an eleven-mile front from Lessines to four miles east of Ath, a town near Brussels. Here, caked in mud and blood, the wheels of my grandfather's mobile pathology lab come to rest for the last time. In the minutes after the eleventh hour of the eleventh day of the eleventh month, the dry, depersonalized statistics of the Great War are rounded off into a numb abstraction: over 15 million lives lost. Out of a population of 8 million, Canada had rallied a largely volunteer army of 600,000—a quarter of the nation's young men. Over 66,000 Canadians were killed and 170,000 scarred and crippled. Statisticians fail to do justice to the anonymous thousands forever marked by trauma, haunted by the memory of the ceaseless onslaught of bullets and bombs, mustard gas and shell shock, syphilis and meningitis, trench fever, vermin, and lice.

Still, public health doctors have something to celebrate: In past wars, eight soldiers died of disease for every one killed in battle; on the Western Front, the Canadians have pulled off a spectacular reversal—only one soldier died of disease for every twenty killed in battle. Producing one-fifth of all the serums used by British armies, the Connaught Laboratories are hailed as novel and unique, combining teaching, research, and production in a non-commercial, altruistic manner. Indeed, the eclecticism of FitzGerald's idea—an open-minded hybrid of tradition, innovation, and experimentation—typifies the Canadian brand of strategic thinking that characterized the nation-building military victory at Vimy Ridge.

Then, on returning troop trains, Nemesis delivers the Spanish flu to the city of Toronto.

Only two weeks before the Armistice, the death toll peaks when eighty-seven citizens perish in a single day. Moaning, writhing bodies overflow into hospital corridors, rank with blood and excrement. As faces and feet turn a dark brownish purple, it is hard to distinguish white men from negroes. Pockets of air leak through ruptured lungs, making a bizarre crackling noise, as blood oozes on the linens. The flu leaves no internal organ unravaged, including the brain. In the end, gasping for breath, bloody froth and mucus gushes from noses and mouths in angry jets, leaping several feet. As lungs fill up with thick red fluid, as the last human voice of consolation fades into a whisper, the victims drown.

Robbed of the wild revel they justly deserve, the conquering veterans encounter hunched figures scuttling through deserted city streets, draped in white gauze masks and black armbands, trailed by parentless children. Far more than the war, the terror of the flu causes suicide rates to spike; people agonize whether to risk going outdoors to help others or stay home. Parades, parties, and public gatherings are banned; schools are closed. In courtrooms, people taking oaths are no longer asked to touch the Bible with their lips. Funerals are limited to fifteen minutes; in church, no one dares infect a crucifix with a kiss or breathe their sins through the window of the confessional.

For two months, the virus rampages through the city, then retreats, having exhausted the supply of susceptible hosts. Thanks to the stringency of Charles Hastings's existing public health measures, Toronto achieves the best record of all Canadian cities with a relatively low death toll of 1,800. In contrast, in Philadelphia, a city of comparable population, 759 people die in a single October day and over 11,000 that month.

In Gerry's absence, Dr. Defries has been struggling to concoct a vaccine made from eighteen New York flu strains, working day and night in the lab. The Connaught distributes the experimental vaccine nationally free of charge to help stem the tide of death that will eventually claim anywhere from 20 million to 100 million people from every country in the world except Australia. As doctors stab hypodermics, big as darning needles, deep into hip muscles, the huskiest of men faint dead away. It is wrongly thought that the flu is caused by bacteria; not until 1933 will scientists prove it to be a virus. But the appearance of the medicine helps build morale and temper mass panic, among the normally sanguine public health authorities as well. In the minds of Canadians, the name of Connaught is slowly earning a place of trust. Meanwhile, outside the nursery window of 186 Balmoral Avenue where my eighteen-month-old father dreams behind the slats of his crib, innocent of masks and microbes, needles and knives, skipping children sing a popular ditty:

I had a little bird
Its name was Enza
I opened the window
And in-flu-enza.

EIGHTEEN

The Sugar Sickness

> Ever tried. Ever failed. No matter. Try again. Fail again. Fail better.
>
> SAMUEL BECKETT
> *Worstward Ho*

As my grandfather disembarks from the SS *Canada* into the city of Halifax, the arc of his career neatly dovetails with momentous national and international events. On his arrival on February 15, 1919, he learns that on the previous day, President Woodrow Wilson, a model of prim Irish Protestant rectitude, tortured by migraine headaches not unlike his own, has drafted the Covenant of the League of Nations at the Paris Peace Conference. Horrified by the madness of mass slaughter, the elites of my grandfather's generation are binding together in an effort to turn back the eternal recurrences of history; the new global body is attempting nothing less than the redemption of the sins of the Western world.

Article 23 of the League Covenant, calling for the establishment of a health organization, instantly catches Gerry's imagination. Devoted to the prevention and control of disease on an unprecedented international scale, the non-political body will draw upon the expertise of public health doctors from nations all over the world. Soon my grandfather will number among them.

By the spring of 1919, a confluence of postwar events—the epic sweep of the flu pandemic, the spread of venereal disease, the problems of immigration, together with the springing up of lobby groups like the National Committee of Mental Hygiene—spurs the Canadian government to create the Department of Health in Ottawa. With the federal invasion of provincial jurisdictions, a progressive new health act marks an historic shift in responsibility. All the same, Gerry harbours no illusions about the uphill battle ahead. In wartime, everyone pulls together against a common enemy; in peacetime, citizens quickly revert to the old complaints about public health officials, those nagging, meddlesome busybodies, wagging their fingers, holier—and healthier—than thou.

When Gerry's old ally, John Amyot, is made the first deputy minister of health, the appointment necessitates the handing over of his part-time academic duties at the University of Toronto to Gerry, who is promoted to the position of the first full professor in the department of hygiene and preventive medicine. Dr. Defries and other Connaught staff members are appointed to the teaching staff, the postings formally linking the department with the Connaught Laboratories. The lab and the department of hygiene are now closely related yet independent units of the university, uniting the functions of teaching, research, manufacturing, and public service.

With the financial resources of the Connaught Labs, Gerry is now able to aggressively promote both undergraduate and postgraduate instruction in public health, ushering in a golden age of preventive medicine. From the start, Gerry does not hesitate to lay on the workload with a trowel. He overhauls the Diploma of Public Health, formalizing intensive courses to extend over a winter session of eight months; the undergraduate course in medicine is extended from five to six years. He clearly sees the fundamental difficulty in drilling preventive medicine into the skulls of medical undergrads: it is dead boring. He feels that the old system of didactic lectures is next to useless, lulling doodling students to sleep with the monotonous recitation of not-so-vital statistics. Trying to lift the fog of lethargy and indifference, he draws on the resources of the city, taking his charges on mandatory "observational visits" to local clinics to study tuberculosis, VD, and mental illness. He breaks up classes into groups of twenty; each student is told to give a short lecture on an aspect of preventive medicine, which is then debated by the class. Not only is attendance compulsory, but students must endure a three-week summer extension course to allow for the inspection of schools, factories, restaurants, dairies, and water filtration and sewage disposal plants.

The innovation nearly sparks a rebellion. Students protest that their summer earnings will be cut, making the tuition fees for the following year unaffordable; besides who wants to drag themselves to class at nine o'clock on a Saturday morning in June? After a few weeks of unrest, Gerry steps into the lecture hall one morning and, measuring his words as sharply as an apothecary's knife, announces: "In spite of the views of the conservatives, the diehards, and the knockers, this course will proceed."

Anti-vaccination rally, Toronto City Hall, 1919

That same fall of 1919, a severe outbreak of smallpox tests the mettle of Toronto's public health system like never before. When Charles Hastings calls for the mass inoculation of the entire population of the city, threatening to expel school children whose parents refuse to comply, his imperious attitude triggers a public backlash. On November 13, 1919, the small but vocal Anti-Vaccination League stages a public demonstration against compulsory smallpox vaccination on the steps of city hall. Protesters wave placards emblazoned with the slogan: "Compulsory Vaccination German Born—Down With Compulsion!"

Veterans complain that Prussian-like coercion is antidemocratic, unacceptable to the men who have risked life and limb for the cause of freedom. Half a million Canadian soldiers were vaccinated during the war, but now many are attacking the very people who had saved them. For the sake of the cause, Gerry stands before his captive audience of undergraduate medical students, dramatically rolls up his sleeve and injects the vaccine into his bloodstream; then he does the same to the entire class. Through the coming winter, the city records 2,800 cases of smallpox and eleven deaths. When no deaths are attributed to the vaccinations of over 200,000 people, the dispute dies down.

Simultaneously, Gerry assumes a leadership role in a newly created advisory body reporting directly to the minister of health. The Dominion Council of Health is designed to coordinate the three levels of government and various private organizations on myriad health matters ranging from quarantine and pasteurization of milk to medical inspection of immigrants, child welfare, and drug addiction. In the years to come, he will dutifully travel to Ottawa each spring and fall to the DCH meetings, tramping past the wrought-iron gates and grey-stone walls of Parliament Hill, the Gothic revival spires and sloping copper roofs redolent of the massive asylums of his not-so-distant past.

With a population of eight million, Canada is witnessing the transformation of public health from an isolated grassroots movement to a modern, specialized, state-run bureaucracy, stretching like a gigantic safety net from the Atlantic to the Pacific oceans. Gerry FitzGerald finds himself exactly where he wants to be: perfectly positioned on the ground floor, the right man in the right place at the right time, tightening his network of political, scientific, and academic relationships, his fledgling, self-supporting lab an engine of national influence. All nine provinces are freely distributing its nine products; a thriving export market is reaching countries as far away as New Zealand, the Caribbean, and China.

To cap a triumphant year, Gerry is made a fellow of the Royal Society of Canada, a national body of distinguished scholars. How could my grandfather possibly suspect that something even bigger, even more exhilarating, is looming on the horizon?

═

In the summer of 1920, recovering from his war wound, Fred Banting retreats to London, Ontario, where he sets up a private general practice. Muscular and ruddy-cheeked, Banting stands six feet tall, slightly stooped, his bright blue eyes peering from behind owlish, gold-rimmed spectacles. Alternately pugnacious and shy, boorish and witty, he hates all sham, all formality, all niceties of social convention. A farmer to his bones (he will never own more than one suit), his searing honesty—some called it rudeness—cuts through the airs of the pretentious like a scalpel. Like Gerry FitzGerald, he is prone to the hero worship of immortal, patriarchal figures like Louis Pasteur, masking feelings of inferiority and a secret fear of failure. His pious father, William, had steered Fred, his youngest, towards the ministry; for two years, the son had tried hard to please the father, taking arts at Victoria College. When Fred switched to the secular religion of medicine, his father found it a bitter pill to swallow.

To boost his paltry income, Banting lectures at the University of Western Ontario; tense and bored, possessed of a volatile artistic temperament, he is encouraged by an artist friend to relax by painting landscapes and soon splotches of paint join the traces of blood on the young doctor's smock. One day, while preparing notes for a talk on carbohydrate metabolism, he comes across research on diabetes mellitus, the mysterious "sugar sickness" that has killed millions since the days of the Egyptian pharaohs and Greek philosophers. Banting finds that little is known of the bizarre and horrifying affliction, known colloquially as "the pissing disease," which attacks the pancreas, the mysterious bean-sized organ behind the stomach that helps the body digest food. Scientists do know that diabetes tends to more often afflict the prosperous and well-nourished rather than the poor; that it is most often found in Germany and the United States. Gripped by unquenchable hunger and thirst, victims suffer from headache, slurred speech, and staggering gait, as if emerging from a night of hard drinking. Growing grotesquely thin, they fall prey to boils, carbuncles, and blindness; the stabbing pain of neuritis often leads to addiction to morphine. The skin turns as dry as parchment and hair falls out in handfuls. When limbs turn gangrenous, doctors have no choice but to amputate; simply submitting to anaesthesia means putting your life on the line. Sapped of their power to fight off invading microbes, diabetics routinely perish from recurrent infections like pneumonia and tuberculosis. Strict starvation diets are almost impossible to follow; and so comes coma and the release of death. Children, mercifully, die faster than adults. Ironically, as deaths from infectious disease are being steadily reduced by vaccines and better sanitation, the prevalence of diabetes is rising.

Banting spends the weeks of October boning up on the pancreas. On the Sunday night before his lecture, he stays up reading into the early morning hours. It is Halloween—the ancient festival of the Celts who believed the spirits of the dead crossed the barrier into the land of the living—and a full harvest moon shines through his window. Chain-smoking, he paces the linoleum floor of his simple, ten-foot-square bedroom; slowly, the germ of an idea forms in his mind. At 2 a.m., he climbs out of bed and jots in his pocket notebook: "Diabetus. Ligate pancreatic ducts of dogs. Keep dogs alive till acini degenerate leaving Islets. Try to isolate the internal secretion of these to relieve glycosurea." Banting's ignorance of diabetes reveals itself in his misspelling of its name, but spelling counted for little now.

On November 8, Banting ventures to the University of Toronto to meet with Professor J.J.R. Macleod, chair of the department of physiology and the reigning expert on diabetes. A brisk, cultured Scotsman, Macleod doubts the theory that the pancreatic islets of Langerhans, cells of insulin-producing tissue, produce a mysterious

"Hormone X" that burns sugar, but he listens. Banting stammers his way through his pitch, intimidated by the credentialed research scientist staring back at him with glacial detachment.

"I don't mean to be condescending, Dr. Banting, but will you please outline your previous training and experience in research?"

Banting is crestfallen that he has botched the presentation of an idea of almost hare-brained simplicity; when he confides the situation to his Irish landlady, she urges him to write it all down and return the next day.

═

On May 16, 1921, Fred Banting climbs the stairs to the second floor of the university's twin-towered medical building on King's College Circle, trailed by a fourth-year medical student, Charles Best, assigned as his assistant. Having reluctantly approved Banting's proposal, Macleod has given him Room 221, a cramped, disused, dingy anteroom with sickly, yellow-papered walls, grimy windows, and a complement of ten caged dogs. Down two flights in a mouldy basement lab only slightly better equipped, Gerry FitzGerald works side by side with his growing band of Connaught colleagues, grinding out vials of vaccine and antitoxin; when his sensitive ears pick up the sound of muffled barking above his head, my grandfather has no way of knowing he is hearing the sounds of history in the making.

Fred and Charley make an odd pair: Banting, twenty-nine, the quirky loner, profane, down-to-earth, temperamental, impatient, readily offended; Best, twenty-one, handsome, blue-eyed, clean-cut, tall, fair, athletic, popular, and gregarious. Best has his own motives for attacking diabetes: the disease killed his favourite aunt. As a boy, Best used to accompany his father, a Canadian GP based in rural Maine, on calls to remote farmhouses; often his father operated on kitchen tables while Charley dropped ether on a cone to keep the patient anaesthetized. Young Charley loved horses, charging wildly across a frozen lake with his father, exhilarated by the sound of the iron-shod hooves chipping the ice. Yet no matter how fast he races through life, he will not best the depression that dogs his family's bloodlines.

═══

June 1921: The staff of the Connaught Labs assembles for a company picnic on a sunny Sunday afternoon at the farm. The photographs I have pulled from a family album show my grandparents standing over their two small children, Jack, four, and Molly, seven, while serving tea. A staff member picks them up, one by one, and perches them on the back of a chestnut hackney mare. Young Molly is thrilled, for the horse is named after her. Receiving more needles than a haystack, Molly, the mare, has proven herself a national heroine, for her blood has been used to produce all the antimeningitis serum for the entire Canadian population for the past five years.

My four-year-old father is too young to understand the strangeness of his father's world—the caged animals, the thrusting needles, the earnest men in bone white coats. He has not yet shaken the big hand of his godfather, Charley Best, nor can he grasp the enormity of insulin and its magical, godlike powers.

He strains to make sense of the fragmented voices of the grown-ups, lifting and falling in excited tones. What are the horses for? Why are they so important? If children must be saved, what must they be saved from? Even as he grasps the sweaty mane of the white mare, as he feels the bony flanks of the dumb beast under his groin, even as he starves for a tender human touch, my father dreams of outracing the night-mares of the Balmoral nursery, charging headlong into bigger, better, happier worlds.

═══

Banting's ignorance of the physiology of the pancreas will ironically prove an advantage; he will later confess that if he had tried to digest the complexities in the literature, he never would have lifted a finger. He is mainly following his intuition, taking a blind stab in the dark, trying to "bridge a spark gap between two remote ideas." Over the sweltering summer months, he and Best methodically cut open the bellies of chloroformed dogs and tie off their pancreatic ducts with catgut; whimpering in pools of blood, the dogs die of shock, infection, and overdoses of anaesthetic. Banting spends the last of his cash to procure a fresh batch of mongrel dogs from the back alleys of Toronto's slums, fastening his tie around their necks and leading them back to the lab. Failure follows failure, making it all the more blissfully sweet when the breakthrough comes. When an injection of a crude extract of degenerated pancreas—Hormone X—lowers the blood-sugar levels of a snow white terrier, the two men know they have witnessed a species of secular miracle.

Ever the cautious, fastidious scientist, Macleod declines to share Banting and Best's enthusiasm over their promising findings, a withholding attitude that will eventually pitch a tenuous alliance into a scandalous feud. Staking everything on the work, Banting has given up his house and private practice in London; there is no turning back now. He demands a salary, a bigger lab, a boy to look after the dogs. When Macleod coolly replies that other researchers will suffer, Banting explodes. "I'll show that little son of a bitch!" he rants as he storms onto College Street. Macleod soon relents, but the Scotsman has cut Banting to the quick, opening a wound that will fester for the rest of his life.

On November 16, 1921, two days after Banting's thirtieth birthday, a shaft of inspiration once again shoots up from the depths of his unconscious mind. Again, around 2 a.m., restless with insomnia, a chain of thoughts leads him back to boyhood memories of the cattle he helped slaughter on his father's farm. Knowing from his reading that the pancreatic islets of Langerhans are more richly concentrated in newborn infants, Banting deduces they must be even richer in embryonic mammals before birth. Because there are no digestive juices to poison the internal secretion, their pancreases should be composed of almost pure islet tissue.

At first, he thinks of inducing abortions in dogs, but the task would be too cumbersome. Then he remembers that if a farmer wants to fatten cattle for slaughter, it is best to breed them first, for pregnant cows are more voracious feeders. On November 19, Banting and Best drive to a slaughterhouse in northwest Toronto, not far from the Connaught Farm, where they slice open the stomachs of nine pregnant cows. Carefully, they cut out the three-month-old calf fetuses, carry them to the car, and rattle back to the university.

Best mashes up the bloody glands and makes a quantity of extract. They are right: The fetuses are free of trypsin, the powerful digestive juice that impedes the isolation of the hormone; the oversized islets of Langerhans prove a rich source of crude extract. No longer needing to sacrifice dogs, they are freed from the dilemma of killing the healthy to save the sick. Vibrating with excitement, Banting and Best have successfully isolated the active principle of Hormone X, which they christen "isletin."

Before Christmas, a shy, sensitive Ontario-born biochemist, James Bertram Collip, twenty-nine, on sabbatical from the University of Alberta on a Rockefeller Foundation fellowship, joins the team. He is a key addition—historians will argue Collip is the best pure scientist of the four. Working into the early hours of the morning, Collip devises a way to purify the crude isletin, using, of all things, a prohibited substance: alcohol.

Gerry seizes the moment. Offering the team the use of his basement lab, he advances five thousand dollars to cover the costs of preparing purified extract for clinical trial, and if successful, to make mass quantities for millions of the world's suffering diabetics. On January 1, 1922, he appoints Bert Collip, assisted by Charles Best, to direct the production operations of the Insulin Division of the Connaught Laboratories. Professor Macleod, who has coined the simpler term "insulin," can no longer deny the amazing truth: For thirty years, the dark mystery of death-dealing diabetes has stymied the best scientific minds toiling in the finest labs in the civilized world; yet, here in insular, puritanical, joyless, bigoted Toronto, two green Canadian rubes, working on a shoestring, have cut open the heart of the beast, emerging with bloodied hands, covered in glory. And they have done it in a mere nine months.

Gerry's fledgling lab is a mere seven years old, the same age as his little freckle-faced, red-headed daughter, Molly. Can he handle the awesome responsibility of shooting an unproven, experimental wonder drug into the sickened bloodstreams of the world?

On January 11, 1922, an experimental insulin extract is carried into the diabetic ward at TGH where a hopeless charity case, fourteen-year-old Leonard Thompson, emaciated and sheet white, lies dying. Doctors observe only a slight improvement from the injection; the trial is premature. Intense pressure falls on Collip to prepare a purer batch, for the boy is plunging headlong into the jaws of death. Collip works furiously, tinkering and mixing and blending and stirring assorted random mixtures of fresh whole beef pancreas ground up in concentrated alcohol, injecting it into rabbit after rabbit. Amazed, he watches as the animals contort in violent convulsions, their heads snapping back, eyeballs bulging, and limbs stiffening. They toss themselves from side to side, then collapse into a coma, breathing rapidly. Their legs flail as if they are being chased by a mad dog; the slightest nudge sets them off again. The seizures recur every fifteen minutes until, at last, they die.

Charles Best and Fred Banting on the roof of the University of Toronto medical school, 1921; insulin revenues helped vault the fledgling Connaught Labs into a world leadership position

Taking a leap of intuition, Collip injects the rabbits with glucose and, miraculously, they recover. He has stumbled across the phenomenon of hypoglycemia, or insulin shock, occurring when the blood sugar falls below a certain level. A witness later wrote: "It all looks simple now but it was the most thinking per square metre that I have seen."

Late on the night of January 16, Collip achieves the eureka moment—the isolation of pure insulin. "I experienced then and there all alone in the top storey of the old pathology building perhaps the greatest thrill which has ever been given me to realize."

Soon Collip drops into Banting and Best's lab. (He likes to call them "B-2.")

"Well, fellows, I've got it."

"How did you do it?"

"I've decided not to tell you," comes the nonchalant reply.

When Collip says he intends to quit the team and take out a patent on the pure pancreatic extract in his own name, the poisonous brew of mistrust and suspicion explodes into flames of pure paranoia. Furious, Banting clenches his fist and swings a wild haymaker, knocking Collip to the floor, then pummels him where he lies. Astonished, Best grabs Banting's shoulders, pulling him off Collip with all the strength he can muster.

"I may have helped to save millions of diabetic lives," Best later recalled, "but I know of one life I saved for certain—Bert Collip's."

This time, the injections produce a dramatic drop in Leonard's blood sugar level. The boy's listless eyes gleam brightly, his parchment skin assumes a rosy hue, he rises from his bed and walks. It is an epic moment, equal to Louis Pasteur's saving of the Alsatian boy, Joseph Meister, from death by rabies. Here in dingy Toronto, a city as dull as ditch water, in the killingly frigid month of January, a select handful of doctors can, for a moment, drop their poker-faced clinical personas and stand in awe of the near-supernatural. It was as if, before their eyes, they saw Lazarus resurrected from the dead. Yet even in the flush of triumph, even as insulin holds the potential to deliver diabetics into ripe old age, burning resentments and recriminations threaten to strangle the infant discovery in its crib.

The role of the cool-headed peacemaker falls on Gerry FitzGerald; he knows he must rein in the rearing stallions that threaten to gallop out of the barn in all directions. If the team is to co-operate with the Connaught Laboratories in making pancreatic extract, he insists they adhere to two key conditions. On January 25, only days after the inglorious bout of fisticuffs, Banting, Best, Collip, and Macleod sign a contract in which each agrees not to take out a patent with a commercial firm during the working period with the Connaught; in addition, no changes in the research policy will be allowed unless first discussed among FitzGerald and the four collaborators. A colleague of Banting's, Lewellys Barker, will say it best: "There is in insulin glory enough for all."

═

Down in the dank sub-basement of the medical building—the same lab where Gerry lined up his bottles of horse blood back in 1914—workers begin installing special production machinery designed to churn out bottles of insulin by the truckload. But then disaster strikes: Bert Collip, an absent-minded note-taker, is unable to recapture the breakthrough of mid-January. He simply cannot duplicate the original conditions of time and temperature.

Incensed by such unforgivable sloppiness, Banting grabs Collip in the corridors of the medical building and threatens to kick his ass all the way to College Street if he doesn't rediscover the method in a goddamn hurry. Passersby yank them apart. For the next two months, only tiny, fitful doses dribble out of the lab; only a handful of the most desperate, terminal cases are treated. Tragically, Toronto is mired in an insulin famine.

Working with the crudest of equipment, Charles Best and his colleague David Scott scramble to resurrect the secret. They evaporate alcohol or acetone from the extracts in open trays, using a bizarre, makeshift wind tunnel driven by an old exhaust fan, bristling with hot, unprotected wires. The voracious condenser in the metal still burns two tons of hand-delivered ice per day; wading through floods, Best and Scott step over skittering wild rats, engulfed by the reek of acetone, a dangerously volatile chemical. One day, the rattling motor driving the exhaust fan shakes a gigantic glass bottle of acid off the shelf. In horror, they watch it shatter on the floor and, for a split second, expect to die. But the insulin team is blessed by yet another miracle—the acetone fails to spark a fire and blow the Connaught Laboratories to hell and back.

Heart-rending letters scribbled by dying children clutter Banting's desk. From all over the continent, diabetics are pouring into Toronto by ship, train, horse and buggy—the insulin rush of 1922. In the oppressive summer heat, people camp out at the doorstep of the medical building or they seek out Banting's office at 160 Bloor Street West, beseeching the great man himself to inject the insulin, not just any run-of-the-mill doctor. Meanwhile, Bert Collip is mysteriously neglecting his duties. Impatient with the famine and the ongoing feuding, Gerry tells Best and Macleod that somebody better start doing the job that Collip doesn't seem capable of handling. And another dark prospect now looms: losing the discovery to competitors. The thorny issue of patent rights needs ironing out.

On April 12, FitzGerald, Banting, Best, Collip, and Macleod write to the university president, Robert Falconer, proposing a defensive tactic to block unscrupulous drug companies from making the extract and racking up massive profits. When the details are published, anyone is free to prepare an extract, but no one can secure a monopoly. The University of Toronto forms an Insulin Committee and the Connaught is set up as a testing centre. Insulin must first be assayed by the university in order to guarantee its reliability; a nominal fee is levied for testing insulin outside Canada. After covering the lab's expenses, the rest of the fee is divided into two parts: one half goes into a public trust to maintain patent rights and finance future research on insulin; the second half is divided into three shares to underwrite future research projects by Banting, Best, and Collip. By patenting, it is possible to control the production of insulin so only reputable firms that adhere to the highest standards will be licensed. Not a penny of personal profit is to be made from insulin.

Meanwhile, during the weeks of April and May, all four men—Banting, Best, Macleod, and Collip—work feverishly to regain the lost secret of the pancreatic secretion. By mid-May, they find it.

Buckling under the pressure to meet the surging demand, Gerry and the Connaught finally are forced to admit they need help as mass production is fraught with difficulties. Founded in Indianapolis in 1876, Eli Lilly is a family-owned drug company that sells only to doctors and pharmacists. With 1,100 employees and five million dollars in annual sales, the American firm attracts Gerry because of its high ethical standards and progressive, enlightened ties with fledgling academic research communities. After several days of meetings at the King Edward Hotel with Eli Lilly, the grandson of the firm's founder, Gerry and the Toronto team hammer out an agreement that will prove far-sighted. The university will eventually license North American drug firms making insulin for sale and collect royalties for research. Meanwhile, the Canadians will grant a temporary, exclusive licence to Eli Lilly. For Eli Lilly, the pact is a potential gold mine.

One day that same spring, Dr. Joe Gilchrist, a diabetic classmate of Banting's and self-described "human rabbit," leaves the lab after volunteering to take yet another injection of experimental insulin. Strolling the streets of Toronto, he is suddenly overcome by a feeling of impending doom. Sweating profusely, his body trembles and his legs buckle; his mind swoons into a dark fog. Gilchrist is the first human being to experience insulin shock.

A passing policeman, patrolling the pavement of Toronto the Good, promptly leads the doctor, dizzy and weaving, back to the station where he books him on a charge of public intoxication.

In the spring of 1922, Gerry sails to London to forge a separate insulin patent agreement with the British. They are skeptical of miracle cures emerging from the backwoods of North America, but soon enough two large English chemical firms are gearing up to supply doctors and hospitals with the Canadian discovery.

For months to come, the transition from small- to large-scale insulin production will prove difficult, erratic, and expensive in both Toronto and Indianapolis. In its effort to produce a consistent insulin standard, Eli Lilly rips through a staggering hundred thousand rabbits over the first six months of work. Colonies of cottontails slip free of their cages, hopping through the corridors for their lives, chased by frantic technicians. Passing godlike life-or-death sentences, clinicians on both sides of the border agonize over which patients deserve the preciously few doses of life-saving hormone. Not until the end of 1922 will enough extract be produced for use by Canadian diabetic specialists and clinics outside Toronto. The city continues to rely on the Americans until the early summer of 1923; not until the fall will sufficient quantities of insulin find their way into drugstores.

Despite the infuriating production hitches, all agree that pure insulin is nothing less than a miracle—"an achievement," according to a Boston diabetic specialist, "difficult to record in temperate language." Elliott Joslin resorts to biblical analogies, likening insulin to the windy breath of God transforming slain corpses into a great army. "By Christmas of 1922," he exclaims, "I had witnessed so many near resurrections that I realized I was seeing enacted before my very eyes Ezekiel's vision of the valley of dry bones."

Jack and Molly astride a mare, named after Molly, that was responsible for producing meningitis serum for the entire Canadian population from 1916–1922

In the fall of 1922, in the heady thrall of the insulin rush that was pumping life into his young lab, my grandfather rushes headlong out of town. It seems an inopportune time to abandon his post, but months earlier he had agreed to travel to the University of California in Berkeley where from 1911 to 1913 he had germinated his original idea for the Connaught Labs. This time, he will spend a sabbatical year occupying the prestigious chair of bacteriology and experimental pathology and renewing his attack on an old enemy—diphtheria. In his absence, he hands over the reins of the Connaught to the loyal Bob Defries.

Gerry and Edna had loved their two-year sojourn in pre-war California; this time, they will take Molly and Jack, now eight and five, along with them. In this same year, my grandfather has published an eight-hundred-page textbook for use in his hygiene classes. Typical of many of his projects, *An Introduction to the Practice of Preventive Medicine* was a collaborative effort, enlisting the help of several academics of various disciplines. Packed with the latest data, the tome is the first in Canada to range over the full sweep of public health and preventive medicine, laying out the latest knowledge before the up-and-coming generation of young doctors—a generation he fully expects his own son, Jack, to join.

In Berkeley, Gerry once again works in the lab funded by Phoebe Hearst, the philanthropist who had perished three years earlier in the influenza pandemic, unprotected by the veil of her fabulous wealth. Gerry will publish five research papers, three with Dorothy Doyle, a colleague of Irish blood, all the while keeping a close eye on the situation in Toronto where the discovery of insulin has created an urgent demand for new and more refined methods of measurement and control. In February, he writes Fred Banting, describing how a local doctor has used insulin to pull a woman out of a diabetic coma in an Oakland hospital: "It is the first time such a thing has been accomplished on this coast and everyone is jubilant."

═

After a day of chasing paper in the Berkeley archives struggling to understand the intricacies of my grandfather's science, I walk out into the serene afternoon sunshine. In my hand I am clutching a black-and-white photograph of my smiling grandparents perched on the front steps of a rented home at 2915 Garber Street, with Molly and Jack at their feet. Hoping the place still exists, I climb a green hill overlooking San Francisco Bay. Suddenly, before me stands the same charming three-storey house fronted by a white picket fence. Nothing has changed in 83 years: The same carved wooden columns flank the front door; the same oranges droop from the bushes; seabirds still skim the tops of the palm trees. With its steeply sloping roof and twin brick chimneys, the place subtly echoes the cool formality of 186 Balmoral Avenue and for a moment I am aware I am holding my breath.

I knock on the front door but no one answers. Just then, a teenaged au pair ambles by, hand in hand with a small boy who, if this were the year of 1922, could pass for my five-year-old father. I show the girl the photo, and like me, she is amazed how the place appears unblemished by time. I ask her to take a picture of me sitting on the front steps, as if summoning the four ghosts to enter the frame with me. As the shutter clicks, the little boy in short pants gazes at me quizzically; I feel a strange mingling of sadness and triumph, of having found something of meaning—the house on the hill, the place in the sun, a forbidden channel of existential joy penetrating the darkness. Yet, for all my small epiphanies, I know all is loss.

═

Back in Toronto, Fred Banting, the reluctant and bewildered hero from Alliston, Ontario, is now stumbling into the bear pit of global celebrity. When not fending off the voracious press or nursing a worsening ulcer, Banting is turning down stupendous sums to lend his name to commercial enterprises. Dr. John Harvey Kellogg of the Battle Creek Sanitarium, the inventor of cornflakes and avid supporter of eugenics, offers to build the young Canadian a special wing and pay him ten thousand dollars a year; George Eastman of Eastman Kodak urges Banting to travel across the lake and work at the University of Rochester. Then, on a rainy night in late August 1922, Banting receives a visit from a mysterious American businessman. Scanning his shabby bachelor digs with disgust, the stranger spits: "You're a damn fool." All Banting has to do is hand over the Connaught patents to a group of Wall Street businessmen and he will receive $100,000 cash, a five percent royalty, and the medical directorship of a glittering national chain of clinics.

"It was not even a temptation," Banting writes in his diary. "It meant that the suffering diabetic would be exploited." Later, he declares: "Insulin doesn't belong to me. It belongs to the world." Some doctors find his ethos of high-minded self-sacrifice excessive; there is, perhaps, too much Methodism in his madness. But his friend, Gerry FitzGerald, as ruthless in the execution of his own altruism, stands behind him all the way.

═

On October 19, 1923, my grandfather greets the prim figure of Wickliffe Rose, a slight, immaculate man in pince-nez and bow tie, as he steps from his Pullman car at the North Toronto station at Summerhill Avenue. Given a blank cheque by John D. Rockefeller, Junior, the sixty-one-year-old Rose, a classical scholar of Tennessee breeding, is now embarking on the first leg of a five-month whistle stop tour through nineteen countries, where he will talk to dozens of leading scientists, spot young fellowship talent, and assess the needs of the international research scene. Insulin has put Toronto on the map; and Toronto is Rose's first port of call.

Four years earlier, the Rockefeller Foundation had donated a gift of five million dollars to support medical education across Canada, one million of it earmarked for the University of Toronto. The heroic age of North American scientific philanthropy is now radically transforming medical education, making the full-time teaching system a reality and keeping Canada in step with European and American medicine.

Under the Rockefellers, Rose had adapted the tradition of the "circuit riders," the early disciples of John Wesley, the English founder of Methodism who had planted his rigorous brand of religion in eighteenth-century Georgia. The method was used brilliantly to tackle an epidemic of hookworm, an energy-sapping parasite that was plaguing over two million people in the US south. Emulating the evangelical Wesleyites, Rose's circuit riders—a combination of fact-finders and missionaries—no less zealously spread the gospel of science through the illiterate rural South. It was the largest public health crusade in American history; and it wiped out hookworm.

Rose then set up the machinery to internationalize the work he had begun in the American south, pushing for the creation of an International Health Board and the building of training schools in public health, first at Johns Hopkins and Harvard, then in cities abroad. Rose knows that the prestige of the discovery of insulin in Toronto had been further enhanced when the Connaught Labs exemplified the ideals of international cooperation in the drug's improvement and distribution. He also knows the city's public health system, under Charles Hastings, is firing on all cylinders. The chairman of the Rockefeller Foundation, George Vincent, is so impressed with Toronto he is sending his fellowship doctors, hand-picked from various countries, to study in the city to gain the best kind of practical experience. Dr. Leon Bernard, the respected professor of preventive medicine at the University of Paris, adds his voice to the international chorus, praising Toronto as having developed the best system anywhere in the world.

Gerry with his right-hand man, Robert Defries, 1921

In accepting Rose's invitation to join the International Health Board—other members include his heroes William Welch and Simon Flexner—my grandfather seals his ties with the greatest fortune in the history of the world. But some of his colleagues express outrage at being dependent on the whim of philanthropists; Howard Kelly, one of the fabled Four Horsemen of Johns

Hopkins, numbers among the most vocal, citing "the train of suicides which followed Rockefeller's early days as he was building up the Standard Oil monopoly. If we could see all the little white stones which mark the graves for which the Rockefeller and Carnegie interests have been responsible, I wonder if the mountainsides would not look as if a snowstorm had struck them." But if my grandfather harbours any misgivings about staining his palm with blood money, he never says so; he is on the ride of his life.

==

On the morning of Friday, October 26, 1923, six days after Wickliffe Rose's train rumbled out of Toronto, Fred Banting prepares to drive down to the university lab after spending a day with his parents in Alliston. He has tucked the morning newspaper under his arm but has yet to scan the headlines. The phone rings; picking up the receiver, he hears the excited voice of a friend: "Congratulations! Have you read the news?"

Banting's eyes fall on the column of newsprint: He has won the Nobel Prize in Medicine. The first ever awarded to a Canadian, the award comes with a cheque for forty thousand dollars. But the good news is shattered by the revelation that he has to share the prize with, of all people, J.J.R. Macleod. "Go to hell!" Banting bellows, flying into one of his patented tantrums. "Macleod! Macleod! Macleod! Macleod! No mention of Best!"

He storms out, jumps in his car, and speeds to the laboratory, ready to give his arch-enemy a piece of his mind. When he arrives, he finds my grandfather standing on the steps; knowing Banting would be furious, Gerry has come to cut him off at the pass. He takes his friend by the arm, but he resists. His face livid with rage, Banting vows he will not accept the prize; in fact, not only will he not accept, he is going to cable Stockholm and tell that old fogey, August Krogh—the Danish Nobel laureate who nominated Macleod and Banting—to go to hell.

Banting defies Gerry to name one idea in the whole body of research from beginning to end that had erupted from the brain of

Macleod, or to name one experiment that he had performed with his own hands. As Banting rants on, Gerry cannot get a word in. When he finally has the chance, he says, "There is a gentleman waiting to see you in my office." It is Colonel Albert Gooderham.

"The weight of his presence cooled me down," Banting later recalled. "He was one man whose calm and strong personality always reminded me of my father."

Gooderham offers to cover his expenses to travel to Stockholm, but Banting sticks to his decision to reject the prize. Think of Canada, think of science, Gooderham and FitzGerald urge him. You are the first Canadian to be honoured; what will the world think if you fling it back in their faces?

On the spot, Banting decides to share the prize and half the cash award with Best. As Best is in Boston addressing Harvard medical students, Gerry suggests Banting send him a wire about the equal share. Within days, Macleod gives his half of the prize to Collip. Ernest Hemingway of the *Toronto Star*, a future Nobel Prize winner himself, is assigned to ferret out the story, but this driven son of a diabetic doctor cannot rouse himself out of a Toronto-induced depression.

═

One fall day as I stroll across the lawns of Queen's Park to take my regular swim in the basement pool of Hart House, I pause to poke my head inside the baronial Great Hall. How easy it feels to be wafted back to the evening of November 26, 1923, where, under the vaulted timber ceiling, the hanging wrought-iron chandeliers, and the heraldic coats of arms of fifty-one British universities that adorn the oak-panelled walls, I watch the ghost of Fred Banting take his place at the head table to receive an honorary degree before four hundred guests. As he shakes my grandfather's hand and takes his seat, I notice that the detested Dr. Macleod—no photograph of the four co-discoverers together is ever taken—has chosen to sit three seats away, removing himself, like a glass jar of volatile chemicals, from his combustible colleague.

In gilt lettering, a quotation from Areopagitica, *Milton's passionate seventeenth-century tract on freedom of thought and speech, circles the walls of the banquet hall like a halo. Yet Banting feels anything but free; nor does he relish the prospect of making a speech. As he drones on in his patented, self-conscious way, shoulders hunched, he is hounded by a feeling that his fame was not properly earned. The playing of jazz on Hart House pianos is banned, a restriction that will not be lifted until 1957; the presence of women is also forbidden, an irony in a place where principles of personal liberty are exalted. But on this special occasion, an orchestra is allowed to beat and blow its seductive rhythms across the crowded hall as fair-skinned, bejewelled young debutantes line up to meet the eligible bachelor cloaked in celebrity—"a Lothario long starved" said one biographer, as any bony diabetic. Fearing the intensity of his own feelings, he will callously drop woman after woman after leading them on. As another frigid winter grips Toronto, he composes an article quoting Sir William Osler. Like his hero, Fred Banting believes that aspiring young doctors grasping for the brass ring of success must pay a price: they must sacrifice their libido, sublimating their animal sensuality into the holy pursuit of science; in short, they are obliged to "put their affections into cold storage."*

NINETEEN

The Four Horsemen of the Apocalypse

It is part of the Canadian genius to reduce the heroic to the banal.

JAN MORRIS

I am spending a day in the library of my grandfather's fabled lab, gathering and photocopying reams of archival material. I am feeling oppressed by the sheer weight of the task I have set for myself, the dense puzzle I feel compelled to solve. Most mornings I awake exhausted and depressed, feeling as if I were an ancient Druid priest hauling a slab of stone over endless hills and plains to Stonehenge. Why am I a slave to such punishment? Then it hits me: I am feeling what my grandfather must have so often felt, yet was able to stave off year after year—a feeling of being *totally overwhelmed.*

It is on this day, poking around a dusty closet, that I make a startling discovery: a decrepit wooden filing cabinet stuffed with over five thousand glass-lantern slides that my grandfather used in his public lectures all over the world. Transfixed, I sift through the treasure, holding random images up to the winter light that pours through the window, pulling me back into the early decades of the century. As if jabbed with a shot of adrenalin, I sense I am unpacking three lifetimes crammed into a narrow pine box. And one of those lives feels stolen from me.

The dreamlike pictures evoke worlds within lost worlds. First come the medical buildings, the statistics, the profiles of famous, bearded scientists, safe as pasteurized milk; then the faces of sheer wanton, infectious disease, lurid and grotesque, the shock approaching the pornographic. The germ of diphtheria strangles the ghost-pale child to death in its crib; the victim of smallpox, "the speckled monster," is consumed by a fire of searing pustules and lesions; the sad, syphilitic sexual sinner, his face and hands and genitals disfigured by oozing sores, gazes back into my gaze, as the insidious, avenging spirochete attacks the brain, the heart, the bones with pitiless force.

Gerry's massive glass lantern slide collection educated the masses on the perils of uncontrolled sex.

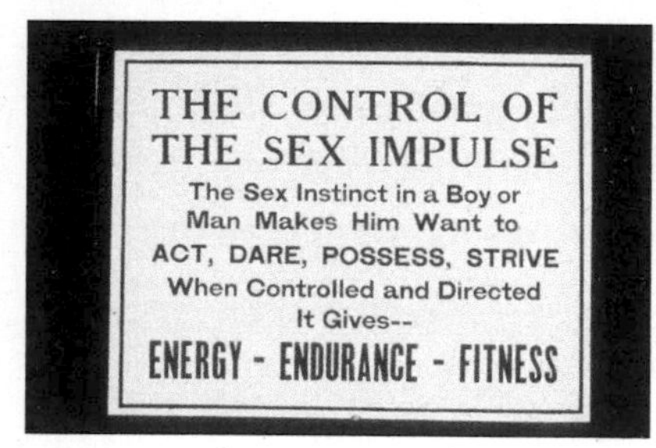

In counterpoint to this medieval tableau of tortured freaks and blinded babies, dozens of peppy public health slogans, vignettes, and moral fables charm with a puritanical quaintness: "Take no liberties with any girl that you would not have another man take with your sister"; "The girl who would yield to you has probably had relations with another man; the chances are she is diseased." A man on horseback, reining in his rearing steed, stands as a metaphor for "the control of the sex impulse." Repression is exalted as a virtue, one that bestows energy, endurance, and fitness. Immersed in the hand-painted images of a forgotten era, I am edging one step closer to understanding what makes Gerry run.

Institutions are more important than men, my grandfather never tires of saying, and in the spring of 1924, he will at last come to see his rarefied scientific ideals embodied in bricks and mortar. At his desk at 61 Broadway, John D. Rockefeller, Junior, a migraine stabbing his forehead, raises his fountain pen and with a practised stroke cuts a cheque for the sum of $650,000. "Riches," he will one day sigh, "breed but sin."

Over the next two years, a stately red-brick, four-storey School of Hygiene will rise on the southern edge of the University of Toronto campus adjacent to the twin-towered medical school, at 150 College Street. After years of ceaseless toil, Gerry FitzGerald, the hypersensitive, forty-one-year-old son of an obscure, melancholic village druggist, is heaving the key pillar of his vision into place. Gerry stipulates that his new educational institution be linked with the Connaught Labs, a move lauded as a brilliant strategy during a time of paltry government funding. Successfully competing with hospitals and clinical practitioners for scarce financial resources, he has managed to create a unique, integrated organization, set on a university campus and completely self-sustaining, the first of its kind in the world. Gerry persuades the faculty of medicine to add a mandatory course in preventive medicine to the undergraduate curriculum, yet convincing the best minds of an aspiring generation of students to join his specialty will prove a tougher sell. Focused on slow, incremental, long-term change, preventive medicine lacks the immediacy and glamour of traditional medicine, the adrenal rush of emergency surgery on damaged organs and limbs, the equivalent of plucking the hapless from burning buildings; somehow, my grandfather must make noble the pedestrian act of plodding from house to house, installing the fireproofing.

That summer of 1924, Gerry seizes an opportunity to forge another scientific first for the Connaught Labs. To this point, the traditional multiple injections of diphtheria antitoxin have imparted only temporary protection, but now a new French-made "toxoid" is bestowing lasting immunity. And it is safer, cheaper, and easier to make. After six months of maddening trial and error, the Connaught records the first successful preparation of diphtheria toxoid in North America. Then, in typical fashion, Gerry drives the advance one step further: mass immunization of unprecedented precision and efficacy.

Working with the city department of health, the Connaught Labs launches a five-year field trial of the toxoid, travelling from school to school in a district of Toronto suffering an outbreak of the disease. A team of needle-wielding doctors and nurses move through lines of pupils, one hundred per hour, consent forms in hand, sleeves nervously rolled up. Gaining the confidence and co-operation of school principals, teachers, and parents, not to mention the GPs who object to the infringement on their services, takes political savvy. To allay parents' fears, Connaught doctors give speeches on radio and write newspaper articles, reminding the public that the toxoid is both free and safe. For children, Gerry assembles a series of the hand-coloured lantern slides that I will find stuffed in a dusty cabinet decades later, each image targeted to a specific disease. Grasping a teddy bear, a toddler gazes up at an enormous white stallion on the Connaught Farm, "Seeing her true friend." In another image, a sweet little girl bedecked with blond curls perches on a swing surrounded by fluffy rabbits; the caption reads, "All well again!"

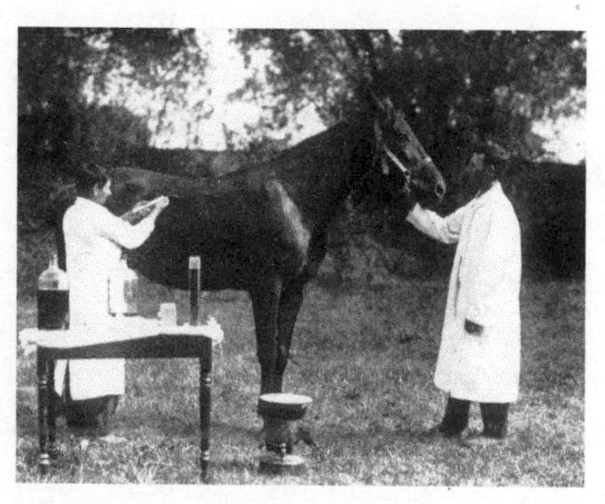

Bleeding of a horse for preparation of diphtheria anti-toxin

The Connaught will eventually publish a classic study of the incidence of diphtheria in thirty thousand immunized Ontario schoolchildren and twenty thousand un-immunized controls, finding a ninety percent reduction of disease in the immunized group. Not only is Connaught the first lab to mass produce millions of units of the toxoid, it has made history by statistically demonstrating for the first time the value of a non-living vaccine in preventing a specific disease—the second great immunization advance after Jenner's smallpox vaccine. At first, the diphtheria breakthrough is not well known outside the country and the Americans and British remain skeptical of their colonial cousins; only when Gerry and his colleagues personally present their results abroad does the ground start to shift. When the Connaught campaign proves far more effective than a similar one in the United Kingdom, the Toronto lab at last draws international attention. Canada is now acknowledged as the world leader in overcoming the difficulties of producing and distributing on a large scale a safe and effective preventive medicine. The innovative Canadian lab stands as the model for all subsequent campaigns to control a wide range of contagious diseases in countries around the world.

═

If the taut net of science can arrest the scourges of infectious disease, my grandfather never ceases asking himself, why not the emotional travails of childhood?

In the winter of 1925, a team of mental hygienists descends on Toronto's Regal Road Public School at Davenport and Dufferin where hundreds of unwary elementary school kids play as they have always played, but now in the role of guinea pigs. The children have already been immunized against diphtheria, but something new is afoot.

The team is led by Bill Blatz and Gerry FitzGerald, commissioned by the Rockefeller Foundation to conduct a five-year study of "the mental hygiene of childhood." Only months earlier, Blatz had been appointed head of the St. George School for Child Study, the first of its kind in the world, occupying a renovated home at 47 St. George Street. The ninth and youngest child of German immigrant parents, the thirty-year-old Blatz is a short, wiry man with a toothbrush moustache and round spectacles, impeccably attired in a three-piece suit, starched collar, bow tie, and gold watch-chain. A spellbinding speaker, he routinely fills Eaton Auditorium and Massey Hall, shocking stiff Orange Toronto sensibilities with his wildly innovative ideas on the raising of children, "a general all-round disturber of the intellectual peace" according to the press. One colleague sees Blatz as mercurial: Humane and compassionate one minute, he can quickly turn "sharp and witty, with a cruel intelligence which would stick a knife in you, for he didn't suffer fools gladly."

Blatz's philosophy rests on the commonsensical notion that kids thrive if their needs for security are properly met. But after reading the entire works of Freud, he is determined to prove him wrong. Blatz believes long-term therapy can be used as a way for people to talk around their problems and escape reality. "The unconscious was groomed for the unwary, the credulous, and crafty," he writes, echoing the sentiments of C.K. Clarke, C.B. Farrar, Clare Hincks, and the vast majority of the Canadian medical establishment united in their antipathy for the Viennese charlatan. "Human beings behaved peculiarly, so a scapegoat or whipping boy had to be found; the 'unconscious' was in the offing. Once it was brought into the open (a paradox indeed), one could behave as one wished; the unconscious was in operation, and any kind of behaviour could be excused, if not explained."

Jack and Molly at a birthday party at the Connaught Farm

At Regal Road School, Blatz diligently classifies sixty-five types of misbehaviour from bullying to truancy to swearing, stealing, eating in class, untidy clothes, or messy books; each act is carefully coded and numbered. Over the ensuing years, Blatz will amass a mountain of data on 1,400 kids but take only a few stabs at analyzing it. Gerry files his own brief report, but the raw material is never fully tracked or interpreted. Such is the fate of many such over-ambitious studies: few are completed or submitted to scientific scrutiny. Critics complain that a general formulation of mental hygiene remains amorphous and incomplete; the work suffers from one great, overriding weakness—by trying to do too much, it does not do much good at all.

In time, Blatz's "nursery school of the laboratory type" will expand to Windy Ridge, a three-storey house on Balmoral Avenue, next door to my grandfather's house and the epicentre of my childhood dreaming. Whether marked by brains and/or wealth, Toronto's elite families cheerfully submit to Dr. Blatz's progressive principles of "scientific child rearing" with its abiding faith in fact over intuition, happily untainted by talk of the unconscious mind.

═

Fred Banting had delayed accepting his Nobel Prize for lack of travel money, but by the summer of 1924, he is ready to go to Stockholm. He has recently married Marion Robertson, a vivacious, blue-eyed X-ray technician at TGH and the daughter of the first fully trained doctor in the Scots-blooded Ontario town of Elora. Gerry and Edna happily accept the Bantings' invitation to accompany them on a meandering six-week vacation through the Continent en route to Sweden to accept the prize. Jack and Molly, eight and eleven, are packed off to summer camp and on August 5, 1925, the Bantings and FitzGeralds sail from Quebec City on the luxury liner *The Empress of Scotland*, a seven-hundred-foot, eleven-deck, German-built ship that was, briefly, the largest in the world. Among its two thousand passengers stroll English aristocrats and well-heeled American businessmen; the sprightly twenty-seven-year-old Marion Banting is in her element. The tangle-footed Fred declines to dance; every night, as his glamorous bride spins across the floor of the baronial stateroom, he slips away to study alone in his cabin.

As the graceful liner slices through the Strait of Belle Isle between Labrador and Newfoundland, Banting glimpses a field of icebergs and sets up his easel on deck. There he meets the Yorkshire sculptors Sydney and Vernon March, a clever and charming pair of brothers returning to England after installing the Samuel de Champlain monument in Orillia. Banting is much taken with the artists and the men sketch together. At some point on the Atlantic voyage, Gerry commissions Sydney March to create a bronze bust of himself destined for the library of the new School of Hygiene. And he likes the idea of March making separate foot-tall plaster statues of each of his two small children.

On the first leg of their summer idyll, the couples motor through Normandy in perfect weather. In Caen, they wander through cathedrals built by William the Conqueror—Fred's and Gerry's fathers shared the name of William—then breeze through quaint villages nestled in the rolling hills. "Words fail," writes Marion in her diary as they reach the Alps. "It's the loveliest spot I've ever seen . . . we all wanted to just stay put." In Paris, the foursome occupies a grand three-bedroom suite at the Hotel Westminster. The City of Light is crackling in radical artistic ferment, a wild cultural experiment in jazz, psychoanalysis, interpretive dance, and surrealism, a rabble of post-Victorian bohemians cavorting till dawn, drenched in cheaply bought alcohol and opium and sex, hurtling towards some nameless, impossible destination.

On August 19, "the boys"—for so their wives call Gerry and Fred—board a train down to Geneva for a medical convention, unattended by their spouses, who think the side trip too rushed. When they return from Switzerland, the Bantings and FitzGeralds transfer to an exclusive club at 33 Faubourg Saint-Honoré, the former residence of the aristocrat Henri de Rothschild. Banting feels ill at ease in the *haut monde* of palatial grandeur, impeccable manners, and hovering valets. He loves telling the story of how his hero, Louis Pasteur, a guest in the palace of Napoleon III, vanished from a tea party to study a microscopic drop of royal wine.

On August 22, Gerry and Fred tour the Canadian battlefields of Arras and Vimy Ridge, worming through the flower-strewn cemeteries and the vast carpet of bone white crosses. Outside Lille, they retrace their harrowing paths in the war, tramping across a mass of moonlike craters, risking stepping on one of the countless undetonated shells buried in the mud, haunted once more by the memory of the groans and mangled bodies of the poisoned, the wounded, and the dying. During a day in Brussels, Gerry reminisces about his working honeymoon at the Pasteur Institute back in the innocent, pre-war days of 1910; moving on to Ostend, Brugge, and Blankenberg, the FitzGeralds buy the Bantings a luncheon set of delicate, handmade Brettonne lace for a wedding present. In the Hague, the couples peruse old bookstores and art galleries together; then the FitzGeralds press on separately to Copenhagen, Oslo, and Trondheim, arranging to reconnect with the Bantings in Sweden. Arriving in his Stockholm hotel room, Fred finds a bouquet of roses and a heartfelt letter from a grateful Swedish mother whose child was saved by insulin. After a round of meetings, receptions, and dinners, Fred and Gerry, the ambitious pair of small-town Ontario boys, don their dress suits and on the evening of September 15 climb the stone steps of the Royal Caroline Institute. In his trademark halting monotone, Fred Banting delivers his Nobel Prize acceptance speech, his mouth as dry as ashes, recounting in tedious detail the medical discovery of the century.

Back at the hotel, Marion Banting scribbles in her diary: "Mrs. FitzGerald and I are holding hands nervously at home. Ladies don't go, it seems."

On December 18, 1925, the new five-storey, red-brick Toronto Psychiatric Hospital at 2 Surrey Place admits its first patient. Only half a block east of the School of Hygiene, the building embodies the fulfillment of the long-delayed dreams of C.K. Clarke who, on his deathbed the previous year, had anointed none other than C.B. Farrar, Gerry's American friend, as his successor.

C.B. had happily reconnected with "Fitz" after the war, both men rising fast in psychiatry and public health, each struggling to wrest professional autonomy from mainstream medicine. Less happily, Farrar informs my grandfather that he has paid a steep price for his intense immersion in his work: His marriage has broken up. With their two small daughters, aged four and one, Evelyn Farrar flees back home to Richmond, Virginia. Although permanently separated by hundreds of miles, the couple will never divorce.

Farrar had spent three years working under Henry Cotton, the controversial superintendent of Trenton State Hospital in New Jersey. Cotton convinced himself that he had found the single, root cause of insanity: infected teeth and bowels. He believed that a poisonous toxin, generated by bacteria, leaked from the teeth and travelled through the bloodstream straight to the brain, driving a person insane. So began Henry Cotton's train of "bacteriological surgery" that will run unchecked for decades to come. In time, the Trenton surgeons were ripping out tonsils, gallbladders, spleens, stomachs, testicles, cervices, and ovaries—but especially colons. Cotton argued that surgery was vastly preferable to the quackery of psychoanalysis; the Freudians, he jeered, often blamed their therapeutic failures on the patient's "resistance" to the talking cure. "We offer no such excuse for our work," wrote Cotton, "because patients who are resistive and non-co-operative can be given an anaesthetic and the work of de-septization thoroughly carried out." Appealing to men of mere literary training, Freud was a scientific fraud: "More in truth than jest, psychoanalysis will in time be superceded by gastric analysis." When one of Cotton's sons, named after his revered mentor Adolf Meyer, showed signs of a stomach infection, Cotton cut open the boy's gut and resected his colon.

Cotton saw himself as the new Lister, leading a revolutionary vanguard of surgical shock troops, waving the gleaming swords of scientific certainty. The elites of British and American psychiatry agreed with him, readily embracing the heroic work that was saving vast sums of money for the state of New Jersey. Meyer endorsed Cotton's methods as "an outstanding contribution to twentieth-century medicine"; yet in this very year of 1925, he is compelled to commission an investigation into the suspiciously high rates of cure that Cotton has claimed. Although Meyer admits that Cotton's claim of an eighty-five percent cure rate is "preposterously out of accord with the facts"—the report finds nearly half of his patients have died, over one thousand from the intestinal surgeries alone, most by post-operative infections; thousands more are maimed for life—Meyer suppresses the damning evidence and the press drops the story. Not until 1960, decades after Cotton's death (and the suicides of his two sons), will the relentless extractions of infected teeth grind to a halt at Trenton State Hospital.

History does not record whether C.B. Farrar actively joined Henry Cotton in his reckless campaign of "exodontia." But he did choose to work by his side in Trenton for three years before heading north to Canada; and, like most of his peers, Farrar failed to raise his voice in protest as, throughout the Roaring Twenties, body part after body part, corpse after corpse, was routinely wheeled from the operating theatre of the Trenton State Hospital.

═

In his new institution at 2 Surrey Place, Farrar parades his gowned patients like zoological specimens before his classes of medical students; he wants the clinical atmosphere to mirror an ordinary hospital. In a soft tenor voice tinged with a mid-Atlantic accent, his lectures and demonstrations flow with cultured phrases. Like C.K. Clarke, Farrar holds the burning conviction that only physicians are qualified to treat the mentally and emotionally distressed; the dark threat of that mystical pseudoscience, psychoanalysis, must be stopped dead in its tracks. Ever since Ernest Jones fled Toronto in 1913 under a storm cloud of puritanical scorn, Freudian thought has yet to penetrate the northern shield of Canadian sensibilities. Nor will it for decades to come; into the 1930s, the name of the founder of the talking cure will fail to infiltrate Toronto's medical school calendar. In fact, the works of Freud will remain quarantined in the science library under lock and key; students may gain access to the books only upon receipt of a note from the professor.

Early in his career, Farrar had been intrigued by Freud's *Interpretation of Dreams*, reading it in the German original, but gradually he lost interest—a strange contradiction for a well-advertised connoisseur of the depth psychology of Sophocles and Shakespeare, Wordsworth and Proust, the poetic precursors of Freudian thought. Yet as the new head of the Toronto Psychiatric Hospital, he now openly despises the "fairy stories" of Freud, condemning psychoanalysis—together with Catholicism, Communism, and Unionism—as "the Four Horsemen of the Apocalypse." In Farrar's materialist world, the child is *not* the father of the man; all that matters is matter alone. He believes the words of Emil Kraepelin, not Freud or the poets, should be engraved in gold and hung on the wall of every psychiatric clinic: "Every attempt to understand the mental life of another person in its inner workings is fraught with manifold sources of error."

Farrar claims to reject all intellectual dogma, whether biological or psychological, yet to the end of his life he will aggressively champion the hereditary origins of all psychiatric illness—and the final solutions of eugenics and euthanasia. He is convinced that the greatest single contributor to social and economic wastage is mental defectiveness, and the greatest single cause of mental defectiveness is an inescapable, tainted heredity. The wards of the state, the mentally handicapped, delinquents, and criminals—all are wallowing in vice, illiteracy, and squalour. Economically useless, they are in constant danger of committing anti-social acts. Most are promiscuous—the base factor of mental inferiority. Lamentably, the humanitarian spirit of society—including the free distribution of Gerry FitzGerald's vaccines—is increasing the numbers of these useless, incompetent, undesirable, and dependent classes.

Farrar believes that anyone who protests the eugenic solution of sterilization, an essential cog in the wheel of the public health movement, is clearly sentimental; it was not even necessary to discuss objections raised on "quasi-religious grounds." The benefits far outweigh the risks, Farrar insists: "Along with ninety-nine potential defectives whose propagation is prevented by sterilization, possibly one potential genius will be lost to the world."

The aggressive lobbying of Farrar and Hincks will lead the province of Alberta to adopt the Sexual Sterilization Act for mental defectives in 1927. Within a few years, the Eugenics Society of Canada will call for the sterilization of not just the feeble-minded—the root cause of all social evils—but all people with inherited physical disabilities, from deaf-mutism to hemophilia to manic-depressive psychosis. Farrar steps up pressure on Ontario to follow Alberta. But despite heavy lobbying, the provincial government will not be moved, fearing the wrath of Roman Catholic voters.

Does Gerry FitzGerald believe in eugenics as passionately as his old friends Farrar, Hincks, Bates, and Clarke? In his heart, does he fear a mass descent into race suicide? Does he think of his thirteen-year-old Irish immigrant grandfather John, the grandfather he never knew, terrified to the marrow of his bones, risking disease and death, clawing his way to reach the New World? Does he think of his own father, Will, cold and aloof, cracking on the anvil of a punishing life? In a tentative moment, as Gerry rises to shave each morning, as he lifts the razor to his throat in the bathroom mirror, does he doubt, for a split second, that superior blood runs in his veins?

In his professional writings and public speeches, at least, my grandfather keeps mum.

═

As I sift through a pile of yellowed newspaper clippings, I come across an item in the Toronto Telegram, *dated April 19, 1926. The headline reads: "A Peep Behind the Curtain of Time." Under a picture of my forty-three-year-old grandfather runs a list of his multiple credentials, his honorary degrees, his club memberships, his mountainous adult accomplishments; next to the photo is an oval inset of baby Gerry, swaddled in Victorian frill, his brow furrowed, the cosmic mysteries of infancy reflected in the tinge of sadness in his eyes.*

That night, a desolate image arises out of my dreams, as clear as a crystal spike of ice. I see an abandoned newborn boy, wrapped in a blanket. He is lying on a vast snowbank, infinite as outer space. From a great distance, his tiny, narrow face stares at me, like a silent accusation. The blankness, the whiteness, the emptiness feels overpowering in its purity. I feel as if an avalanche had swept the baby away, but incredibly, he survived the disaster; incredibly, the tiny being had, inch by inch, clawed his way back home through the frozen wasteland, willful, stubborn, and impossibly alone.

═

In the spring of 1926, Fred Banting writes Charles Best in England: "The new public health building is almost ready for roofing. It looks great. Fitz is as pleased as 'Punch' and wears a smile not unlike the historic gentleman." In the letter, Banting scratches out a black-ink caricature of Gerry in craggy Irish profile, beaming broadly.

That fall, following an unhappy month at Camp Temagami in the wilds of northern Ontario, my nine-year-old father, Jack, tied and blazered, walks two blocks south from Balmoral to an artist's studio on Edmund Avenue. Here he begins his formal sittings for Sydney March, the English sculptor Gerry had befriended on the Nobel Prize trip to Stockholm. This is the year Jack has been accepted into Upper Canada College, a short, inexorable walk due north of the house on Balmoral. Jack's maternal grandfather went to the school and so would he; it is the done thing. Gerry shoots a quick postcard to Edna, who had extended her summer vacation in Seal Harbor, Maine: "Well darling, just a line to say that our son and I went over the top together at UCC this morning. . . . We all waited around, Dads and sons, looking rather foolish for some time. No one seemed to be on hand. Finally 'the Dean' appeared coming along on his stick. We fathers, those of us who had remained, were told our presence was not absolutely essential to the opening ceremonies, so we departed. My friend Wallace said, 'They didn't seem to do it much better at UCC than they do at the university.' That was some consolation. Jack and I are sharing our room. Last night he asked, 'How many days till Mother gets home, Dad?' I said, 'I suppose she will be leaving any day now.' This is to indicate that your boys miss you awful. Hope you have real Seal Harbor days in the meantime. Love and love and love from your Jeremiah."

That Christmas of 1926, Banting presents Gerry with a gift—a landscape in oils, painted in the frigid tundra of the Yukon, which tacitly mirrors his own barren emotional life. For years, *The Midnight Sun* will hang above the fireplace in the living room of Balmoral, passed down as an heirloom from Gerry to Jack to James, from eldest son to eldest son to eldest son.

═

I am alone, a child of six. I am gazing up at a dark Arctic scene, framed in gold-lacquered wood. No one has told me the story of the painting on the wall or the man who made it, so its meaning is left to my imagination. Below the image, on an antique table, stands the foot-tall plaster statue of my father as a boy, only slightly older than I am now, and who looks so much like me. I know I must not touch it, yet I surrender to my impulse to rub my finger over his hard head. I rub and rub, like it's Aladdin's Lamp; but try as I might, the boy's lips remain sealed, unyielding, as final as the lid of a coffin.

TWENTY

Mens Sana In Corpore Sano

> An old suspicion has it that no building is sound whose foundations have not cost a human sacrifice.
>
> SIGMUND FREUD

On June 8, 1927, the University of Toronto's School of Hygiene, born of British traditions, American money, and Canadian innovation, officially opens in a ceremony in Convocation Hall. Following the speeches, a line of gowned dignitaries strolls the few hundred yards eastward to 150 College Street and glides up the semicircular steps to the front entrance of the new four-storey building of neo-Georgian brick and stone. Over linoleum and hardwood floors, my grandfather leads the way, flanked by his wife and sister, the Bantings, and the Bests. They cut through the inner courtyard, fitted with pineapple pediments, then peek into Room 103, a brightly lit lecture hall whose acoustics are considered the best in the university. Next come the labs; among the rows of incubators and refrigerators and test tubes, the visitors watch the technicians perform their secular rituals, the sharp smell of sterilizing chemicals stinging the nostrils. Here is one chemically treating the blood of a horse with sodium nitrate; here is another filtering the egglike yellowish white mixture through a glass funnel; here is yet another pouring it into vials, capillary tubes, and syringes. Poking their heads into the packing and shipping department, they spy rows of women, robed in clinical whites, busily sterilizing, testing, measuring, labelling, and packing the clinking bottles of medicines. In a Ford-like assembly line, the Connaught is now pumping out serums, vaccines, and antitoxins for sixteen different diseases, shipped by the trainload across the nine Canadian provinces and overseas to nineteen countries from New Zealand to the Caribbean, from China to the Irish Free State. And, of course, here too is the first factory in the world dedicated to the production of insulin, extracted from tons of pancreases ripped from freshly slaughtered animals and ground fine in a giant meat chopper.

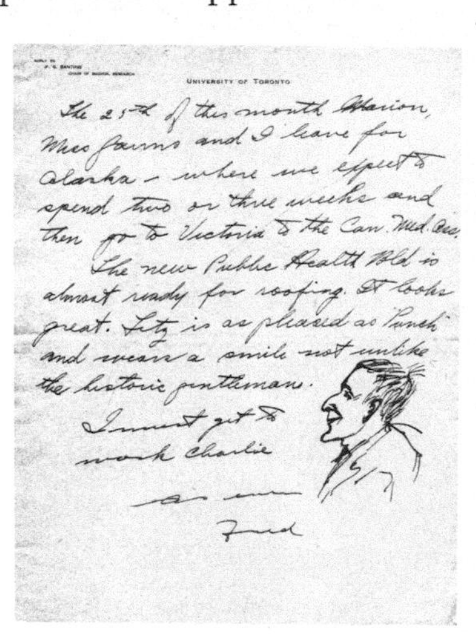
UNIVERSITY OF TORONTO

The 25th of this month Marion, Miss [illegible] and I leave for Alaska – where we expect to spend two or three weeks and then go to Victoria to the Can. Med. Assn.

The new Public Health Bld. is almost ready for roofing. It looks great. Fitz is as pleased as Punch and wears a smile not unlike the historic gentleman.

I must get to work Charlie

as ever

Fred

Banting writes Best about the new School of Hygiene building, 1926: "Fitz is as pleased as Punch . . ."

Climbing the stairs, my grandfather passes his palm over the black banisters of ornamental iron. He inspects the second-floor library with its panelled walls and shelves of Circassian walnut, finely carved stone fireplace, and three great windows gazing westward towards the setting sun. In the corner, on a pedestal, stands his newly finished bronze bust, the penultimate destination of his mortal remains. On the third floor directly above the library, Gerry's office is similarly furnished with soft green chairs, wood panelling, and a fireplace of stone; over the mantel hangs a portrait of the venerated Albert Gooderham, and on his desk is propped a framed photograph of Molly and Jack, the children he rarely sees in the flesh. Finally, Gerry presses heavenward, towards the roof. Inlaid in a wall of red brick, a series of classical stone balustrades ring the top of the school like an ancient parapet. Here, in my grandfather's symbolic "room at the top," experimental dogs and rabbits and guinea pigs twitch in their cages, awaiting their sacrificial fate.

In just over a decade, Gerry's brainchild has catapulted from a humble horse barn with three employees to a sophisticated

Built with a Rockefeller Foundation bequest, the University of Toronto School of Hygiene opened at 150 College Street in 1927

national training centre of 125; the School of Hygiene is now "analogous in scope" with the Lister and Pasteur Institutes in London and Paris. Distinguished by its interdisciplinary character, the school is staffed by a versatile team of experts in the emerging medical sciences of bacteriology, immunology, microbiology, biometrics, parasitology, virology, epidemiology, environmental health, nutritional science, pathological hygiene, industrial hygiene, sanitary engineering, public health administration, and clinical instruction in communicable diseases. A photo in *Maclean's* shows Gerry's beaming face superimposed on a backdrop of bottles and needles; the School of Hygiene is portrayed as "a perfectly functioning peacetime munitions plant where science fashions its weapons for the war against the armies of disease." The writer describes the tall, lean, bespectacled Canadian as "a reticent gentleman of detached mind, but assured of an understanding and sympathetic audience, he is capable of speaking his opinions with clarity and eloquence."

On this warm June afternoon, as the tour of the School of Hygiene winds down, Gerry finds himself in the foyer by the front entrance, bidding adieu to the dwindling knots of dignitaries. He stands on the black-and-white floor tiles arrayed like a chessboard, a wrought-iron chandelier hanging overhead; looking up one more time, he silently reads the credo of his hero, Thomas Huxley, inscribed in a stone tablet embedded in the wall below a clock, noble words that infuse his life's work:

> "That man . . . has had a liberal education who has been so trained in youth that his body is the ready servant of his

> will and does with ease and pleasure all the work that as a mechanism it is capable of, whose intellect is a clear cold logic engine with all its parts of equal strength and in smooth working order, whose mind is stored with a knowledge of the great and fundamental truths of nature and the laws of her operations . . . one who, no stunted ascetic, is full of life and fire, but whose passions are trained to come to heel by a vigorous will, the servant of a tender conscience who has learned to love all beauty, whether of nature and art, to hate all vileness and to respect others as himself."

Finishing up his first year of boarding at Upper Canada College, my father, Jack, just turned ten, is absent from the day's festivities. Never as strong a student as Gerry, he is ever-eager to please his busy father, the tall, blurred figure who calls him "Sonny Boy" and comes and goes with such unpredictable suddenness. Only days after the opening of the School of Hygiene, Jack mails a postcard destined to fall, half a century later, into my own hands:

> Dear Dad,
>
> As you can see, I stood fourth this term. I will tell you the grand news, that I *beat* ted [McMurrich], he stood sixth. I am sure you are pleased, and so am I! I think I am going into 4b but I don't really know.
> Love, your son, amen.
>
> P.S. Mama got me nice hunting knife.

Fred Banting's lab research, meanwhile, is going badly. Even the serial improvements of insulin have delivered a cruel irony. Under the guiding hand of Gerry and his university colleagues, the cost of the average dose has been steadily cut down from the original

186 BALMORAL AVENUE
TORONTO

Dear dad,
as you see I stood 4th This term, I will tell you the grand news, that I beat ted, he stood 6th
I am sure you are pleased, and so am I.!! I think I am going

into 4b but I dont really know.
Love your son aman.
P.S.
Mama got me nice hunting knife:-
blade 4¾ in.

═

level of $1.50 per day to 10 cents; the League of Nations has recently standardized an international formula to ensure insulin of a uniform potency. But insulin has significantly increased the lifespan of diabetics and thus the bottomless hunger for the drug; combined with the general increase of life expectancy by vaccination and other public health measures, countless new cases of diabetes are now surfacing in adults. The era of the coma has been supplanted by the era of complications: Diabetes now poses more medical problems—blindness, kidney disease, circulatory ailments—than existed *before* the discovery of insulin. One doctor proposes that insulin be given free to the poor—on the condition that they never have children. FitzGerald, Banting, and their colleagues expect the roaring demand of the 1920s to taper off, but in fact insulin production will double every five years. In its first year of mass marketing, Eli Lilly raked in one million dollars in royalties; in the decades to come, insulin will transform the private American firm into a pharmaceutical leviathan. The University of Toronto, meanwhile, ploughs its pot of gold back into research.

To escape the punishing demands of work and an increasingly unhappy wife, Banting takes long, outdoor painting expeditions with his friend, the Group of Seven artist A.Y. Jackson. Banting wants an obedient and unquestioning partner who will treat him with "Airedale trueness," but the cultured Marion is no lapdog. Fastidious and refined, she is nauseated by her husband's slovenly habits, particularly how he cuts the roast in the same clothes he wears while cutting into his lab dogs. She hates his old threadbare dress suit; he retorts: "No one has ever had an idea in a dress suit." Burning a short fuse that might suddenly ignite some nameless rage, Canada's most famous doctor is now slipping into a string of clandestine affairs. Then, one day, it just happens: He hits his wife.

Banting often finds sanctuary at the FitzGerald dinner table on Balmoral Avenue. He loves children and sometimes plays charades with Molly and Jack, even as the charade of his marriage is crumbling. But mostly he unburdens himself to the ever-attentive Gerry; talking with friends, he believes, is "like letting pus out of a wound." Balmoral serves as a hard bowl of wood and glass where accumulating emotional poisons can, for a time, be ground down, like pestle to mortar, by the smiling civility of the social circle.

Year by year, assorted dinner guests crowd in, medical and familial, their rising and falling voices leaking psychic vibrations that embed themselves in the curtains, the paintings, the furniture, the floorboards, my body yet to be born. The more Gerry listens, the more they unburden, often straying into veiled and grievous tales of cheating spouses, drug addiction, and suicide.

Cutting across the carpet comes Gerry's maternal cousin, Bill Woollatt, a hard-driving businessman of restless energy whose craggy profile and Cheshire cat grin mirror their shared bloodlines. Addicted to speed and action, Bill founded the Canadian Greyhound bus line and runs Thorncliffe Racetrack in the Toronto suburb of Leaside, revitalizing the sport of harness racing in the city. Known for his overriding optimism, he knows everyone from the stable boys to the jockeys to the executives. An innovator, he allows airplanes to land in the racetrack enclosure to entertain the crowds. And yet his manic charm masks a darkness within, one that must inevitably infect his son, the fourth William in the Woollatt line, who despairs at pulling his high-flying father down to earth; for why else would he, only in his twenties, choose to destroy himself?

C.B. Farrar himself, being human, tells Gerry of his own troubles. Over the years, the two friends have formed a kind of interdependency, helping settle each other down. C.B. laments the plight of his estranged daughter Evelyn, who, coincidentally, at eighteen, is the same age as Molly FitzGerald. Living in Virginia, Evelyn is drinking to excess and picking up strange men; her mother is at the end of her rope. Evelyn will marry four times, perpetually seeking the father she never had.

Then, of course, there are Sidney and Bill, each bone-thin bachelors encroaching on middle age. Working half-heartedly as a teller in a downtown bank located mere yards from the headquarters of the mental hygiene movement, Bill is stifling the pangs of a failed romance. A quiet, quirky character, Sid has recently returned from Dublin where he spent a few years indulging his passion for family history even as a savage civil war raged around him. Mistaken for a Brit, the IRA burned Sid out of his home and he returned to Toronto to toil at a law firm at 689 St. Clair Avenue West that bore the Dickensian name of Dobbin, Froom, & FitzGerald.

Both brothers admire Gerry's breathtaking conquests in the world of medicine, knowing they will never exhibit similar prowess in law and banking. Both try to find the strength to confess their frailties to their accomplished elder brother in private; as his brothers' keeper, Gerry sympathizes keenly, but even he has his limits; best that they direct their deepest travails to C.B.

Alone with his quiver of troubled thoughts, Sidney habitually shuffles up Bathurst Street, northward from his one-bedroom apartment. Sometimes, he pauses in the middle of the Cedarvale footbridge; sometimes, he stares at the hand-holding lovers strolling the dirt path that cuts through the wooded ravine one hundred feet below; sometimes, he dares to imagine a better world for himself.

But mostly he feels the shame and terror his culture attaches to the open expression of masculine fragility. Such events do not appear in the carbon black columns of newspaper print or insinuate their way into intimate, candlelit dinner conversations, or even find satisfactory redress in the consulting room of the esteemed Dr. Farrar. Such events do not happen at all. Perhaps it is for the best; perhaps it is best that as he silently treads the icy sidewalks of Bathurst Street, Sidney FitzGerald keeps his truest feelings entirely to himself; even, perhaps, *from* himself; as if he had never felt them at all.

═

In September 1929, following another lonely month of summer camp, my twelve-year-old father, Jack, walks the three blocks north from Balmoral Avenue to Upper Canada College, entering another year of boarding. His narrow, curtained cubicle in the spartan dormitory seems like a second home to him now, for in the course of a year he seldom sees his preoccupied parents. On long weekends, he bicycles up the twelve miles of dirt roads to the Connaught Farm with a friend, playing in the woods or wading up to his thighs in the Don River or watching the herd of diphtheria horses graze in the long grass, a boy innocent of his privileged place in the world and his distant father's rising star. Sometimes, he retreats alone into the dark womb of the nickelodeon theatre on Yonge Street to escape into the flickering silent images; sometimes, dawdling back along St. Clair Avenue, he heads towards the newly dedicated statue of Peter Pan that stands opposite a black, cast-iron fountain, one side used by horses, the other by humans. A replica of the original *puer aeternas* ensconced in Kensington in 1912, the bronze statue rises twelve feet high; for a split second, it makes Jack think of the quiet, twelve-inch plaster boy, himself, perched on the mantelpiece at Balmoral, under Fred Banting's austere Arctic landscape. As he gains a foothold on the hard, carved heads of woodland nymphs, toads, and doves, Jack thinks, for a split second, of the nervous, long-eared white rabbits he sees caged in his father's lab; as he gazes up at the frozen face of the boy statue who never grew up, my father, for a split second, cannot escape feeling what its creator, the childless Scots playwright, Sir James Barrie, once felt: "It is everyone for himself in this world and no one knows this better than a child."

The Connaught Labs is not immune to the effects of the catastrophic stock market crash a month later and the onset of the Great Depression. In a nation of ten million, every third Canadian worker will lose his livelihood—no family allowance, no medical plan, no welfare net will catch his fall. Families will disintegrate; philanthropy will dry up. The lab is forced to cut back its expenditures, salaries, and new appointments, yet the work presses manfully on, cleaving fast to its lofty ideal of public service. Ironically, the austere decade of the Depression will coincide with a golden age at the Connaught. Seizing the words of the socialist Victorian educator, Charles Kingsley, my grandfather embraces, yet again, a life infused with a "divine discontent . . . the very germ and first upgrowth of all virtue."

═

On May 5, 1930, Gerry travels to Washington, DC, for the first-ever International Congress on Mental Hygiene, a high-water point in the history of the movement. Here he once again shakes the hand of its charismatic leader, the crusading ex-asylum patient Clifford Beers. Organizers are expecting 1,500 people for the five-day event, but over 4,100 delegates—mobs of Russians and Japanese, Africans and South Americans, fifty-three nationalities in all—crush the doors of the Willard Hotel. Beers delivers a witty keynote address, drawing a thunderous ovation that he will remember as the high point of his life.

In his trademark straw hat, he threads through the jammed lobbies and meeting halls, flitting from receptions to lunches to field trips. "I am never happier," he declares, "than when I am busy." Beers is thrilled his once-mad ideas are now considered sane, yet privately he remains forever ashamed of his fragile condition. He watches a string of leading psychiatrists, psychologists, psychoanalysts, social workers, and public health doctors deliver dozens of scholarly papers in English, French, and German; Hincks, Blatz, and Farrar bear the Canadian banner, while Adolf Meyer and William "Popsy" Welch move famously among the throng. Speakers of Freudian tint—Abraham Brill, Karl Menninger, Helene Deutsch, and Otto Rank—also take their places behind the lectern. In the ceaseless tug-of-war between the armed camps of nature and nurture, the soldiers of the unconscious are pulling hard from their end of the rope. Freud had recently published *The Question of Lay Analysis* in which he insists artists, writers, philosophers, teachers, and the intelligent layman can grapple with the subtle mysteries of the unconscious mind as well, or better, than "the man with the syringe." Doctors, he says, "fight every new truth in the beginning. Afterwards, they try to monopolize it. . . . To leave psychoanalysis solely in the hands of the doctors would be fatal to its development." For their part, the influential doctors Welch and Meyer fear the mental hygiene movement is at risk of seduction by the "cults of psychoanalysis," seeping like a virus into the collective consciousness.

At the congress, Ernst Rudin, director of the Rockefeller-funded Kaiser Wilhelm Institute for Psychiatry in Munich, speaks on "The Importance of Eugenics and Genetics in Mental Hygiene." As far back as 1911, Rudin has been pushing for the culling of "the extraordinarily large number of inferiors, weaklings, sickly and cripples." He was not alone, even on the left side of the political fence: Margaret Sanger, G.B. Shaw, and H.G. Wells actively endorsed eugenics, while G.K. Chesterton's searing 1922 polemic, "Eugenics and Other Evils," stood as a virtual voice in the wilderness.

When Hitler is elected to power in 1933, Rudin will fanatically embrace the creed of Nazism, draw up the Third Reich's compulsory sterilization laws, and substitute the term "racial hygiene" for "mental hygiene." He was inspired by the landmark 1927 ruling by the eighty-six-year-old US Supreme Court Justice Oliver Wendell Holmes, Jr., who ordered the forced sterilization of a feeble-minded seventeen-year-old Virginia girl conceived in a madhouse and now the mother of a seven-month-old daughter. "The principle that sustains compulsory vaccination is broad enough to cover cutting the Fallopian tubes," Holmes intoned. "Three generations of imbeciles is enough." The ruling has emboldened the agenda of European and North American eugenicists with C.B. Farrar aggressively leading the way for Canada; during the post-Holocaust Nuremberg trials, Nazi war criminals will cite Holmes in their own defence.

Beers has recently decided it is high time to surrender his onerous duties to his Canadian friend Clare Hincks—"the only man in the world suited to the job"—and move on to international work. Only after Beers offers Hincks the medical directorship of the American arm of the movement will he learn of the Canadian's own eerily similar mental history. Beers has carved out a career from brilliantly publicizing his own flights of madness, yet he sees no contradiction in urging Hincks to cover up his own.

Three months after the wildly successful international convention in Washington, Beers's brother William, a gregarious Yale alumnus and prosperous businessman, is floundering in a depression in the wake of the stock market crash. In August, he is found hanging from a tree on the serene lawns of his mental hospital. Months later, the body of another brother, George, is discovered drifting in a river. George had resisted treatment because of his earlier, unhappy experiences in the Hartford Retreat, the same asylum whose transgressions had inspired his brother's famous book; when George did finally rouse himself to seek help, his psychiatrist said he didn't require hospitalization. Beers seems outwardly unaffected by the suicides of his two brothers in brutally quick succession, coinciding with the cresting of his own success; it is, he simply notes, "an irony of fate that members of my own family have been unable to receive benefit from the work I started."

═

That summer of 1930, Gerry FitzGerald is riding the wave of his own hard-earned success. Accepting an invitation to join the Health Committee of the League of Nations, his career is, like that of Clifford Beers, taking an even wider international turn. With sixteen-year-old Molly heading for finishing school in Switzerland and thirteen-year-old Jack boarding at UCC, my grandfather decides to lease the Balmoral house and move into the Alexandra Palace, a posh, seven-floor residential hotel at 184 University Avenue, directly across from the Toronto General Hospital, which Molly comes to call "the Ally Pally." The place drips Edwardian class: smooth classical columns, ornate wrought-iron balconies, and a gliding staff of porters, doormen, and chambermaids of Asian blood—an exotic sight in bland, WASP Toronto. Distinguished residents include Sir Adam Beck, chairman of the public utility, Ontario Hydro; Premier Howard Ferguson; TGH chief Dr. Duncan Graham; and assorted well-heeled businessmen. For a monthly rent of $250, the FitzGeralds occupy a six-room furnished suite as their *pied-à-terre.*

The move out of Balmoral has a seismic effect on one of Gerry's emotionally dependent younger brothers. On June 23, 1930, Bill FitzGerald shuffles up the steps of the Toronto Psychiatric Hospital. Perhaps, as Gerry suggested, a consultation with Dr. Farrar can help him cope with his ceaseless bouts of insomnia, nervousness, and depression.

Opening a fresh file, Farrar uncaps his fountain pen and invites the knife-thin, round-shouldered Bill to unfurl the story of his forty-five years. Incapable of making the simplest of decisions, Bill feels overpowered by feelings of inadequacy. He has been working as a branch manager of the Dominion Bank in Hamilton for ten years; the daily stress has burned a duodenal ulcer in his stomach, leading to an operation and five weeks of hospitalization. Since going under the knife, he feels possessed by an unnamable blackness. Five years earlier, he had allowed a woman he loved to slip out of his life and he brooded about her constantly. In October 1929, the week of the stock market crash, he was transferred to the bank's Toronto head office as a supernumerary. He spent long, tedious hours doing nothing, occasionally relieving other employees in other branches. With no fixed address, he stayed for a while with Gerry on Balmoral, then moved in with his older brother Sidney, a few blocks away on Vaughan Road.

Only days earlier, Gerry—the person on whom he most relied—had left for British Columbia for a two-month working vacation, to be followed by a month in Europe, where he will take up his new post with the League of Nations. When Gerry decides to lease the family home on Balmoral, Bill feels as if the sky has fallen. The place has served as a kind of refuge, and now it is gone.

He worries that he is turning into his father, a man of perplexing personality—sometimes quiet and sensitive, sometimes a callous bully—who had become extremely dependent on his adult children after the death of his wife. Bill is worried he is now repeating the pattern, clinging to his three siblings. His sister, Hazel, tells Farrar that Bill has always been seen as the "odd or weaker member of the family"; Farrar, in turn, recommends that Bill move in with Hazel. The reasons for Bill's various travails, Farrar concludes, lie in "constitutional characteristics of inadequacy." Commenting on the patient's regret for not having married the woman he loved, the doctor adds: "It is possible that with a wife to lead in responsibility, he would have settled into a fairly normal life." Then, with a few quiet words of encouragement, Farrar sends

Bill off to the School of Hygiene, only half a block away, to provide samples of his urine and blood.

═

In late September 1930, Gerry FitzGerald's train rumbles into Geneva, Switzerland, the cosmopolitan, perpetually neutral city state in the heart of Europe. In the glass-roofed council chamber of the Palais Wilson, Gerry meets his boss, Ludwig Rajchman, a dynamic Polish bacteriologist born into a prominent family of free-thinking political activists. Under Rajchman's inspired leadership, the Health Organization will prove the most effective of the League of Nation's non-political agencies in the postwar years, forming a cordon sanitaire from the Baltic to the Black Sea and establishing a strong presence on every continent. In the rarefied, exacting altitudes of Calvinist Geneva, my grandfather finds himself in his element, beetling from meeting to meeting, doctor to doctor, language to language, thick dossiers stuffed under his arm, a moving cog in a vast social machine forging an even vaster moral vision. He will attack a diverse set of problems: syphilis in Bulgaria, leprosy in Rio de Janeiro, radium treatment of cancer of the uterus, deafness and dumbness, quinine, yellow fever, rabies, distribution of vitamin standards, and reform of medical education. The League's naive attempts to arrest the global opium trade will prove, alas, a task beyond human reach. As he works, my grandfather never forgets the sublime words of Thomas Huxley, which are carved on the wall of the School of Hygiene, the credo that exalts the virtues of the harnessed intellect—"a clear cold logic engine with all its parts of equal strength and in smooth working order."

On a Saturday morning in October, Gerry breaks away from his meetings and boards a train north to Lausanne. He is planning to surprise his daughter, Molly, long-legged and red-headed as himself, now installed in Les Fougeres, a swishy finishing school, where she is learning German and French. Precociously worldly at sixteen, flashing a lethal wit, Molly had left Toronto's Havergal College after showing up drunk to her grade eleven exams; despite collecting a gaggle of friends, she hated the place, her mother's alma mater, often running away. Rumours, ultimately unverifiable, rumble through the Connaught Labs that the real reason the daughter of their leader had been packed off to Switzerland was to elude the judgments of Toronto society on a pregnancy out of wedlock.

Not allowed to speak English in her new school, Molly pours out a stream of letters home, recounting her weekly round of tango lessons, tennis matches, concerts, and skiing. Rattling off her marks, she also reports her trips into town, where she hands out picture books, toys, chocolate, and berets to children of the poor. She rooms with Helen Oakley, a wealthy Toronto girl who has been educated by a succession of English governesses. An unusual future awaits Miss Oakley: A passionate jazz fan, she will meet and be mesmerized by the genius of Duke Ellington. Rebelling against the life of a dutiful chatelaine laid out by her family, she will become a jazz journalist, contribute to the fledgling *DownBeat* magazine, organize "Battles of Swing," and pioneer interracial collaborations with Benny Goodman and black musicians. But for the moment, the two Toronto teenagers are busy writing girlish letters home.

"My darling angel," Molly writes her mother. "Daddy has arrived! He sprang onto me without any warning on Saturday morning and oh my, was I excited! You bet your boots I was!"

Father and daughter go to lunch, shop, and take a boat tour of the lake. "We didn't stop talking, except to eat, for the whole afternoon." At school, she is allowed only two baths a week, so she uses her father's hotel room to take another. They dine at the luxurious Beau Rivage where Molly "nearly danced with a lot of dirty Spaniards." At the interstices of historical coincidence, father and daughter remained blithely unaware that in the very same hotel, over their heads in an upstairs suite, F. Scott Fitzgerald, spawned from the loins of the same medieval Norman knight, is furiously scribbling his latest work-in-progress, *Tender Is the Night*, while knocking off a fifth of gin. At the best private asylum in

Europe, the nearby Les Rives des Prangins, Fitzgerald's flaxen-haired, thirty-year-old wife, Zelda, recently diagnosed as schizophrenic, is chafing under a cruel regimen of forcible restraint and solitary confinement; within months, she will be chafing under the Germanic rigidity of Adolf Meyer at Sheppard Pratt. As ever, Scott Fitzgerald's latest work of art—the tragic story of Dick Diver, a doomed socialite psychiatrist whose exhausting devotion to his invalid wife plunges him into a feeling of emptiness "too shameful to be divulged"—is drawn from life.

After her father returns to Geneva, Molly fends off stabs of homesickness with a chatty letter. "Daddy darling," she writes, "many thanks for your letter from the boat. I am at Mlle. Chambert's table this week, sitting beside the American girl who came over on the *Europa*. She is beautiful but very dumb. The other day, I told her that you came out on the *Europa* and shared a suite with Henry Ford. I also told her that you were president of the Bank of Nova Scotia and had a ranch next to the Prince of Wales! She thinks you are the cat's pyjamas and looks at me with absolute awe."

Yet for all Molly's worldliness, deep under the pale skin of the child of privilege, spoiled as she is neglected, unutterable feelings of shyness and anxiety lie hidden.

═

On a brilliant sunny May afternoon in 1931, my grandparents host a garden party on the lawns of the Connaught Farm. Under a white tent shaded by elm trees, dozens of members of the Royal Society of Canada—the cream of Canada's scientific and artistic elite—cluster around the tasteful white tablecloths, set with tea and crumpets and vases of tulips, to chat informally. Together with their bonneted wives, the array of high-voltage male minds—the kind of superior bloodlines from which the newly formed Eugenics Society of Canada measures its standards—pose for a panoramic photograph destined to hang on the wall of my grandfather's study. Gerry has often spoken of the social value of vaccines; for his wife, the phrase has taken on a double meaning.

On the far left stand Gerry and Edna, on the far right Fred and Marion Banting, near a beaming Charles and Margaret Best, proud parents of a newborn son, Sandy, destined to inherit his family's brilliance and depressive temperament. In the middle stands the nation's pre-eminent psychiatrist, C.B. Farrar, distinct with glinting spectacles and black toothbrush moustache. In the city of Toronto in the depressed year of 1931, breadlines stretch for blocks, longer than the bloodlines of the embryonic Canadian aristocracy; 111 suicides will be recorded, the highest number ever. As he moves among his friends, Gerry is carrying distressing news unfit for polite company: Only weeks before, Sir Andrew Balfour, the fifty-seven-year-old head of the London School of Hygiene and Tropical Medicine and Gerry's British counterpart in the Rockefeller-funded system, an obsessive perfectionist whose sanitary work in the Sudan set a standard for the whole African continent, had fallen precipitously into a ferocious depression. Only months after being knighted, he hanged himself on the grounds of a mental hospital. Initially his colleagues were able to keep the gruesome details quiet, but soon the story was widely publicized.

But on this spring afternoon on the sylvan outskirts of the city of Toronto, the afflictions of the wider world seem, for the moment, far away, for the happy quartet—Banting and Best, Farrar and FitzGerald—are each riding the steeds of public acclaim. Farrar has recently beaten out the radical psychoanalyst Harry Stack Sullivan for the editorship of *The American Journal of Psychiatry*; C.B.'s sensibility will dominate the journal for the next thirty-four years. After several years of work at the School of Hygiene, Best has prepared a liver extract for the treatment of pernicious anemia, hailed as a great clinical advance. Only months earlier, the Banting Institute, a six-storey block of red brick and stone trim, opened one hundred yards east of the School of Hygiene on College Street; built by one million dollars of government and university money, it now stands as the home of Canadian medical research. A household name, Fred Banting's every move is chronicled like a Hollywood celebrity, a striking anomaly in an egalitarian country historically suspicious of hero worship. Prickly and aloof, he feels the adulation is undeserved, silently accusing himself of being a flash in the pan. Throwing himself into countless new experiments, Banting delivers far more failure than success; no matter what he does, he can never escape thoughts of that mythic summer of 1921; the harder he works to bottle the elusive moment of inspiration, like a dream, the faster it melts away.

As he moves among his peers that spring day, Gerry attracts a flurry of handshakes and words of congratulation, for they know the Rockefeller Foundation has just appointed him one of the six scientific directors of its International Health Division–the first Canadian named to the position. Gerry's appointment is a crowning triumph in a chain of crowning triumphs. In addition, the foundation has donated a further $650,000 to double the size of the School of Hygiene building. Gerry's friends say he seems to be living a charmed life, holding the world by the tail; some call it the luck of the Irish.

Just before Christmas 1931, my grandfather announces to his staff that he has been offered and—"with no few misgivings"—has accepted an appointment as dean of the Faculty of Medicine, effective January 1, 1932. Here is yet another feather in his cap, the third major appointment within the space of months, hard on the heels of the jobs with the League of Nations and the Rockefeller Foundation, all in the depths of a world depression. The deanship is seen as a clear endorsement of Gerry's vision, a recognition by the faculty and university of the growing importance of hygiene and preventive medicine in the medical curricula.

Gerry, Dean of Medicine, 1932

After my father's death, I discover among his spare possessions a crumbling manila folder that I realize contains a flood of congratulations for his father's latest appointment. I leaf delicately through the aging papers as if examining the Dead Sea Scrolls. The dean of the University of Glasgow medical school pens a letter typical of dozens: "I wish I knew your secret for attracting money. In comparison with Toronto, the equipment of Glasgow can only be called pitiful." Another colleague declares, "As a disciple of the preventive medicine cult, I am pleased to know that our high priest is in a position where he can suggest policies along these lines, and certainly knowing you as I do, I feel sure that they will be made effective."

Then I find a note scribbled by seventeen-year-old Molly, writing her mother, Edna, from her new school in Munich: "I am awfully proud of Daddy being a Dean. But his poor, humble daughter hasn't had a letter for six weeks!"

TWENTY-ONE

The High Priest of Quality Control

> In an age of advanced technology, inefficiency is the sin against the Holy Ghost.
>
> ALDOUS HUXLEY
> *Brave New World*

In the grip of the Great Depression, J.G. FitzGerald and C.B. Farrar, defenders of their respective scientific faiths, now preside over Canadian public health and psychiatry with alphalike authority. Gerry enjoys an edge; over the decades, public health has made inroads further and faster than psychiatry, the bastard child of medicine he had abandoned over twenty years earlier. Both nascent specialties, of course, continue to find common ground in the mass movement of mental hygiene; and now, having slipped past the door marked "Dean of Medicine," Gerry is poised to aggressively push his preventive-minded crusade against all forms of mental and physical affliction.

Psychiatric treatments remain virtually unchanged from Gerry's Edwardian days at 999 Queen Street West; there are few effective medications, only derivatives of codeine and morphine used to subdue the unruly. The mad are seen as a lost cause, the odds of recovery low. Over the first three decades of the century, the number of North American asylums has more than doubled; one notorious state hospital in Milledgeville, Georgia, confining over ten thousand patients, has been reduced to a Dante-esque holding tank of society's refuse. Doctors are reduced, too, serving as mere "managers of despair"; little can be done except hope for a spontaneous remission—or more aggressive eugenics policies.

Impressed with Dr. FitzGerald's breakthrough with the diphtheria toxoid, Dr. Herbert Bruce, the lieutenant-governor of Ontario, draws a parallel with mental defectives. Sterilization, he says, is "damning up the foul stream of degeneracy and devaluation, which are pouring pollution into the nation's lifeblood." If the present rate of increase of insane continued unchecked, he warns, half of North America will be confined to insane asylums by the end of the twentieth century.

Farrar publicly endorses Bruce's views. He believes tainted bloodlines remain the first cause of insanity, yet he also believes the disturbed act of suicide, for example, can sometimes stem from sane motives, particularly in respectable members of the higher classes. In 1932, seventy-seven-year-old George Eastman, the inventor of the Kodak camera and generous funder of eugenicist programs, the man of impeccable pedigree who had tried in vain to lure Fred Banting to Rochester with promises of personal glory, calmly shot himself in the heart minutes after attending to the final details of his will. In a note, he wrote: "My work is done. Why wait?" C.B. opined: "The facts of Mr. Eastman's life were so well known, his mind and character were of such fine quality, and his value to the world has been so great, that his last courageous act is beyond cavil."

Like Gerry, C.B. is struggling to attract ambitious, young doctors to a budding specialty with a limited appeal. For his part, Gerry is sparking a wholesale revamping of medical education in the midst of a disastrous economic downturn. In time, recruitment to his cause of public health will exceed the rate in the US, but the struggle will remain Sisyphean. Critics gripe that Gerry is engineering an elite breed of "super doctors," striving to attain an impossible state of omniscience. As the Great Depression deepens, many GPs fear nurses will take over their work, cut into their income, and encroach upon their relations with their patients and families. Gerry responds that nurses, when properly educated, are invaluable allies, playing a key role, for example, in the delivery of rural health services. He argues that it is the general practitioner fighting on the front lines of curative medicine who is best positioned to impart the principles of disease prevention, assuming the responsibility of urging millions of people to assume responsibility for their own health.

But even some members of Gerry's own inner circle fail to respond to his upbeat health ethos. Each day, Fred Banting smokes several packs of cigarettes and downs ten ounces of rye whisky to blanket the stress of his crowded life. Through a persistent hacking cough, he croaks a stark repudiation of Dr. FitzGerald's philosophy: "Exercise is for the most part a fad, like all this talk of diet." When he discovers his wife, Marion, is having an affair with Donat LeBourdais, the handsome young educational director of the Canadian mental hygiene movement, Fred goes wild and assaults the brazen cuckolder, finding no double standard in the fact that he has been indulging in an affair of his own. Intimate details of unpardonable sexual sin are shouted from the street corner newstands of stodgy, stick-in-the-mud Toronto—sweet revenge for the press corps whom the Nobel laureate had treated so shabbily.

Stung by the scandal, Banting is becoming a curmudgeonly misanthrope, depressed and aloof, plagued by hangovers and insomnia. "My heart is broken," writes Marion, "because you are so different to the man I idealized." Fred plunges back into his work and his hobbies, obsessively collecting Canadiana and searching in vain for a cure for cancer, even as the medals and honours continue to pile up. In many ways, he feels like a drowning man, so perhaps it is not surprising that one day he will draw inspiration from the musings of his nine-year-old son, Billy, who asks how ducks could eat underwater without drowning. Fred thrusts flailing rabbits and rats underwater, then tries to revive them by tracheotomies or injecting belladonna into their hearts. The fickle press prints stories about the imminent arrival of another insulinlike miracle serum and fanciful scenarios of heroic teams of researchers rushing to accident sites and magically raising the dead from the watery depths. Meanwhile, to toughen up his young son and heir, Fred ushers Billy Banting into the lab and places a scalpel in his hand; then he makes the boy cut off the squealing heads of rodents, their blood squirting on his hands.

That same summer of 1932, eighteen-year-old Molly FitzGerald returns home to Toronto after a final, hormonally charged year of schooling in Germany. In Munich, she shared a flat with Anna Mongeolais, a worldly Russian aristocrat (after whom she will name her firstborn child), danced with Prussian officers, and tasted the forbidden fruits of sex. In late October, together with her friend, the jazz-crazy Helen Oakley, Molly "comes out" at a debutante party to Toronto society, introduced by her mother to a line of guests at the Alexandra Palace, a traditional ritual designed to attract suitable husbands for nubile daughters of the well-to-do. "A fair, slim aristocrat in an enchanting Norman Hartnell model of honey lace and burnished brown satin," gushes a society columnist, "met their guests with dignity and charm." Molly is squired about to parties, cotillion balls, receptions, and supper dances in the Crystal Ballroom of the Royal York Hotel. Described as "a golden girl, lacquered hair and frills flaming round her golden skirts," she poses for the obligatory Ashley and Crippen portrait, published in the *Toronto Telegram;* a headline reads: "Eager Young Feet Dance Into Land of Sweet and Chic Elite." Unlike her brother, Jack, the quick-witted Molly is not destined to shuffle off to university.

Gerry with his wife, sister, and daughter, Connaught Farm, 1930s

For Gerry, the year is fuller and more stressful than ever, racing back and forth from country to country, bearing the cross of his many cross appointments, nailing down his various obligations like the flapping corners of a tent in a high gale. First it is the canyons of Manhattan for meetings with the International Health Division of the Rockefeller Foundation. Then comes the mountain air of Geneva and the semi-annual congress of the Health Organization of the League of Nations; then prosaic Ottawa and the Dominion Council of Health's three-day meetings. In public, Gerry keeps up his sunny persona but increasingly within the family circle, the smile drops and his face turns blank and remote, even taking on an occasional Banting-esque surliness—the street angel turning home devil.

Ironically, Gerry and his idealistic colleagues at the League of Nations have been trying in vain for years to suppress the worldwide traffic of opium and heroin, stoked by Prohibition; ironic, for Gerry grew up in a house where the handmade narcotics of his father's apothecary shop had dulled, but not cured, his mother's unspeakable pain; doubly ironic, for now the next generation of FitzGeralds is following suit. Gerry himself uses drugs to quell his insomnia and migraines; pushing fifty, Edna is suffering menopausal "vapours" and routinely resorts to medicinal solutions while stout-hearted Hazel, the family glue, takes care of her. Isolated in the boarding house at Upper Canada College, fifteen-year-old Jack is struggling with his studies, retreating into the solace of his jazz record collection; when he comes home on holidays, he pilfers his mother's sodium amytal tablets to settle his nerves.

The invisible gods of synchronicity, meanwhile, keep pressing their shoulders to the wheel. In London in that same year of 1932, Ernest Jones, the high priest of the Freudian cause, is anointed president of the International Psychoanalytical Association. Its swelling membership has now reached ninety-two analysts of assorted nationalities, with Canada a notable exception. On a parallel course, Gerry FitzGerald has graduated a similar number of acolytes from his School of Hygiene; in his contemplation of the achievement, he has no reason to regret the choices that have driven him to the peak of a noble profession.

═══

It is a Saturday in late January 1933, the dead of winter, and my grandfather strides briskly westward to Convocation Hall to deliver an upbeat public lecture on the quickening pace of medical research. Franklin Roosevelt and Adolf Hitler have recently taken office as the respective, duly elected leaders of the United States and Germany; embodied in the avatars of the New Deal and the Final Solution, the historical forces of liberal democracy and fascism are digging in for global Armageddon of unprecedented savagery. My grandfather cannot possibly suspect that politics and technology and medical science are perversely fusing into a mutant ideology of hate and fear, as voracious and nihilistic as an infectious disease; and perhaps it is no historical coincidence that this same month of January 1933, marks the official announcement of the first of a quartet of aggressive somatic treatments that will thunder across the psychiatric hospitals of the world during the darkest years of the Great Depression, straining to legitimize psychiatry as a true son of science. Together, the Four Horsemen of Psychiatry—insulin coma, Metrazol, electroshock, and lobotomy—will forever brand the face of the profession; and the leading steed, ironically, is charging out of the city of Sigmund Freud.

After six years experimenting on animals in his own kitchen, a strange and withdrawn thirty-two-year-old bachelor delivers a momentous report to the Vienna Medical Society. Manfred Sakel, a Polish psychiatrist of misanthropic temperament who claimed to be a direct descendant of Moses Maimonides, the medieval Jewish rabbi and physician, describes an encounter with a famous actress who suffers from diabetes as well as morphine addiction; when he gives her an accidental overdose of insulin, she swoons into a mild coma. When she awakes, her craving for morphine subsides and she becomes less restless and agitated. Excitedly, Sakel begins to treat all addicts with insulin; after they are revived with doses of glucose, he notices some seem calmer, more co-operative, and gain weight.

The gaffe-prone Sakel then accidentally gives an overdose of insulin to an addict who is also psychotic, and he falls into a deep coma. When the man awakes, he is lucid—for a time—and the doctor mistakes a temporary remission for a cure. He makes the grandiose leap that if he deliberately induces an insulin coma in schizophrenic patients—a dangerous, life-threatening procedure at the best of times—he might permanently eradicate their madness. Boldly, he sets out to test the idea on asylum inmates.

In the early morning hours, patients are rolled into a semi-darkened ward, where Sakel injects insulin into their buttocks. For four or five hours, the patients' pulse and heartbeat race at high speed; they sweat and drool, toss and turn, moan and weep; they vomit and lose control of their bowels; they twitch with violent tremors and muscle spasms. Those given the highest doses fall into violent convulsions, biting their tongues and breaking their teeth. Some fracture a femur, arm, jaw, or spine; others tear hip bones out of their sockets.

After falling into a coma, the patients are left anywhere from twenty minutes to two hours, nurses closely monitoring their pulse and respiration. Then, godlike, Sakel pulls them back from the brink of death with a dose of glucose solution. If the coma persists too long, the patient dies; in the years ahead, many will.

If the patients do awake from the coma, their ordeal is not yet over. Bewildered and disoriented, they regress to a primal, infantile state, reaching out for the nurses and doctors and kissing their priestlike hands; famished by sugar depletion, they cry out for their mothers and suck their thumbs. Witnesses who happen to walk into an insulin coma ward testify that the place resembled a medieval tableau of torture ripped from the pages of Dante's *Inferno*.

An exultant Sakel claims cure rates as high as eighty percent, yet he is at a loss to explain the effect. Perhaps, he reasons, trauma itself—the threat of imminent death—is the healing agent. Years later, some doctors will speculate that remission of symptoms was simply a case of the unusual degree of attention lavished on the patients, for the procedure requires a minimum of fifty sessions and a small army of doctors and nurses to perform. In any event, the North American press aggressively champions the new "bedside miracle" and hails Sakel as the "Louis Pasteur of Psychiatry."

In Toronto, Sakel's method naturally piques the curiosity of Fred Banting and C.B. Farrar. Clearly, anything that holds out hope for the mad should be tried. In those wintry days of early 1933, Gerry FitzGerald is as yet unaware of the novel use of insulin his Toronto lab has famously mass produced. A newspaper editorialist has recently praised him for "his genius for making intricate health topics understandable by the lay mind" and on March 7, he sits down in a Toronto radio station and delivers a talk on preventive medicine to a general audience. Leaning into the microphone, Canada's leading medical educator intones: "In no other group of men and women is character and integrity more essential than in those who are to practise the healing art of medicine."

No recording of my grandfather's voice has survived.

═

On December 11, 1933, Dr. Alan Gregg, the Harvard-educated director of the division of medical sciences of the Rockefeller Foundation, visits my grandfather in Toronto. After lunching with Gerry and his inner circle at the School of Hygiene, Gregg writes in his diary: "It is amusing to note that FitzGerald and his staff take three-quarters of an hour of active exercise playing deck tennis after lunch; they all keep fit by so doing. It is the only instance I have ever seen of a hygienist trying to live a hygienic existence."

Gerry remains ever-mindful—even superstitious—about dates and anniversaries. December 11 falls two days after his fifty-first birthday, a birthday he shared with his mother, Alice, who died at fifty-one; as his old mentors age and die off, his thoughts are increasingly turning to his own mortality, especially on the eve of a dangerous trip abroad. That very week, emulating the Wesleyan circuit-riding tradition of Wickliffe Rose, Gerry is scheduled to set sail on a rigorous three-month, Rockefeller-sponsored survey of the troubled health conditions in India, Ceylon, and Egypt. The fetid jungles of the Near and Far East stink and crawl with ancient diseases like cholera, malaria, hookworm, and bubonic plague; he knows, all too well, the risks he runs. And so my grandfather draws up a "Memorandum of Wishes," to be executed in the event of his death. He stipulates that his remains be cremated and the ashes deposited inside his bust in the library of the School of Hygiene. In the event of his death abroad where cremation cannot be undertaken, he desires to be buried where he died and his body not returned to Canada. In the event of his death during "the legal infancy" of his son, Jack should be given the opportunity of a university education, preferably at McGill University, the University of Cambridge, and/or the University of Toronto. He is to be encouraged to acquire a broad training in natural science, preferably in Italy, Germany, France, and the United States. That said, Jack is to be "given complete freedom in the choice of a profession or occupation"—a freedom my father will never truly exercise, bound up in the subtle forces of paternal coercion disguised as persuasion.

Flanked by a pair of crack American and English hygienists, F.F. Russell and W.W. Jameson, my grandfather will cover 24,000 miles in ninety-three days by train, car, and boat. Landing in Cairo via Paris, they visit assorted hospitals, institutes, and labs. In India, a cauldron of anti-British unrest, the Rockefeller doctors have struggled to eradicate hookworm—the centimetre-long parasite that bores through the bare soles of the feet, attaches to the wall of the gut with cutting "duodenale" teeth, and inflicts a litany of ills from anemia and circulatory disorders to emaciation, stunted growth, and mental retardation. Wickliffe Rose's American hookworm campaign had brilliantly succeeded, but in India, a combination of apathy, lack of funds, and stark cultural differences has foiled the best-laid plans.

For my grandfather, a citizen of a cold, vast, thinly-populated land of only ten million, the concentrated filth and chaos of tropical India seem as dreamlike as an alien planet. India's dense population of 350 million—one-fifth of the world's total—is mostly rural and rife with bubonic plague, smallpox, leprosy, flu, and rabies. In 1928 alone, 350,000 Indians died of cholera; yet malaria is its single greatest scourge, feebly combatted by two quinine factories. If disease does not carry you away, monsoons, tigers, or deadly snakes likely will. The staggering mortality rates dwarf the losses suffered back in Canada; here, religious resignation consoles the masses that a better world lies beyond. Here, death is a way of life.

Over several weeks, Gerry covers five thousand miles, trudging through fifty villages and inspecting scores of huts made of mud and palm leaves. Then, on a hellishly hot day in January, the delegation arrives in Benares, the holiest of all Indian cities.

Hundreds of temples, shrines, and statues snake along the banks of the Ganges, the 1,500-mile river the Hindus consider divine; a supreme brilliance shines here, lighting the way to salvation. Every day, thousands come to drink and wash in the filthy water, risking infection from typhoid, dysentery, and cholera. Yet the faithful believe the river cleanses the body and soul, having the power to wash away sin. Here, in this beautiful and lethal place, no sin is so terrible as to be thought unpardonable.

Within days, 500,000 pilgrims are expected to flood into "The City of the Gods" and Gerry watches transfixed as the ancient ceremonies unfold on the banks. Hindus believe that at the moment of death, the soul discards its worn-out body and takes on a new one until, through countless reincarnations over millennia, it achieves perfection and unites with the divine. Karma, the present product of one's past actions, defines one's future. Cultivating mental balance, stoical resignation, and detachment from the endless horror and suffering surrounding him, the eternal optimist fends off suicidal despair by performing good actions to shape his future. True liberation comes from spiritual self-realization.

My grandfather, the alien secular man of science performing good acts of his own, cuts through the timeless air laced with the fragrance of incense and decay; the murmurings of pilgrims, sensuous and trancelike, sift through the acres of blood red saris, drying under the sun by the river. Around him swarm thousands of foraging monkeys, deemed sacred and untouchable; this does not prevent surplus animals being bought up by the laboratories of the West, including the Connaught, for the development of experimental vaccines. The Hindu faithful have come searching for moksha, the tradition that holds that those who die here and are cremated on the ghats, the wide steps lining the shore, are liberated from the eternal wheel of life, death, and rebirth. In the death houses lining the banks, Gerry sees people caring for their expiring family members; he watches them hoist their remains on logs piled two-storeys high, and set them ablaze. The bodies must be reduced to a fine ash before being cast into the river, which will carry their souls to paradise. The corpses of those unsuitable for cremation—pregnant women, lepers, children, and victims of cobra bites—are tossed like rotting logs into the same waters.

Donning a pith helmet, Gerry is taken out on the Ganges on a government launch, accompanied by the medical officer of health of Benares, a young Hindu who had been trained at Johns Hopkins. Gerry has heard of the ancient legend that the Ganges, fouled with rotting bodies and raw sewage, could so miraculously purify itself that a mile below the source of pollution, its waters flowed free of malignant bacteria. When Gerry asks about the legend, the westernized doctor shakes his head. "I'm sorry, Dr. FitzGerald, but the story is not true."

═

In June 1934, Fred Banting is made a Knight Commander of the Order of the British Empire by George V; the King happens to be a diabetic. The Conservative government in Canada had lifted the suspension of granting of titles to its citizens, a controversial move in a democratic country. Like the Nobel Prize, Banting nearly refuses the royal distinction, but in the end he kneels and feels the tapping of the sword blade on his shoulders. When Gerry sends his congratulations, Banting replies: "Many thanks for your letter on the honour that has come to me. So far my friends are having more fun about it than I am." In the months to come, he will routinely growl: "The next person who calls me 'Sir' will get his ass kicked." The knighthood helps mend Banting's damaged reputation and the embarrassing debacle of his highly publicized divorce. But he is increasingly adopting the life of a recluse.

My grandparents have now quit the glamorous Alexandra Palace, moving into an elegant buff-brick house at 14 Prince Arthur Avenue, a quiet residential street on the northern fringe of the university campus; perhaps it is no coincidence that the street is named after Prince Arthur, the Duke of Connaught. They continue to rent out the Balmoral house of which Edna is so fond, but plan to return there one day.

In September 1934, my seventeen-year-old father, thin and shy, leaves for McGill University in Montreal, having completed grade twelve at Upper Canada College. Eight years of boarding, leavened by short Christmas and summer breaks, have turned Jack in on himself, a hair-trigger Irish wit deflecting sudden, unwanted intrusions. He makes friends easily, yet often saunters off alone to the reassuring, womblike darkness of the movie house. He is drawn to older girls, but mostly he spends every last available dollar on the latest lacquered seventy-eight, diverting himself with the ecstatic pleasures of Louis Armstrong's horn, the genius father of jazz, born in a New Orleans brothel to an alcoholic prostitute and sent to work at seven, who now treats the stings of life with daily intakes of marijuana smoke.

My father has chosen McGill, yet no matter where he might go—Canada, the United States, Europe—he can never escape the looming shadow of his father. Gerry is delighted by the choice of McGill, for his friend Grant Fleming, the medical director of the Canadian arm of the mental hygiene movement, is dean of medicine there. Gerry has built an empire from the systematic accumulation of contacts, and here is one of countless men of influence that Jack, if so inclined, can use.

But Jack is not yet ready to follow his father's carefully phrased "memo of wishes" to the letter. A middling student, he enrols in economics, the dismal science, for which he has no aptitude, and it only makes him feel dismal. He joins a fraternity, Alpha Delta Phi, where he slouches and frolics with the influential gods of booze and sex and jazz; he once drives the three hundred miles from Montreal to Toronto and back in a single day and night to attend a party. As all sons must, my father is taking a rude stab at rebellion. The pressure to head to medical school—to please his father—can be staved off for a time; for a time, the sensuous freedom of the black man's hip-swivelling music will mock the cold marble pillars of science.

But in his heart, my father knows he is putting off the inevitable.

═

With Jack leaving home for university, his unmarried forty-six-year-old aunt, Hazel, is tacitly liberated from her duties as nanny. She has spent her life as a selfless caretaker, first for her invalid mother in Harriston, then her father in his declining years at the foot of the Avenue Road Hill, then her growing nephew and niece. Finally free to pursue her own interests, she buys the second-floor dress shop at 3 Charles Street West, at the corner of Yonge Street, which she calls "Gerald's."

Behind silk screens, a clutch of seamstresses sew dresses while stylish Toronto ladies of Rosedale and Forest Hill lounge by the tall bay windows, cluttered with cooing pigeons, and gaze down Charles Street East. On the hardwood floor, wardrobes are set up in a circle where the ladies can step inside for fittings. Thin and chic, perpetually dressed in black, Hazel smiles her charming smile, puffing her du Maurier cigarettes, and attends to their every whim. The Toronto Ladies Club stands one block north, on the northwest corner of Yonge and Bloor, across from Britnell's bookstore; there she often lunches with her niece, Molly, or sister-in-law, Edna, both of whom admire and envy her, for not only does she possess style and grace, she *works*.

Diligently cultivating contacts with couturiers in Paris and New York, Hazel travels by train and boat nearly as frequently as her medical brother. She is the first to import costume jewellery into Canada, and during World War II, her shop will be known as the only place in Toronto where a woman can find a pair of silk stockings. Over the twenty-year life of the shop, Miss FitzGerald will compile *Gerald's Book of Brides*, a scrapbook stuffed with society-page clippings and photographs of the steady stream of young women whose satin-smooth gowns she cut. Yet Hazel will never become a bride herself, despite three offers. One suitor proposes marriage while crying in her lap; she finds him unmanly and turns him down. Hazel cites the catastrophic losses of the *Titanic* and the Great War as the reason for her lack of suitable suitors. Over the span of her eighty-eight years, she will roundly reject the use of the word "spinster."

Hazel opened her dress shop, Gerald's, on Charles Street West, in 1934

On the morning of September 11, 2001, a virus creeps silently into the computer that houses my accumulated research on my grandfather; in the dawn of the new millennium, viruses seem more epidemic in cyberspace than in the bodies of humans. I lug my hard drive up the stairs of a newly opened repair shop at Yonge and Charles Streets, run by two young men with thick Arabic accents; one is wearing a red baseball cap with the word "Canada" emblazoned over a maple leaf. Then I realize the coincidence—the shop occupies the same space that my great-aunt Hazel used as a dress shop during the Depression years. My eyes fall on the same wooden floors, the same tall bay windows, the same flapping pigeons and, for a moment, time stands still.

The TV is on. As I place my computer on the hardwood floor, my eyes fall on the dreamlike horror of the smoking twin towers of the World Trade Center. Only later do I learn that at the pinnacle of the 105-storey monument to global capitalism stood the stock brokerage firm of Cantor Fitzgerald, occupying the top three floors, my grandfather's symbolic "room at the top."

As I walk home, I recall a dream I had almost exactly ten years earlier, just weeks before my father's death. I leaf through my dream diary, and sure enough, it's there:

"I am watching a passenger jet rising into the sky after takeoff. Then it slowly twists on its back, turning back on itself; something is wrong. Slowly, slowly, it crashes in flames into a high-rise building, smashing through the screaming people waving frantically from windows as I gaze up from the street, stunned as a statue. The plane seems to represent my father and the building my grandfather—the drawn-out, entrail-pulling agony of the eternal paternal drama, sons crashing into fathers, fathers into sons, the millennia of overreaching Icaruses straining for godlike perfection in the airless heavens. And here I am, the psychic detective, down below, the grandson rooting around blindly for the black box buried in the toxic rubble of his unconscious, poised to transcribe for the record the last cries of mortal men."

TWENTY-TWO

A Monument More Lasting Than Bronze

Success, noun: the one unpardonable sin against one's fellows.

AMBROSE BIERCE
The Devil's Dictionary

At its zenith in the 1930s, the Connaught seems to emanate a quasi-mystical vibration. A first among equals, the founding father appears to know exactly where he is going, urging others to rattle down the same path, possessed by the same passion. A gleaming, gliding train is the metaphor that most leaps to mind when people think of the labs. FitzGerald designed the locomotive of the public health system, they say, and Defries makes it run on time, supported by an unflagging, dedicated crew.

Mild-mannered and self-effacing to a fault, Bob Defries perseveres as the efficient, day-to-day manager behind the scenes while the nomadic Gerry blows around the world like a kite. In his free time, "Curly" Defries tends his flower garden on Cumberland Avenue where he lives with his long-widowed mother. The devout bachelor has earned a reputation for his quiet paternalism; each Christmas, he drops an envelope containing a few dollars on the desks of the all-female clerical staff, all extremely underpaid. A photography buff, Defries buys a 16mm Eastman camera and shoots hours of colour footage of the endless stream of distinguished international visitors; but most of the film is lost. Some whisper that Defries is linked romantically to a lab employee, but if there are clandestine affairs at the Connaught, they are generally intellectual, not carnal.

Donald Fraser acts as a balance between FitzGerald and Defries. The son of an Alsatian mother and a legendary professor of romance languages at the University of Toronto, Fraser is a handsome war hero and zestful Renaissance man who speaks fluent French and German and retains a smattering of Latin and Greek. Level-headed and shrewd, his unswerving honesty makes "Ask Fraser" a solution to any problem baffling the occupants of the School of Hygiene. Full of wisecracking repartee, Don's dispensing of encyclopedic erudition never seems pretentious or pedantic. His short, blunt sentences come with a smile, for he possesses an unfailing good nature; he embodies the ideal of the cultured man: one who knows everything about something and something about everything. "His own serene philosophy of life," one colleague observes, "kept him largely free of worry."

Fraser's home at 7 Wychwood Park serves as a social hub for School of Hygiene staff and other members of the university. Mortified by the death of true science in Hitler's Germany, he supports refugee scientists fleeing the Nazis; a believer in complete intellectual freedom, he not only helps them secure university positions, but aids their composition of scientific articles for English-language journals. Everywhere he goes, Don Fraser preaches the gospel of prevention by immunization; much to his wife's horror, Fraser tests the experimental vaccines and antitoxins on himself and his three small children; he says he wouldn't advocate anything as safe that he wouldn't give to himself or his own blood.

In March 1935, after reading the annual report that includes a review of the lab's twenty-year history, he writes Gerry: "I realized, with not a little reflected pride, that I myself had been associated, even in small measure, with you and your great achievement. A quotation from Horace—a fleeting shadow, but the only substance left from a vain attempt to veneer a classical education on a warped base—comes to my mind:

> "Exegi monumentum aere perennius
> Regalique situ pyramidium altius."
>
> "I have completed a monument more lasting than bronze and far higher than that royal pile of pyramids."

The son of tough Scots immigrants, the epidemiologist Neil McKinnon is a horse of a different colour, a forceful iconoclast and contrarian of Celtic sensibility known for his intense loves and hates. At the Connaught, McKinnon is a martinet for whom nothing matters but facts, logic, reason, truth; when medical students float lofty theories, he brusquely packs them off to the library to find the data to support their ideas. Explosive, tactless, argumentative, and sometimes abusive, he loves playing the devil's advocate; his blunt, critical approach is calculated to force students to think for themselves.

To McKinnon, a speaker of Gaelic, his fellow Scot Don Fraser is "the greatest fellow who walked the earth." He feels Defries, however, is too controlling and has a sole virtue: He is nice to his mother. "She is a better man than he is," he quips. When McKinnon, newly hired in 1925, suggested to Defries the idea of measuring the effectiveness of the new diphtheria toxoid—a landmark achievement in the annals of public health—Defries turned white with anger. Who was this brazen upstart? But later that day, when Gerry gave the project the green light, McKinnon felt vindicated. "It was the freest place," he recalled in his retirement. "That's the way FitzGerald wanted it. If you got an idea, you didn't ask anybody, you just

Promotional slide of Connaught's global reach, 1920s

went ahead and did it. At the Rockefeller Institute, they felt guilty if they broke a pipette."

In the summer of 1932, McKinnon is made director of the Connaught Farm. He installs his wife, Hollie, and three small daughters, all under six, in a newly built house close to the apartment the FitzGeralds used on weekends for entertaining. Since its birth during the Great War, the Farm has grown to seventy-five acres, gobbling up land vacated by beleaguered farmers. Insulin royalties have ensured strong, long-term financial stability; intermittent infusions of endowment money spawn new barns and labs, stables and operating rooms.

Practising what they preached: Gerry with Earle McHenry, pioneer of Canadian nutritional science; physiologist Charles Best; and epidemiologist Neil McKinnon, Hart House track, University of Toronto

At the farm, McKinnon patiently listens to everyone's personal problems, offering quotations of poetry and practical horse sense as solace. His most frequent confidant is Gerry, a man he unabashedly worships, a Horatio to his Hamlet; making the excuse that they are surveying the property, the two men disappear for long walks in the woods, together with the FitzGeralds' Kerry Blue terrier, Rhapsody in Blue; "Rhap" is forever tearing into the flesh of the McKinnons' smaller, weaker Wirehaired terrier. For hours on end, the men immerse themselves in conversation, as if, in their restless searchings, they might find with the setting of the sun a shining bottle of truth serum lying in their path.

"We loved every inch of that place," McKinnon recalled in his anecdotage. "It was sacred to us. We never wanted to leave."

Except, perhaps, his wife. Prone to bouts of melancholia, Hollie McKinnon is a cousin of the lab's benefactor, Albert Gooderham. Like her husband, she earned a medical degree at U of T, but gave up a career to raise their three girls. Not fond of animals, Hollie tends her garden or drives downtown to the symphony; often, she takes long, solitary strolls through the woods without telling her husband, a habit that triggers his famous temper, for he fears she may harm herself. During the Second World War, she will work as an industrial physician for the lab but never shake the coils of the depression that stalk her into middle age. One day she will attempt to fling herself from the Bloor Street viaduct but will be saved by Lorne Hutchison, the Connaught comptroller; finally, in 1950, she will consummate her darkest wish, swallowing a lethal dose of Nembutol.

Like Neil McKinnon, Charles Best is a magnificent equestrian who knows how to break and train horses; he loves to thunder about the property in all weathers on the back of his favourite mount, Priest, an imposing gelding seventeen and a half hands high that was given to him by the Toronto chief of police. Best and McKinnon had both served with mounted units during the Great War and they love to talk more about the horses than the fighting. Best is currently riding high, having discovered the enzyme histaminase in 1929 and one of the B vitamins three years later; he has recently purified the anticoagulant Heparin, revolutionizing heart operations the world over.

On weekends, my grandfather chooses not to ride with Best or the McKinnons, contenting himself with petting the noses of the horses and feeding them cubes of sugar. One such weekend, Alan Gregg of the Rockefeller Foundation visits the farm; as McKinnon dismounts after a ride with his daughters, the American takes him aside.

"Do you know how you could hurt Dr. FitzGerald?"

The question startles him.

"Why hurt him? I would never want to hurt Dr. FitzGerald."

"Well, you could."

"No," McKinnon insists, "I would never want to hurt him."

"You better not stop riding horses then," Gregg replies, "because if you did, you'd break his heart. He's talked about you and the children riding far more than the lab work. He is so proud of the way you enjoy the horses. He wants everybody to be happy and he thinks you are, supremely. So you would break his heart if you ever stopped riding."

═══

From a stack of crumbling press clippings saved by my grandmother, I shore together fragments of my grandfather's personality as seen through the eyes of others; reading between the lines, I gather up my own intuitions. By the age of fifty-three, his life seems uncommonly rich, animated by world travels in the company of fascinating companions, the serial triumphs of the teaching and lab work, and the coming of age of his children. The only damper is the recent *Globe* article on the death of his seventy-three-year-old benefactor, Sir Albert Gooderham, who was stricken after playing a round at Rosedale Golf Club, where he was president. A streptococcus infection from a sore throat poisoned the blue blood that ran in his veins; his beloved lab worked night and day to devise an antitoxin, but he cannot be saved.

While sitting on an international commission on biological standardization, Gerry is able to triumphantly declare to a press conference: "Ten units of insulin now have the same potency whether purchased in Tokyo, Copenhagen, Buenos Aires, or Toronto." Improved production capacity has sharply cut the Canadian retail price of the drug from ten dollars per one hundred units in 1922 to only twenty-five cents by 1936, making it affordable for the working class diabetic. On a 1936 trip to San Francisco reconnecting with his old Berkeley colleagues, a reporter poses a hypothetical question: "If you had $100,000, would you care for one hundred tuberculosis patients or let them perish and devote this fund to preventing thousands of others from contracting tuberculosis?" Ever-wedded to the facts, Gerry replies: "That is speculative." Then he adds with a wink: "Confidentially, I would keep the $100,000." In Vancouver, he modestly denies any personal skill in original bacteriological research, deflecting reporters' questions away from himself to the work of a promising young Englishman, Claude Dolman—recently lured away from Alexander Fleming, the discoverer of penicillin—whom he had appointed head of the new Western Division of the Connaught Labs. Gerry is among the first to admit he stands on the shoulders of giants. "I merely sit in an office," he muses, "and try to keep remotely in touch with the great men."

As I read *The Birth of Penicillin and the Disarming of Microbes*, the memoir of another English scientist drawn into the Connaught, I gain a deeper appreciation of the quirky day-to-day culture of my grandfather's maturing lab. Arriving in the spring of 1936, Ronald Hare, a self-described "inhabitant of the curative camp," is surprised to find himself quickly converted to the preventive side led by FitzGerald—"quite as remarkable a man as any of those I ever knew." Hare has long assumed that his native Great Britain was far ahead of the rest of the world in the practice of preventive medicine. In fact, the UK is still averaging five thousand cases of diphtheria every year while in Canada the disease is nearly eradicated. A key factor was a sophisticated annual campaign of public education, using modern advertising techniques, that is now attracting international attention.

"I was naturally taken aback," Hare wrote, "at the demonstration that a nation generally considered to consist of backwoodsmen or prairie farmers had, so to speak, stolen a march on us and succeeded in doing something that we had not even seriously considered. I did not know it then, but I was soon to find out why. Canada had a FitzGerald, we had not."

Practising what he preaches, Gerry pushes regular medical checkups for the entire staff and stocks a fully serviced cafeteria—one section for men, one for women—directed by a trained dietician, a reflection of his lifelong fascination with nutrition. In a male-only "Officer's Mess" in Room 423, Gerry holds court to his inner circle: "Dr. FitzGerald reserved his lunch time talk for the hierarchy; and it was an unwritten rule that it was only they who sat at his table," noted Hare. "Even they had to be careful because no matter how eminent they may be, none of them must sit in the seat reserved for the Chief, even when he was away. Strange as it may seem, this minute attention to diet did not save a high proportion of the older members from gastric troubles, who compared notes on their symptoms at the table in revolting detail."

Gerry also runs a mandatory program of recreational sports, including an annual golf tournament and regular games of deck tennis on the windy roof of the School of Hygiene; singles are taboo, as they do not encourage team spirit. Starting in 1930, the doctors compete for the FitzGerald Cup, a silver, foot-tall trophy with the names of annual winners carved into plaques; Best and FitzGerald, the keenest sports enthusiasts of the group, will become the only two-time winners. Whenever they can, the men pound around the track at Hart House, stripped to the waist, then swim as many lengths of the pool as time allows.

Gerry, right, playing deck tennis with a future deputy minister of health, Don Cameron, on the roof of the School of Hygiene, 1930s

Like all leaders, Gerry serves as a blank screen on which his staff projects their subjective views. Many testify to his courteous manner: His treatment of all persons, irrespective of rank or station, is gentle and considerate. He loves animals to the point of sentimentality, even as he sacrifices thousands to the greater good. "Tall, slim, quiet, a complete master of his emotions and in full possession of all the facts, FitzGerald never hectored or bullied," Hare observed. "He merely suggested, without appearing to do so, a line of attack that not only seemed logical but almost inevitable." Others see him as a study in contrasts—boyish, spontaneous, and easy-going one minute, forceful and authoritarian the next. Obsessive about bacteria swimming in unseen underworlds, he sometimes steps aside to let his companion open a door for him; otherwise, he pulls out his handkerchief and wipes the doorknob. Yet almost in the same breath, he might say it is acceptable to eat a piece of meat that had fallen to the floor; a little dirt is sometimes good for you.

"Not only so detesting disease that he made its eradication his life work," Hare records, "FitzGerald detested noise even more; so much so that the tap of heels on the hardwood floors that paved the institution was anathema to him, for which reason every newcomer was warned to get himself or herself shod with rubber heels before arriving. Loud laughter, and above all, singing, would have been the signal for instant dismissal. Nevertheless, provided one avoided Rabelaisian humour, which he did not like; the making of loud noises; played deck tennis with becoming frequency and golf once a year; and taught preventive medicine the way he wanted it taught, FitzGerald was an almost ideal Chief. He was always interested in what one was doing and kept himself up with progress by dropping in about once a week for a chat. He had known most of the pioneers, and although by the time I had got to know him he had given up lab work, he knew enough about it to sympathize with one's problems. To me, he was a very great man and Canada was all the poorer when he died."

As I ingest these vignettes of Gerry's crowded life, I sense the ceaseless push and pull inside his body, the toothy, FDR-like optimism biting back at the Great Depression. It is mentally he seems most free, lifted like a bird by the spontaneous gusts of ideas, his own and others', the civilized and the cruel locked in tight Platonic embrace, stoking the furnace inside him; for under his skin, beneath his controlled, genteel facade, blazes a fireball of frenetic energy. Through it all, year after year, Gerry FitzGerald never tires of repeating his favourite saying, the distillation of his purest wish: "Play to win or don't play at all."

═══

After two years at McGill resisting the dismal science of economics and playing the grooves out of his swelling stack of jazz seventy-eights, Jack FitzGerald succumbs to his father's gift for friendly but firm orchestration of the lives of others and agrees to enter medicine. Although Jack is a middling student, Gerry is not above pulling strings to have his only son enrolled in pre-med in the elite enclave of Trinity Hall, Cambridge; nor is Jack prepared to decline and risk compromising his father's love. The unspoken implication hovers in mid-air: If he works hard, he will be groomed for a spot in the Connaught (naturally starting at the bottom and rising through the ranks); then, perhaps, one day, having proven himself worthy of the laws of primogeniture, he might succeed his father as its head.

But first, Jack is allowed to indulge in what he called a "post-exam spree"—a four-month, fourteen thousand–mile motor trip from Montreal to Chicago, Vancouver to New York, an archetypal adolescent rite of passage reserved for the privileged, emotionally impoverished sons of the upper-middle-class. But big band jazz, not travel, remains Jack's mad passion. In mid-September, he swings down to Manhattan to catch two nights at the Riley-Farley Round and Round swing club, intoxicated by the all-night jam sessions, talking and drinking and smoking with the glamorous black-skinned jazzmen, mimicking their habits of dress, their silk-lined slang, their tilted postures of cool eroticism.

Back in Toronto, he stands in the living room of 14 Prince Arthur Avenue, wildly waving his arms like a conductor, raving to his father about the power and poetry that bursts from the divine trumpet of Louis Armstrong. It is as if, in this fervent and naive soliloquy, he might awaken the conservative professor of hygiene to the wonders of an exotic universe utterly foreign to his own. Jazz musicians possess discipline and drive, too; yet he may as well have been speaking Chinese to Gerry, who scorns all loud music, no matter its aesthetic shape. Yet, politely, he feigns interest, nodding his head and declaring: "Yes, yes, now isn't that something?"

In the same living room stands twenty-two-year-old Molly, who is casting spells on a gaggle of circling suitors; a young Upper Canada College teacher of Irish blood, Al Harris, who not so long ago caned the backside of her brother, Jack, for the sin of swearing at an opposing football team, numbers among them. Molly's latest conquest is Andrew Hazeland, an architect who is designing a special explosion-proof building at the Connaught Farm for the handling of volatile chemicals. Molly can pass for a volatile chemical herself; to enliven a dull Toronto, she spends her nights dancing the hours away with her several beaux at the Palais Royale by the lip of the lake, drinking and smoking and laughing with an inherited Irish passion; then, for a lark, she heads to Bowles, a downbeat downtown eatery, to gawk at the rows of shabby down-and-outs slurping their soup in the darkness before the dawn. Molly FitzGerald is playing out the drama of her generation, a drama that obsesses F. Scott Fitzgerald in his writings and in his tortured relationship to his wife, Zelda: the thwarted longings of intelligent daughters for their remote, eminent, perpetually busy fathers.

Family in Limerick, Ireland, 1935

This day, Gerry has news of his own to impart. His six-year stint at the League of Nations is now over. In addition, Gerry has resigned as dean of medicine with two years remaining in his second term and is taking a leave of absence from the university in order to take on a plum, year-long, all-expenses-paid project for the Rockefeller Foundation: Starting October 1, he will tour eighty-five medical schools in twenty-seven countries in North

America and Europe, assessing the fledgling state of undergraduate teaching in preventive medicine, together with Edna and an American colleague.

He sets out on his trip riding a swelling wave of public and professional accolades: A British Medical Commission surveying medical schools in North America in that year of 1936 has found Johns Hopkins and the University of Toronto the two most outstanding on the continent. The *New York Times* reports the Connaught Laboratories, in its research and production of preventive medicines, is "now recognized as the finest in the world."

═

On October 1, 1936, delivered by the hand of Fate—or more properly, his father—my nineteen-year-old father, Jack, a hesitant, slim-as-a-stick colonial encased in the naïveté of youth, arrives at the ancient campus of Cambridge University. After a two-year detour through the frat house frivolities of McGill, he is poised to face the subtle spectre of great expectations his father had placed, like the tendrils of invisible vines, in his path. As he clicks open his steamer trunk in his undergrad digs at 8 Portugal Street, my father could be forgiven for not fully divining the history-encrusted grandeur of the place where he will pass the next three years of his life.

When I was a prep boy, sick with the flu and installed in my parents' queen-size bed at Dunvegan, I pulled one of my father's tassel-spined photograph albums from the shelf and curled up with my prize. As I turned the thick black pages, I struggled to make sense of my father's soundless past, eerie and compelling as a dream, his curlicue jottings and jokes, in white ink, speaking to image after image, pose after pose. In anatomy class, I found him ripping open a dog's rib cage as he mugs at the camera with a mock-predatory grin. In a barren quadrangle of flagstone, he sat beside his peers in their double-breasted white jackets with crescent-moon crest, rimmed with dark trim, arms folded, faces as heavy as night. Under the pictures, my father's scribblings offer clues: He was a member of the Trinity Hall Boat Club, a charming row of Victorian boathouses lying along the River Cam. There he was, straining at an oar, his face contorted in pain; tradition dictated that he must row hungover. He seemed a fanatical, if inconsequential, rower, a lightweight of 140 pounds on the Fourth Lents Team; with his delicate, nervous temperament and the slim build of a coxswain, he was best suited for cheering others on—"Stroke! Stroke!"—with a masturbatory urgency

Perhaps, even then, I intuited it all in the black-and-white snapshots. There he was, frozen in time—the wavy, russet hair, the hazel eyes shielded by sunglasses, the crossed legs, the fingertip-dangling cylinder of Silk Cut; this, I guessed, must have been the place where my father found and cultivated the paradoxical, loose-limbed persona of cool that seemed to emit a double message: "Yes, I am irresistibly intriguing—but keep your distance." Yet in the subtle corners of the images, traces of vulnerability leaked out. I saw him pulling back his round shoulders, puffing up his bony, hairless chest, heaving at his oar like a galley slave, striving mightily to be what the wiser world said was a man. As he moved self-consciously among the gentlemen of elegance and grace, the intelligent, the gifted, the chosen few, the born geniuses of past, present, and future, the men whose deeds hold the power to both inspire and dispirit, he can only worship the best from the Canadian fringe, inflated with a grandiose identification with greatness that camouflages an unbearable feeling of incompetence and inferiority. In this ancient alma mater of the likes of Newton and Darwin, Milton and Wordsworth, what can possibly rub off on him? What, except a burning sense of entitlement?

This is the mythic world my father never spoke of as I grew up; the privileged, unflinching world of his father he has not yet earned himself; an intimate yet remote corner of tradition, taste, and civility where he can hide, in the highest of styles. It was there for the taking and he took it: not a father, but a fatherlike world, an airless sphere unto itself, one that will perfectly serve as a slender shell of selfhood, which Jack FitzGerald will silently bear, year after year, forward into the future, driven by the horses of the past.

═══

My father's first week in Cambridge neatly coincides with his own father's arrival in England, the first leg of his nine-month European tour. After Gerry inspects the curricula of the Cambridge Medical School—and finds it sadly wanting in its preventive teachings—father and son briskly perform the familiar hello-and-goodbye ritual they now have down to a science.

With the coming of the Christmas holidays, Jack spends three weeks travelling through Europe playing in a hockey tournament. Stocked mostly with sturdy Canadian legs, the maroon-shirted Cambridge Eskimos whisk across Belgium, Germany, Austria, Switzerland, and France—the same countries, coincidentally, Gerry FitzGerald is now charging through with single-minded energy. Jack skates on the third line, a fringe player a bit out of his depth, yet he is fiercely competitive, complaining about his lack of ice time. Here in the year 1936, I find my father and grandfather criss-crossing the continent, caught in the crack of history between a Great War and a Great Depression, each in his own way chasing the gods of perfection. For Gerry, it is an impossible world of universal health, free of killer disease; for Jack, it is the sounds of swinging big-band jazzmen jabbing junk in the veins that pump the blood that blows the hot mad music of sweet salvation.

"There's always room at the top," Gerry FitzGerald never tires of saying. Gazing heavenward, his one and only son, beating down the alien Irish upsurgings of rage, terror, and longing, feels the power to choose his own path slipping fast away. No choice now but to take a deep breath, as if it were his first and last, and embrace the long, slow climb, exquisite and preordained, upward into the ever-darkening clouds of divine discontent.

PART IV

TWENTY-THREE

Hell of a Good

> People occasionally fall ill precisely when a deeply rooted and long cherished wish has come to fulfillment. It seems then as though they were not able to tolerate their happiness; for there can be no question that there is a causal connection between their success and their falling ill.
>
> SIGMUND FREUD
> *Those Wrecked By Success*

On a June day in 1937, my grandfather steps onto the platform of Union Station, returning home from his Rockefeller-sponsored, circuit-riding mission to convert the world's medical schools to the gospel of preventive medicine. Colleagues are quietly shocked by his haggard appearance; in the space of a nine-month trip living out of a suitcase, he seems to have aged twenty years. Pressing through a tightly organized itinerary of passenger liners, train stations, hotels, and cabs, Gerry had touched down on dozens of magnificent cities—Cambridge, London, Belfast, Dublin, Cork, Paris, Brussels, Berlin, Stockholm, Athens, Moscow, Leningrad, Prague, Warsaw, Strasbourg, and Vienna—yet he could spare little time to drink their cultural riches. Critics worry the Rockefeller Foundation is casting the net of its healing resources too randomly, thousands of miles wide yet mere inches deep. A similar concern is sinking into the conscious mind of Gerry FitzGerald, who reflexively brushes aside the suspicion that his aging body is governed by the immutable law of diminishing returns. A man who is told he has reached the ne plus ultra of his profession may see no more mountains to climb; or he may see far too many more.

═

Gerry's closest friend, C.B. Farrar, meanwhile, is busy scaling his own professional peaks. Radical waves of "furor therapeuticus"—insulin shock, Metrazol, lobotomy—are sweeping across the ocean from Europe and storming the wards of North American psychiatric hospitals. "Do something! Do anything!" is the clarion call of the families of the mad and Farrar leads the ranks of mad-doctors ready to oblige.

Back in 1935, a Hungarian doctor, Ladislaus von Meduna, first used Metrazol, a synthetic preparation of camphor, to induce violent convulsions in mental patients; his idea was motivated by anecdotal reports of cures of schizophrenic patients who developed epileptic seizures. One man described feeling like he was being "roasted alive in a white hot furnace." Even Meduna confesses that his treatment makes "brutal inroads into the organism," likening insulin and Metrazol to the onslaught of exploding dynamite. He is so distressed witnessing the effects of Metrazol that he has to be supported to his office by nurses.

That same year, Egas Moniz, a sixty-one-year-old Portuguese aristocrat, parliamentarian, and neurologist obsessed with winning the Nobel Prize, forms the fixed idea that pathological thoughts became "fixed" in the cell tissues of the frontal lobes of the mentally disturbed. Rotating a leucotome, a razor-thin, pick-like instrument with a wire loop, Moniz severs the guilty brain fibres as if coring the apple of original sin.

One day in 1939, a paranoid patient will enter Moniz's office, pull out a revolver, and shoot him five times. Confined to a wheelchair by his wounds, Moniz carries on with his work undeterred. A decade later, he will win the Nobel Prize he so desperately covets, giving lobotomy a mighty impetus worldwide. Into the 1950s, over fifty thousand psychosurgeries will be performed in the US alone. Dozens will die on the table or from complications afterwards; countless more will suffer severe side effects ranging from seizures to hyperactivity to vegetative inertia. In most cases, the knife merely replaces one set of gross symptoms with another.

As news of the prefrontal lobotomy blows across the Atlantic, C.B. Farrar wastes no time reaching its vociferous American champion, Walter Freeman, a charismatic but heartless Washington-based neurologist and a grandson of America's first brain surgeon. Addicted to Nembutol to quell his own depressions, Freeman has embraced the leucotomy knife with an evangelical fervour, eventually adding his own new twist—the transorbital lobotomy, in which he uses an ice pick to bore through the patient's eyeball and scrape nerve fibres from the brain. Freeman, like Farrar, mocks and despises psychoanalysis: "Insight is a terrible weapon, and few know how to use it constructively," he writes. "When we realize, really get to know what stinkers we are, it takes only a little depression to tip the scales in favor of suicide." To gauge whether his instrument is cutting too deeply, Freeman keeps his crowded assembly line of patients awake with local anaesthesia so they can report their feelings—a grim parody of the method of Freudian free association. To allay—or perhaps heighten—their fears during the ordeal, he instructs them to recite the Lord's Prayer. During one lobotomy, Freeman asks the patient: "What's going through your mind?" After a pause, the man replies, "A knife."

Together with Toronto's foremost neurosurgeon, Kenneth McKenzie, Farrar travels south to observe Freeman in action. Back at the Toronto Psychiatric Hospital, McKenzie cuts open the foreheads of their first set of nineteen patients—all women—scooped from the crammed wards of the provincial hospitals. So it is that my grandfather's close friend and confidant will open the door to the first Canadian lobotomies, spreading from the red-brick walls of his fiefdom to 999 Queen Street West and beyond. As late as 1973, three years after Farrar's death, a psychiatrist in Homewood sanitarium will recommend—though not carry out—a lobotomy as a final solution to my own father's intractable troubles.

But it is for insulin shock that Farrar reserves his most ardent interest. At the annual meeting of the American Psychiatric Association in early 1937, he is impressed when its inventor, the eccentric bachelor Manfred Sakel, claims a spectacularly high cure rate. Farrar starts experiments on twelve young women over a three-month period, working with insulin specialists from the Connaught Labs and the Banting Institute. The excitement generated by the novel use of insulin had naturally riveted Fred Banting's attention; increasingly intrigued by the mysteries of madness, he befriends Clare Hincks, head of the mental hygiene movement, and joins its board of directors. Banting visits mental hospitals and tries to mobilize Nobel Prize winners to pull together and fashion a magic bullet for the mad. After a tour of the wards of 999 Queen Street West, he tells Hincks:

"I found the attitude of doctors and nurses to patients all wrong. They treated the patients as inferiors, not equals, telling them what to do rather than leading them to self-help, self-respect, and independence. When the patients were by themselves with a minimum of supervision, they spoke to each other as equals and were really doing a magnificent job therapeutically for one another. The patients, if given a chance, are the real therapists, not the doctors or nurses."

Yet Banting ignores his own intuitive insight; caught in the thrall of his messianic status, the hero of insulin anxiously presses onward with his search for an aggressive somatic cure. Sitting down with his friends Farrar, FitzGerald, and Best, he proposes a five-year psychiatric research project on insulin coma therapy, first performing experiments on dogs and cats at the Banting Institute, then moving on to human subjects.

With insulin, Metrazol, and lobotomy, psychiatrists are now armed with a formidable one-two-three punch. As the 1930s roll on, North American and European hospitals will rush to embrace the new treatments, as if they hold the metaphorical promise to roust the masses out of a world depression. Desperate for a cure, many asylum patients are actively willing to risk anything, even dangerous and painful methods, to feel the fleeting calm after the storm. No impassioned voices of dissent will arise from the profession within or the public without; no outcry against the broken bones, the damaged brains, the shredded memories; few tears will be shed for the hundreds of lives lost.

In Vienna, meanwhile, an aging Sigmund Freud, tortured by cancer of the mouth, labours to finish his latest work, *Analysis Terminable and Interminable*, soberly holding fast to the unpopular notion that cultivating a lifelong habit of rigorous self-reflection remains the worthy and ethical aim of a civilized man.

═

Returning to Cambridge that fall, my father bashes away at his courses in bacteriology and pathology—the daunting domain of his illustrious father. In his letters home, Jack rates each of his profs, hating himself for pandering for marks. He keeps assiduous accounts of lecture timetables, sports scores, allowances, and ocean fares. He runs practices for the hockey team, the Cambridge Eskimos, admitting, "I make a better coach than player," and feels complimented when he is called "a born slave-driver." In November, his Canadian friend John Hamilton, a demonstrator in pathology destined to become a future dean of medicine at the University of Toronto (and destined to preside over the breaking up of the School of Hygiene in 1975), takes him to see Ibsen's controversial play *Ghosts* at the Arts Theatre. The protagonist, Oswald, had inherited syphilis from his father; the radical theme challenges the sanctity of marriage and the sacred ideal of a son duty bound to honour his father without question. "It was excellent," Jack reports to Gerry, unaware of the irony of the ghostly allegory. "I was tickled about your biography in the medical journal," he adds. "You really are a pretty big guy, aren't you? Hell of a good, as the Cantabs would say!"

At a subsequent dinner, Jack overhears Hamilton speaking to another medical student: "Of course, Ontario is years ahead in public health, due to Connaught. But then, Dr. FitzGerald *is* public health in Canada." Jack is astounded; the veil is falling from his eyes. "That really chilled me!" he writes Gerry. Then he adds: "I don't even know where you are living at present."

Jack dissects a dog in a pre-med anatomy lab, Cambridge University

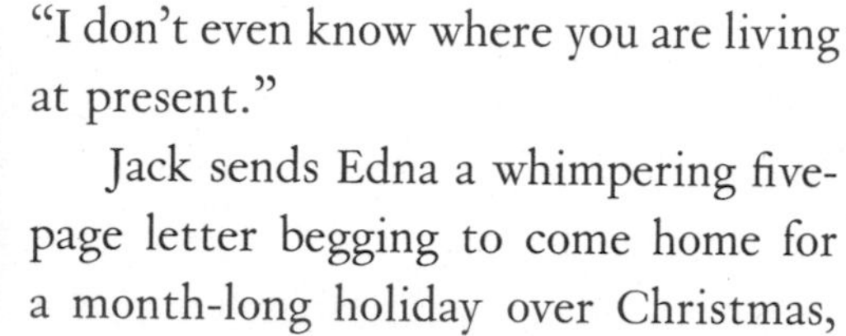

Jack sends Edna a whimpering five-page letter begging to come home for a month-long holiday over Christmas, backing up his plea with carefully worked out expenses down to the farthing. When she agrees, he writes back: "Please don't think I'm too selfish. So overjoyed I felt like screaming. I'm going to whack next term right out of sight! I'm so happy that it is actually a pleasure to work hard!"

Over the holidays, my father, the neglected child of privilege, works at the Connaught, starting as a lowly bottle washer. Amidst the steaming Bunsen burners, he sterilizes rows of glassware, knowing that if he neglects to clean the tiniest speck of a germ, he might spoil one of the experiments in his father's lab. There is, perhaps, some honour to be found in the lowest, most menial of tasks.

On January 16, 1938, days before returning to his winter term at Cambridge, Jack makes his way down to Carnegie Hall in New York City. Molly's classmate in finishing school in Lausanne, Helen Oakley, now a personal assistant to Duke Ellington at the precocious age of twenty-four, has co-organized the first racially integrated jazz concert, a bold breaking of the sound and colour barrier and a temporary antidote to Jack FitzGerald's medical school blues. The show is headlined by Benny Goodman, the twenty-eight-year-old "King of Swing," clad in black tails, with performances by Count Basie, Teddy Wilson, Gene Krupa, Lionel Hampton, Johnny Hodges, and a slew of hot young musicians. During the two-hour concert, Goodman, as obsessive a perfectionist as any lab pathologist, plays a fifteen-minute retrospective, "Twenty Years of Jazz"—aptly, the same age as Jack, born the same year as jazz. The night ends with "Sing, Sing, Sing," bringing down the house. A thunderous standing ovation brings two encores, "If Dreams Come True" and "Big John's Special." Even as my father swoons in the cocoon of his bliss, polite society is properly scandalized; the playing of jazz in the venerable old brownstone concert hall is likened by the press to "letting a whore into church."

On the rocky six-day voyage back to England on the *Europa*, a German liner that once held the world's transatlantic speed record, Jack relieves his seasickness with tablets of Nembutol and Atropine given to him by John Hamilton. "Now I have both a preventative and a cure!" he writes his mother. No sooner has he landed than he races off to a London theatre to see the Quintet of the Hot Club of France, currently tearing up Europe with its own unique brand of jazz. The cult band is fronted by an unlikely pair of Banting and Best–like opposites: twenty-seven-year-old Django Reinhardt, the illiterate Gypsy-vagabond acoustic guitarist, and Stephane Grappelli, twenty-nine, the suave, urbane son of a French philosophy professor, a classically trained violinist of unusual zest and melodic sparkle. Clad in tuxedo and garish red socks, Django dexterously runs his hands up and down the fretboard, his signature heavy vibrato driving him and his partner to soaring heights of improvisation, the exquisite sad-happy sounds shooting spasms of pure, childish joy through my father's nervous system. After the show, he insinuates his way backstage and talks with the band, pursuing his heroes as passionately as his father courts elite scientists.

═

On March 12, 1938, the Nazis invade Austria and execute a swift Ayran cleansing of the native population. In a world willing to pay any price for peace, the virus of fascism meets neither internal resistance nor external intervention from the League of Nations. The invasion triggers a rash of suicides in Vienna, nearly one hundred in ten days; the Nazis issue an edict that only half were attributable to the political situation. When escape seems hopeless, Anna Freud asks her father, "Wouldn't it be better if we all killed ourselves?" Ever defiant, he retorts, "Why? Because they would like us to?" Ernest Jones quickly sets to work obtaining exit visas and mobilizing influential international connections. All flights to Vienna are cancelled, so Jones courageously hires a small monoplane and descends into the clutches of Nazi storm troopers, who promptly arrest him and rain insults on his head; talking his way free, he heads straight over to 19 Bergasse where Nazi thugs have just finished ransacking Freud's apartment. For weeks, Freud has been resisting leaving Vienna, but he is finally persuaded when Jones quotes Charles Lightoller, a British officer on the *Titanic*. When asked why he had not gone down with the ship, Lightoller replied: "I never left the ship, sir. The ship left me."

The day after the invasion, my father writes an ostrich-like letter home from Cambridge: "Mr. Hitler's little game of soldiers in Austria yesterday didn't do very much to disrupt the British calm. The papers talked a lot, but the market was relatively steady. A sure sign! So Hair Hitler can go take a jump in the Rhine."

A few weeks later, Jack attends the Oxford-Cambridge boat race—the gruelling twenty-minute, four-and-quarter-mile course on the frigid Thames—which he regards as a far more important event than the machinations of Hitler and Mussolini. As the waves break into the boats, the Cambridge Eight are forced to stuff football bladders under their seats to keep afloat, a symbolic mirroring of the general sorry state of Britannia. Cambridge loses and my disconsolate father heads down to London, utterly exhausted, feeling "as though I rowed in the boat myself."

That spring of 1938, my father is unaware of a seminal event in psychiatric history that will one day bear down mercilessly on his own head. Five years of radical psychiatric interventions are coming to a head in the fascist city of Rome: first came the Pole Sakel and his insulin coma in 1933; then the Hungarian Meduna followed with Metrazol convulsion in 1935; then the Portuguese Moniz gave birth to lobotomy later the same year. Now, on the verge of a new world war, marches in a hawklike, sixty-one-year-old Italian, Ugo Cerletti, whose brainchild will, in time, sweep all rivals aside.

One day in a Rome abattoir, Cerletti watches the staff grab the heads of pigs with wired pincers spiked with metal points and jolt them with a burst of electricity. He notices the shock does not kill the pigs but simply stuns them, making it easier for the staff to slit their throats; a still-beating heart is needed to pump the blood out. Inspired, Cerletti and his assistants rig up a specially designed electrical device to the temples of a disoriented homeless man whom police had randomly plucked from a train station. When Cerletti jolts the man's brain with a bolt of electricity, his body violently jerks and stiffens, but he does not go into a convulsion. Deducing that they had used too low a voltage, the circle of doctors try it again with a higher dose.

"Not another one!" the man cries out. "It's deadly!"

Cerletti ignores the protest and this time, the man falls into a grand mal seizure. The doctor does not believe it is the convulsions that have produced improvement in the man's symptoms, but the release of hypothetical "vitalizing substances." But even Cerletti is appalled by his creation. "When I saw the patient's reaction, I thought to myself: 'This ought to be abolished!' Ever since, I have looked forward to the time when another treatment would replace electroshock."

In fact, ECT is quickly and widely embraced. The side effects are as severe as those of its forerunners: bone fractures, seizures, and retrograde amnesia. No one dares say it out loud, but electricity is as effective in controlling problem patients in asylums as unco-operative hogs in a Roman slaughterhouse; many patients report no negative associations because they have no memory of experiencing the shock in the first place. Manfred Sakel, meanwhile, peevishly snipes at his competitors like a jealous schoolboy. Insulin coma, he insists, is the first and only true shock treatment.

═

In late May 1938, my grandfather sails to England, a stopover for a planned tour of Scandinavia. He stays at the London home of a physician friend at 20 Seymour Street, an elegant five-storey Georgian building a block north of Marble Arch and Speaker's Corner. The house is only a brisk walk eastward from 14 Prince's Gate, the ornate palace of another ceaseless Irish striver named Joseph Kennedy, the American ambassador to Britain, busy grooming his young sons for political glory.

On Sunday, May 29, Jack comes down from Cambridge for a twenty-first birthday lunch with his father. Gerry gives him a copy of his gold signet ring, imprinted with the crest of the Knights of Kerry—a father-son ritual that will be repeated a generation later.

A week later, on June 6, 1938, Ernest Jones escorts Sigmund Freud and his daughter Anna from Victoria Station to his home at 39 Ellsworthy Road in Primrose Hill. As a condition of his release from Vienna, the eighty-two-year-old Freud must sign a document testifying the Gestapo had not ill-treated him; with characteristic deadpan sarcasm, he writes, "I can heartily recommend the Gestapo to everyone." During the 1930s, Jones has helped many displaced German analysts relocate in England and elsewhere, often at great personal risk; Freud once said that Jones's diplomatic abilities would lead him to be enlisted by the League of Nations. While Freud manages to elude the tentacles of the master race, his four sisters will all disappear up the chimneys of Auschwitz and Treblinka.

Gerry's old connection with Ernest Jones has expired from neglect; yet here they are now, exactly thirty years later, in the same city, mere blocks apart, facing the prospect of an apocalyptic world war. In his private practice, Jones has been busy honing his psychotherapeutic skills, resisting the imperialistic advance of insulin shock, lobotomy, and ECT. One day, a troubled middle-aged man, arriving to see Jones for his third session, unsheathes a knife and places it on the desk, equidistant between the two men.

"Which of us do you think will need to use that?" he asks with sanguine detachment.

The quick-witted Jones replies, "Neither." But the patient persists: "Oh no, one of us has to use it." After a tense silence, Jones senses the power of a negative father transference and answers, "Alright, but suppose we just leave it there for the moment and talk." After inducing him to recount a dream from the previous session, Jones wonders if the man had ever quarrelled with his father. The question releases a torrent of repressed hatreds and the knife is forgotten.

═══

At a June 10 dinner at London's Royal Automobile Club, my grandfather reconnects with his protegé Charles Best, who is in town lecturing on Heparin. The following week, Best is to be made a fellow of the Royal Society; coincidentally, the newly landed Sigmund Freud—he would remain classified as "enemy alien" until his death—is to be inducted at the same ceremony, but he is too ill to attend. In a rare gesture, officials bend the rules and come to Jones's home to deliver Freud the Royal Society

Graduating class of the School of Hygiene, 1937–38. After an exhausting year-long world tour of medical school curricula for the Rockefeller Foundation, 55-year-old Gerry has aged rapidly

book, where his gothic signature joins those of Isaac Newton and Charles Darwin—"Good company," as Freud remarks.

That same week in mid-June, Gerry sets sail for Norway, taking Jack along. With the advent of summer comes the natural tendency of the body to let itself go, a crying out for release from the punishing daily stress; but Gerry drives ahead, as he always has. As father and son hike through a spectacular Norwegian fjord, a sudden, tremendous pain stabs Gerry's stomach like a red-hot knife; it is a bleeding duodenal ulcer, severe enough to send him back to England. He will suffer two hemorrhages in all, spurts of acid eating into the stomach lining, ripping open the vessels and veins, the flow of blood feeling hot enough to burn a hole in a cold marble floor. He is hospitalized in London for nearly two months, the exquisite pain worse in the early mornings, radiating down from his chest to his navel. Back at Cambridge, Jack simultaneously comes down with a case of infectious mononucleosis, as if he were suffering sympathy pains for his father; his wavy red hair comes out in handfuls on his pillow.

Leaving hospital in early August, Gerry recuperates at the Seymour Street house in the heart of London. Yet he realizes that he is losing his ability to concentrate and that his old fear of electrical storms is growing in intensity.

══

Recovering sufficiently from his ulcer attacks, Gerry sails home from London on September 30, determined to relieve Bob Defries who is running the Connaught Labs in his absence. By chance, it is the same day that Prime Minister Neville Chamberlain disembarks from an airplane after a meeting with Hitler in Munich, waving the infamous "piece of paper" that has allowed the Nazis to swallow up Czechoslovakia in exchange for a promise of no further aggression. Most hail Chamberlain as a saviour for achieving "peace in our time," but others, including Fred Banting, denounce him as a cowardly appeaser. "Don't believe it," he growls. "There is not going to be peace in our time. Let's all get ready for war."

Through October and the early weeks of November, Gerry rests at the Connaught Farm, an unprecedented disruption of his long-standing work routine. He has spent his life deftly balancing work and holidays, recharging his body each summer to take a run at the busy fall term. But this fall, it's as if a steady trickle of blood is leaking from the hull of a battered frigate, sapping his waning reserves of strength. He tries to keep himself occupied with rug weaving and other hobbies, but instead waves of apathy and agitation roll over him; each night, he swallows tablets of Nembutol to quell his insomnia. Edna handles his correspondence and postpones meetings.

On the overcast morning of November 23, Gerry rallies himself for a banner day. John Buchan, the Governor General of Canada—the first commoner chosen to represent the British monarchy—is scheduled to tour the Connaught Farm. A thin, wiry Scot of sharp, birdlike features and titanic ambition, Buchan, the first Lord Tweedsmuir, is the Oxford-educated, sixty-three-year-old son of a Calvinist minister and a household name, literary lion, statesman, sportsman, pamphleteer for the League of Nations, speech writer of the king, and confidant of powerful politicians and intellectuals. The progenitor of the cloak-and-dagger spy thriller, Buchan has authored over a hundred books—biographies, histories and "shocker" novels—most notably *The 39 Steps*, made into a film three years earlier by Alfred Hitchcock. Terse and dour, prodigiously self-controlled, the Scotsman speaks in a quiet, feminine voice and rarely smiles; a distinctive scar creases his brow, the traces of a carriage wheel that fractured his skull at age five, lending him a sinister appearance not unlike one of his treacherous fictional characters.

Buchan's struggles with his repressions mirror those of my grandfather. He champions the imperial values of hardiness and grit, nerve, and pluck, even as he stoically denies his own painful bouts of depression and chronic gastritis. "It's a great life," he opines, "if you don't weaken." In his books, Buchan obsesses about men unhappy with their worldly success, tortured by the primal conflict between the calls of selfless duty and personal pleasure; it is no accident that he identified intensely with his close friend T.E. Lawrence, the celebrated Irish-born poet warrior of the Arabian campaign, recently killed in a motorcycle crash. "I cannot but hope," the Governor General declares, "that [the recent accession of George VI] will mean the going out of fashion of this miserable jazz nightclub atmosphere, which has been the worst bequest of the War, and return to something of the old Victorian seriousness."

Gerry escorts Canada's Governor General, John Buchan, on a tour of the Connaught Labs, November 1938. Minutes after this photo was taken, Gerry "went off the deep end."

On that grey November morning, my grandfather escorts Buchan around the labs, proudly holding forth on the conquest of diphtheria and the miracle of insulin. When a press photographer snaps a picture of Buchan and FitzGerald, standing side by side in near-identical top coats and hats, they might have passed for blood brothers. As the Governor General's impeccable, black limousine pulls out of the gates of the Connaught Farm, my grandfather waves a last goodbye. Perhaps it was the feel of the Governor General's firm handshake, the intimidating scar and scowl, a tacitly shared hatred of unmanly weakness, that seem to mock Gerry's quavering emotions; perhaps it is a sudden memory of his druggist father, Will, grim and willfully silent as an Irish ghost, standing on this exact same patch of ground, mere weeks from the grave, at the formal opening ceremonies of the Connaught Farm back in 1917. Or perhaps it is something else again, something beyond shame, that sends my grandfather sprinting to his office where he collapses in a fit of violent sobbing. The anguish gushes and flows, unstopped and unstoppable, fierce enough to pop the steel ribs of a rain barrel. This is no bleeding ulcer; this is something worse, an unforgivable breach of protocol. Gerry FitzGerald has failed to govern himself.

Three days later, as if resisting the psychic coils of his father's descent, Jack steals downtown to Maple Leaf Gardens to see his hero Duke Ellington perform before an idolatrous crowd of thousands. Recovering in Toronto from his long, drawn-out bout of mononucleosis, my father prescribes a dose of swing jazz as his own special brand of restorative. Rocking in his seat, he thrills like a child to the Duke's virtuoso tinkling of the ivory eighty-eights and the wall of wailing horns, lifting the audience into gyrations of ecstasy. Back at the band's hotel, Jack tops it all off with the usual all-night bacchanal, hanging out with the tea-smoking members of the band of ebony-skinned brothers, the illicit tribe of swingers and sinners, sipping the pleasing narcotic of forgetfulness, the coolest of cats landing a clean uppercut on the lantern jaw of Victorian seriousness.

═══

My grandfather's breakdown is cloaked in whispers. The most succinct description comes from the Connaught comptroller, Lorne Hutchison, a man destined to fling himself from the St. Clair viaduct; Dr. FitzGerald, he writes to the university president, has "gone off the deep end."

Neil McKinnon, Gerry's trusted confidant of nearly fifteen years, drives Gerry downtown to the luxurious, state-of-the-art Private Patients' Pavilion at Toronto General Hospital. My grandfather knows the neighbourhood passing his car window as intimately as a lover, yet now it splinters into the random sensations of a waking dream, a perplexing contortion of the familiar. The intersection of College Street and University Avenue stands at the matrix of his professional world, the place of his children's births, an historical crossroads of enduring institutions more lasting than bronze, the visible fruits of decades of hard labour and radiating influence. Flanking the Queen's Park legislature, the School of Hygiene and the Banting Institute lie only a short walk north; the elegantly pillared Alexandra Apartments, his former home, stand directly across the boulevard. Here, everyone knows him and he knows them—the nurses and doctors, the interns and technicians, the administrators and janitors, the dieticians and doormen.

But this time he has come not to instruct, inspect, or inspire. He has come to stay.

To this half-empty "Hospital de Luxe," Neil McKinnon leads his revered boss, the white knight of universal public health care who has dedicated his life to the good of the masses. The two friends pass through the solemn stone entrance of Doric facade and the rotunda of dark walnut panelling, past the walls tinted in tasteful buff, the gleaming fixtures of Canadian nickel, the intermingling smells of the florist shop and the smoking lounges, past the patrons lounging in blue Spanish-leather chairs, heads buried in newspapers. As Gerry rises up the elevator, he feels the force of gravity pull his body down like a lead weight; when he passes through the door of his private room, he barely lifts his head to glimpse the cherry chintz chairs, the elegant mirrored dressing table, the mohair drapes, the vase of cut flowers, the Persian rugs and art deco light fixtures, the scrubbed bathroom with bath and shower, all of the comforting amenities and atmosphere of home, vaguely discomfiting to him now, all gone from his mind as he sinks, overwhelmed, into the coffinlike bed of polished wood finish.

On December 9, 1938, the University of Toronto grants my grandfather a leave of absence with full pay for a period of four months. It is his fifty-sixth birthday. Days later, he slits open an envelope to find a short note from John D. Rockefeller, Junior, thanking him for his seven years of work as a scientific director of the Rockefeller Foundation; his term is due to expire on New Year's Day. Within eyeshot of the wrought-iron balcony of his hospital room stands the roof of the School of Hygiene, laden with the inevitable layers of Canadian snow.

TWENTY-FOUR

The Living Lie

Show me a hero and I'll write you a tragedy.

F. SCOTT FITZGERALD

Like generations of doctors before him, Gerry FitzGerald does not make for a docile patient; he has ideas of his own. No one knows this better than his personal physician, forty-two-year-old Ray Farquharson, a handsome, soft-spoken son of a Scots Presbyterian minister and yet another star scientist to emerge from small-town Ontario. Known as "a physician's physician," Farquharson treats doctors, nurses, clergy, and eminent businessmen and their families. No stranger to tragedy, he lost three older brothers in the Great War and one of his daughters is subject to suicidal impulses.

With Gerry, he does his tactful best with a man habituated to giving, not taking, direction. Before Christmas, he sends him home to continue his convalescence. Alarmed by her husband's unravelling state and worried about money, Edna orchestrates a move from 14 Prince Arthur Avenue into a smaller duplex apartment two doors over at number 18, made of the same elegant Victorian buff-brick, but a visible step down in the world all the same. Frugal to a fault—she never buys her own cosmetics, for Gerry has them made cheaply at the lab—Edna continues to rent out the original family home at 186 Balmoral. Raised under the crystal chandelier of class privilege, my grandmother lives with the certain knowledge that a woman's identity and reputation are intimately interwoven with those of her spouse. She does her best not to admit the unthinkable even to herself—a humiliating loss of status, social ruination, instant banishment from the parlours of the haute bourgeoisie—and bravely puts her best face forward to the world.

Gerry spends much of his time at the farm, which he now wryly calls "The McKinnon Nursing Home." Agitated and despairing, he pours out thoughts of self-destruction to the sympathetic ears of Neil and Hollie. The rocklike Neil stands loyally by Gerry, but privately he weeps for his friend's anguish. Week after week, his three young daughters overhear his voice, words they are not meant to hear, muffled behind the closet door, pouring urgently into the phone: "Promise me you won't do anything, Fitz. I'm coming—promise you'll wait for me." Dr. Farrar, too, visits, night after night for months on end; with his growing estrangement from Edna, Gerry confesses to his friend that in a sudden craving for physical affection, he fell, briefly, into the consoling embrace of another woman while on a transatlantic voyage, an impetuous act that only fans the flames of his guilt. Each time C.B. leaves, he binds my grandfather to an oath: "Goodnight, Fitz. Now, you're going to be here tomorrow night, aren't you? Tell me now, Fitz." Farrar chooses his words deliberately, for he is gambling that Gerry, no matter how troubled, will not dare break a vow to an old friend.

Hazel, Sidney, and Gerry at Balmoral, 1930s

Charles and Margaret Best still often spend weekends with the McKinnon family at the farm. Soon after Gerry's release from hospital, Charley saddles up his favourite horse, Priest, the imposing gelding, and sets off with Neil for a gallop in the snow. Suddenly, Best's horse stops in its tracks, refusing to canter between two trees; after several refusals, Best dismounts and discovers a thin steel wire inexplicably strung across his path. Priest has literally saved his life.

To fill his time, Gerry writes letters; thinking of others is one way to distract himself from spells of morbid brooding; being compulsively helpful staves off his own feelings of helplessness. People continue to be amazed by how such a busy man always finds the time to write letters of congratulation or condolence, remembering birthdays and anniversaries, celebrating their successes or sympathizing keenly with their losses. Yet each day brings fresh news of the fragmenting world situation, medical and political, darkly mirroring his inner life.

In Cambridge, my father is labouring to finish his third and final year of pre-med, needled by the tenacious pangs of homesickness. Every day, Jack feels the ubiquitous shadow of his father's reputation. How could he not, when one of Gerry's favoured protegés, Omand Solandt, a lecturer in physiology, shares his flat; or when his professors never cease to ask after the state of his father's health, to which he does his best to answer in a letter home: "Stay in there, professor, and pitch!"

To Edna, he writes: "I'm just aching to get into hospital. If I don't see a sick person soon, I'll scream. I'm so full of theory that I squeak!" At a Trinity Hall Boat Club dinner that same week, he cavorts with the lads with the usual alcoholic abandon. After dinner, deep in his cups, he scrawls on the front of a

Jack, the self-described "living lie," seated third from right with rowing confreres, Trinity Hall, Cambridge

tasselled menu the stabbing words that, years later, in my searchings for the truth of my own father, I would discover buried in a crumbling file: *John Desmond Leonard FitzGerald—the living lie.*

On a black, frigid night in that same week of early February, shapeless demons swarm my grandfather's fitful dreams. He can stand it no more. Rising from his bed, he swallows twenty grains of Nembutol, the narcotic he habitually uses to quell his insomnia, and falls into a coma, swooning on the cosmic cycles of unconscious family history. Edna finds him and, this time, it is a ghost white ambulance that carries him south to the Private Patients' Pavilion. Emerging from unconsciousness, Gerry makes out the blurred faces of Dr. Farquharson and a row of nurses hovering over him like a surrogate family. "I'm sorry to be such a burden to you all," he rasps. "Next time, I'll do it right."

Over the ensuing weeks—the brutal winter of 1939 is delivering the heaviest snowfall ever recorded in the city—he once again struggles to pull himself together, wrestling with the most agonizing decision of his life. Institutions are a way of life, his lifeblood, his gods; from Buffalo to Baltimore, from Boston to Berkeley to Brussels, he has steeped himself in the accumulated wisdom of institutions and the men who love and have loved them; for decades, he has himself, with passionate commitment, built two great institutions dedicated to the healing of the sick and the protection of the well. And now, as the lean figure of C.B. Farrar whispers the names of highly reputable mental hospitals in my grandfather's ear, the time is edging closer, time to think the unthinkable, time to embrace, yet again, one more institution; for even though he feels as fragile and trembling as an old grey mare, it is time to make the toughest commitment of all.

═

On April 1, 1939, only weeks short of graduation, my father watches his beloved Cambridge Eight—including two of his friends—beat Oxford by four lengths in the annual contest on the Thames. In a letter home, Jack reports the victory with juvenile glee. With a gaggle of hedonistic friends, he races off to the boat race ball in Mayfair that proves "simply grand." Fiercely hungover, he sails for Paris the next morning for three days of Easter holidays, coinciding with a Duke Ellington concert at the Trocadero, followed by a howling, drug-laced all-nighter with the musicians uptown. "Playing good jazz is like an act of murder," the Duke will one day famously declare. "You play with an intent to commit something."

That same month of April, my father writes his father, hoping to pull him out of his gloom:

"I was delighted to hear from Mum that you are actually driving the car. Excelsior! So the long grind is producing results. Let's hope that when I get back, we will be able to do your favourite drives to the 'Big Trees' on Vancouver Island." He reports that the rainy River Cam is overflowing its banks; he went rowing, nonetheless. "It is anything but easy to concentrate on work with the future so unsettled," he adds, "but it seems as though Adolph is not interested in a war."

On May 8, even as her father is sinking deeper into his bed of nails, Molly announces her engagement to Tom Whitley. Days later, Jack writes Edna at Prince Arthur:

> Dear Mum,
>
> I realize, my pet, that you have just about the bloodiest year in your life, this past twelvemonth, and that it has been an awful strain, what with Dad ill, me ill, Molly getting married, and all the rest of it. I'm sure that you are nearly busting under the strain. I know I would, if I were in your shoes. But, the fact remains that I am still trying to leave Cambridge with a good record. And that is going to be next to impossible if I can't keep concentrating on my work. And I certainly can't keep concentrating on my work if I hear any more about the state of utter demolition that has overtaken the family FitzGerald.

> I know that it is a grind trying to make ends meet, my pet. And I also know that I plan every penny I spend. I've even given up buying gramophone records and stopped having drinks in the Hall. And, if you think it's easy to sit here waiting for all Europe to tumble about our ears, *and* get harrowing accounts of how the family is on the rocks at home, well then, you can have another think, on me. The sooner I get out of here the better. I'm feeling enough like a criminal already, having spent all the family income. I should think that with Molly moving out that Dad will have less to worry about financially. I reckon the little angel must be worth a couple of thousand a year! I really feel guiltier every week I stay here. But damn me, I'll be independent this summer, you just wait!

On May 22, King George VI and Queen Elizabeth step from a train at the North Toronto terminal at Summerhill Avenue, specially reopened for the occasion. It is the first visit of a reigning monarch to Canada, part of a 4,300-mile, forty-four-day cross-country tour meticulously organized by Governor General Buchan. The city is ablaze with excitement as hundreds of thousands line Yonge Street to catch a glimpse of the beloved royal couple. Accompanied by Buchan, the king and queen are received on the steps of Hart House by the university president, Canon Cody; my grandfather, however, will miss the opportunity to bow before Their Majesties, for only hours earlier Neil McKinnon had coaxed him into a train at Union Station, heading five hundred miles southeast to the Neuro-Psychiatric Institute of the Hartford Retreat, a place C.B. Farrar regards as the most progressive, congenial—and exclusive—mental hospital on the continent. Patients pay a stiff fee of one thousand dollars per month, but due to his professional stature, Gerry will be granted the charity ward rate of $17.50 per week.

As the train passes through Buffalo, memories of his days at the massive state asylum thirty-five years earlier, working on brain autopsies with Adolf Meyer, flood my grandfather's mind. The nearer he draws to Connecticut, the more agitated and resistant he becomes, begging Neil to let him turn around and take the next train back home. "Oh, poor, poor man," McKinnon recalled decades later. "That's the mind. And there he was, a man who had studied psychiatry, what there was."

What my grandfather cannot know is that he continues to follow an uncanny parallel path to Clifford Beers. The founder of the mental hygiene movement has lost two brothers to suicide; Gerry's two brothers are, like himself, currently mired in deepening depressions. Run ragged by the endless quest for money, Beers had been ordered by his doctors to gear down his speaking engagements and spend more time on holiday in Seal Harbor, Maine, where Gerry and Edna have often passed their own vacations.

And now, this very month of May 1939, tortured by insomnia and flashes of paranoia, Beers is cracking up. Anxious and seclusive, quarrelsome and hypercritical, dictatorial and rude, he has been reading other peoples' mail and rummaging through wastepaper baskets to find negative opinions of himself. He has dyed his hair to maintain a youthful look and refuses to have his rotten teeth extracted. Beers likes to recite his favourite quote—"After all, what the insane most need is a friend"—but by now he has alienated most of his intimate circle.

The Hartford Retreat is suggested as a possible destination, but Beers hates his former place of incarceration—and its present superintendent—with a passion. Beers prefers the Butler Hospital in Providence, Rhode Island, and that's where he will go. En route, highly agitated and suicidal, he berates himself that he is an abject failure and has let everyone down. Within days of his arrival, he will regress to his psychotic state of thirty-six years earlier. His handsome face sinks into a grey mask of mute, cynical mistrust. He refuses medication and eats poorly; he finds sexual symbols in the food. He rants that the hospital is a "colossal fake" and the doctors "impersonators." He pleads for trips to the breezy Maine seaside, but is refused.

The man who has devoted his life to fighting the stigma, shame, and secrecy of insanity has come full circle. A pioneer of the public acknowledgement and discussion of madness, his own condition is kept a secret from a world where failures can still expect no sympathy. In the mind of the public, the association of madness with sin remains as strong as ever, frightening the normal, the decent, the sane, the comfortable.

For some time, the sixty-five-year-old C.B. Farrar, wearing his habitual face of implacable medical authority, has been considering writing a biography of Clifford Beers, but now he quietly drops the idea. Meanwhile, Gerry FitzGerald's train is pulling into Hartford, the insurance capital of America, where, C.B. believes, his friend will receive the best of all possible care.

TWENTY-FIVE

. . . and the Penalty Is Death

> If a patient of ours is suffering from a sense of guilt, as though he had committed a serious crime, we do not recommend him to disregard his qualms of conscience and do not emphasize his undoubted innocence; he himself has often tried to do so without success. What we do is try to remind him that such a strong and persistent feeling must after all be based on something real, which it may perhaps be possible to discover.
>
> SIGMUND FREUD
> *The Question of Lay Analysis*

As Neil McKinnon guides Gerry from the Hartford train station into a cab, the agitation of his esteemed chief mounts by the minute. McKinnon, the hard-headed epidemiologist, has nursed a lifelong antipathy for the pseudoscience of psychiatry; as the two doctors approach the tranquil thirty-five-acre grounds of the Hartford Retreat dotted with majestic maple and elm trees, he hopes he is doing the right thing.

My grandfather knows that the institution uniquely combines the functions of hospital, country club, and university campus—a showcase for the highest of civic aspirations. Together with the McLean Hospital in Boston and Gerry's old stomping ground, Sheppard Pratt in Baltimore, Hartford numbers among the choice ivy league sanitariums that project a silken-gossamer image of east coast, old–New England privilege. A far cry from the dingy, cacaphonous public warehouse of Danvers only one hundred miles distant, here a fleet of ten gleaming Packards and Cadillacs, driven by perfumed female chauffeurs, conducts the two hundred "guests" on excursions through the graceful neighbourhoods of the plutocracy. Gerry knows all this and more; what he may not know is that the Hartford Retreat was founded in 1824, the very year the FitzGerald family fled Ireland for the promise of a better world.

Gerry and Neil are greeted by the psychiatrist-in-chief, Dr. Charles Burlingame, a red-headed, one-man tornado who had touched down in Hartford in 1931. Blue-eyed and barrel-chested, his round, ruddy face distinguished by a moustache and rimless glasses, "Burley" is a prodigiously driven man not so unlike Gerry FitzGerald or Clifford Beers. The son of an insurance salesman of old–New England stock, he was born on the same day as Teddy Roosevelt and in many ways he resembles the former president—the booming laugh, the back-slapping enthusiasm, the incurable optimism, the unrelenting fourteen-hour workdays, the brusque dogmatism that knows nothing of feelings of delicacy, doubt, or despair. Roosevelt once famously declared that cowardice was the one unpardonable sin and the tireless fifty-three-year-old psychiatrist readily agrees.

To the soothing melodies of a four-piece ensemble wafting across the wide lawns, the trio of doctors pass a series of quaint, peaked-roofed yellow cottages complete with a wooden frame bed, flowered wallpaper, oil paintings, a stuffed armchair—all the familiar accoutrements of a comfortable middle-class home. "A mental hospital," Burlingame opines, "should be the most meticulously normal place on earth." Within six short years, Burlingame has erected more new facilities than were built in the previous six decades and has successfully courted a wealthier clientele. Regarding the word "Retreat" as symbolic of stasis—only mediocre and petty people do nothing—he has changed the name of the hospital to the Neuro-Psychiatric Institute of the

Hartford Retreat, draping it in the language of medical authority. Burlingame has also substituted the term "guest" for patient, "psychiatric aide" for attendant, "campus" for grounds, and "psychiatrist-in-chief" for superintendent; like his eminent Canadian guest, he is addressed by his associates as "the Chief."

Burley routinely spells out twelve rules for mental health, celebrating the virtues of work, efficiency, action, and idealism—the very traits that, in better days, imbued the persona of Gerry FitzGerald. Under the commandment marked "Poise," the American psychiatrist has written: "Don't become too intense. Be master of yourself at all times." He fears people are indulging in far too much introspection and brands depression as an affront to the American's inalienable right to pursue happiness. He regards rug- and basket-weaving as good only for the Hopi Indians while one typewriter is worth ten looms as useful therapy. Burlingame, however, does not entirely lack insight: He diagnoses himself as a hypomanic personality, driven by extreme energy and enthusiasm. When he and two doctor friends found a "Hypo-Manic Club," all three claim the title of president.

A confirmed organicist, Burley favours the new aggressive somatic treatments of insulin coma, Metrazol, electroshock, narcotherapy, and lobotomy. When first introduced, the treatments improve staff morale tremendously because *something is being done*; and what must be good for the staff is good for the guests. Impatient with long-term psychotherapy, Burlingame believes the resort to a culture of talking and listening means surrender to a brand of Freudian pessimism. The pioneering psychoanalytical work currently being done by Harry Stack Sullivan and Frieda Fromm-Reichmann on severely disturbed patients in Maryland's Chestnut Lodge means nothing to him; for Burley (as for Farrar), psychotherapy is simply pragmatic personal tutoring aimed at returning the sick to work. For it is education that feeds Burley's primal passion, or more accurately "re-education," both of the sick and their families, and the public at large; over forty teachers present five hundred university extension courses in every imaginable subject. Likening Burley to a medieval knight hell-bent on destroying the infidel, one colleague quips: "He is the only man I've ever known who could advance in a retreat."

Burlingame believes that if psychoanalysis comes to dominate the profession—even worse, if it is conducted by lay people—the result will sow "sterility in the field, followed by intellectual decay." Even as my grandfather arrives in Hartford that spring day of 1939, Burlingame is composing an anti-Freudian screed that will be published in the *Journal of the American Medical Association*; he mocks the "love starvation" school of thought, the idea that a child's unsatisfied craving for affection spawns neurosis. "They would have us dripping love all over the place," he declares, "until the child slips and fractures his future."

Neil keeps Gerry company for several hours but finally he must head back to Toronto. As he leaves, Neil once more urges his friend, a man quavering on the edge of a knife blade, to summon the kind of courage he has shown in the past. As my grandfather surrenders to the probing gaze of scribbling doctors, inescapable mirrors of his days as a budding psychiatrist, the irony stings like a needle.

═

Dr. FitzGerald was admitted in the morning of May 20, 1939. Physical examination revealed a white male, 56 years of age, very well developed but poorly nourished. He appeared to be much older than his actual age. There was no disturbance of his gait. His hair was markedly tinged with grey. Skin somewhat dry.

He was most cooperative to admission care. He appeared, however, to be apprehensive and depressed and stated he had lost interest in everything and was very apprehensive about his condition, and was inclined to worry about things without any apparent cause. To him, however, there really was some basis to these worries. However, during the night, he refused medication and slept very poorly. He was worried about his entire condition and was unable to rest. He received, his first day and the day following, adreno-spermin by intramuscular route and physiotherapy treatment.

He talks in a rather high-pitched voice. He is inclined in his general conversation to refer frequently to his disability and to his many fears. However, his speech is spontaneous and to the point. Occasionally, he appears more reserved and quiet. At other times, when he attempts to do things, somewhat fine tremors of his hands are noted, and although he has not shed any tears, he gives the impression that he might at any time while talking about his condition.

He can give a rather accurate and detailed history of his past life. He states that he has always feared thunder and storms, but that this has become more marked since the onset of his present illness. He states that just prior to his gastric hemorrhage and since then, he has gradually become more apprehensive, has been unable to apply himself and concentrate, has lost interest in everything in general and that he has a definite fear and anxiety about losing his position in Toronto.

His general knowledge and capacity is above average. He is correctly oriented in all spheres. His memory is good for remote and recent events, and there do not appear to be any gross memory defects. He was able to repeat five and six figures backwards and in rather rapid succession. He has a good understanding of the cowboy story and other fables.

He has a certain degree of insight in that he realizes some of his inability to apply himself as he once was able to do. However, he had no insight at all when he begins to explain his inefficiency to do things that he is asked to do and which we have a feeling he is perfectly capable of doing from an intellectual standpoint.

He has been somewhat of a shut-in personality, very much wrapped up in his work and he has not been very active outside of his profession. At one time, he used to play a little golf, ping pong and tennis, but in recent years he has been away from it entirely. He used to play bridge, but he gave it up. No avocation, except maybe a little reading, generally history . . .

No history of nervous or mental disease in his family. As far as he knows, his early childhood was normal.

Each day, my grandfather fires off a chain of letters to his extended family, to the University of Toronto president, Canon Cody, to his closest colleagues at the Connaught Labs—echoing the obsessive scribbling of Clifford Beers, confined to this same asylum a generation earlier. But Gerry's chief correspondent is C.B. Farrar.

May 23, 1939

Dear C.B.,

You are always the good friend and advisor. Were I only an equally satisfactory patient! I found the actuality of being a patient something I hadn't fully realized in advance. The control is of course necessary and in any event I asked for it. But when Neil left on Saturday evening, it wasn't very easy.

I have a wonderful young chap as my daily companion and there are two other women aides on duty in the evening and at night. But all my own limitations and inadequacies are really being revealed in the classes. I can't learn to type or do art work, or even draw. There seems to be no aptitude whatsoever in any of those directions. I seem quite unable to acquire even the rudiments of typing, drawing, woodwork, and even gardening. I am in a group of men who are expert or very proficient in all of these things and I am thoroughly ashamed of myself. I seem or rather am unable to learn, whether because of worry or lack of concentration or just stupidity, I don't know.

The days are all organized and arranged but they can't change the make-up of an individual or live his life for him. I really feel and believe that I am a greater failure here than at home. Hydrotherapy, etc., has no psychic effect on me and because of being a failure, I cannot enter into group activities. I feel that it is a disgrace being so stupid and I only wish I had come as "Mr.," not "Dr." It just doesn't seem possible that anyone as badly equipped and incapable of learning can or could be a doctor. The hospital is so fine I feel I shouldn't be here. I shouldn't inflict a long series of complaints and woes upon you and I apologize for doing so. I am still worried every day and all the time about my University position, which is natural and normal. I have slipped so badly . . .

═

June 4 psychiatric report:

Dr. FitzGerald is very much more apprehensive in the early morning in regard to his condition. He shows a definite inability to concentrate and to apply himself and requires a great deal of reassurance from his physician and nurse to get him into activities. He has a feeling that he is much inferior in his work in his classes than anyone else. He tries to do better than anyone else in the activities, and consequently he becomes very tense and agitated, and during these agitated periods he states that he would be better off dead, and that he is a disgrace to his profession, to his family, and to the Institution . . .

═

June 7, 1939

Dear C.B.,

Everything is splendid here for everyone but myself, I'm sure. I am unhappy and ashamed because I'm so ineffective and useless. It isn't that I don't appreciate all I owe to Dr. Burlingame. He is as you said, a real man. I'm the weak one. I do want to pull myself together. Please don't desert me, C.B. You are my anchor . . .

Neil tells me: "Courage till the whistle blows." You are all agreed that courage and confidence and patience are the prerequisite to recovery. You are agreed, so it must be true. . . . But there is no earthly chance of ever again having the respect or confidence of my dear friends and colleagues. I am further open to shame by asking to leave here, but remaining is so dreadful. Life is really a nightmare. What is left open to me? Nothing but stark ruin and disgrace.

You say in your letter of today, "When you make one of those disparaging remarks about yourself, all your friends feel like rising up to defend you against yourself. We feel that these adverse judgments are not only unwarranted but really morbid."

Well, C.B., I would love to be able to do that but if you really believe what you have written above, you will want to help me get over, if possible, that desire and tendency and feeling. And I can't do it here. Your writing to me is the best medical assistance I can have. Won't you keep it up, C.B., and let me leave Hartford? If in a few months I make no 'headway', then we will have to face disgrace and the prospect of no job—at 57 years of age. I don't want to die if I can overcome whatever it is I have . . .

Will the lab literally turn me adrift? Bobby Defries deserves to be in charge of the work. The work is all important. Institutions are greater than men . . .

═══

July 2 psychiatric report:

Recently Dr. FitzGerald won a golf tournament. However, he felt immediately afterwards that it was a put-up job and that he had really received the booby prize. His tendency to exaggerate any little happening in reference to himself is being deplored and efforts are being made to have him face his illness as such. . . . He often begs that his life might be ended. He states that he cannot commit suicide because he has not the opportunity and he feels that he does not have the innate courage. He is, however, being encouraged in every possible respect.

═

July 12, 1939

Dear C.B.,

The letter from Jack yesterday was absolutely pathetic, poor boy. His own spirit is fine and he appealed to me not to allow myself "to be licked" by my present trouble. But I am, C.B., absolutely I am.

I am not, absolutely not, making headway here at present. Perhaps I never will anywhere, which I fear and dread. It is nearly 75% of my trouble. Will my going away even temporarily mean that I will be forsaken by the University? You know and I know that a number of my colleagues believe I have let the University and my family down shamefully. You say I mustn't suggest such a thing so I won't write it . . .

As the summer heat intensifies, so do my grandfather's fears and pleas. The slightest of remarks or reactions he interprets as expressions of ill will; everyone knows he is a coward and therefore all shun him. Humidity has always oppressed his thin-skinned, high-strung temperament and the manic clattering of typewriters in his daily classes—he had ordered the muffling of the secretarial typewriters at the Connaught Labs, such is his sensitivity to sound—only compounds his mounting distress. Yearning to escape to the coolness of the seaside to restore his spirits, he thinks of his prize protegé, Charles Best, whom he had installed as head of insulin production at the Connaught Labs while still in his early twenties. (Best would later pay homage to his mentor with the words: "There would have been no insulin without FitzGerald.") Best is currently ensconced in Schooner Cove Farm in West Pembroke, Maine, an idyllic seaside enclave of eighty acres he had inherited from his depressed physician father; Gerry writes to ask if he may stay in Maine for the summer, but Best, following Farrar's advice, declines.

July 16 psychiatric report:

> Dr. FitzGerald has been put on extra feedings and is being given small quantities of insulin before breakfast and his evening meal. Upon his first injection, he had a slight reaction because he did not take enough food to cover the small quantity of insulin that was given him. He then began to object about the use of insulin but has since reconsidered the matter and has been extremely cooperative and eager in this effort that is made to increase his appetite, his caloric intake and thus his body weight, presently 127 pounds. It was felt that most of the insulin reaction was associated with a considerable anxiety and fearful state which arose at an evening entertainment.

═

> July 29, 1939
>
> Dear C.B.,
>
> Mrs. FitzGerald in her last letter said that the report from here said I had "ideas." Of course it is the truth. Nothing wrong at all. If I must remain here until I cease to recognize the truth, I would end my life here—meaning spend the rest of my life here. Why not admit the truth. I am not

courageous and the fact has been recognized here. So I am being dealt with as such an individual is bound to be dealt with, inevitably.

The possibility which constantly is in the forefront of my mind is that I may never return as Director of the Connaught Labs and the School of Hygiene. What then? I wouldn't plead as I do were there even the faintest doubt in my own mind. But the lock and key and aide-every-second treatment and the other things to which I have referred make it intolerable. Please come soon, C.B., and give me just a faint trace of hope that a new plan may be worked out. I want to talk to you about Jack and the family, C.B. I am dreadfully worried about their situation. Please don't wait longer, C.B.

═

On August 8, 1939, my grandfather sits at his writing desk, mopping the beads of sweat from his forehead as he composes his daily quota of letters. With a knock at the door, the slight, bespectacled figure of Dr. Farrar appears, dropping in for a brief visit while en route to a meeting in New York. The two old friends have not seen each other since Gerry's arrival in May and my grandfather's haggard face lights up. Months of badgering have finally borne fruit. "What a wonderful man you are!" he gushes, tears welling up in his eyes. "Thank you so much for your thoughtfulness and kindness."

C.B. Farrar, Queen Street, Toronto, 1930

Both men are revered leaders of their respective fields, their identities long defined and driven by professional achievement and recognition. One is collapsing like a house of cards, revealing the hollow shell within, a sight terrible for another human being to behold; the other steadfastly maintains a manly composure, projecting an image of unassailable strength and substance, the untouchable authority of the medical expert.

C.B. pulls out a list of practical suggestions to help Gerry take his mind off his troubles. The list includes negative words he must promise *never* to say or write. With a gentle, paternal pat on the back, C.B. urges his friend to buck up, to be brave, to carry on the fight, to never surrender, as if urging a soldier over the top. Farrar remains unpersuaded by the radically different approach of Sigmund Freud and Ernest Jones—the attempt to explore the meaning of a patient's inner world and the revealing transference of his emotions, originally attached to past parental figures, onto the person of the doctor. No words are considered too trivial, bizarre, or shameful to utter. But C.B.'s conviction remains fixed in marble: no possible good can come of any morbid indulgence of the ugly, unruly feelings of rage and grief, terror and longing that afflict the hospitalized patient. No one notices that the pinched, impassive face of Dr. Farrar bears a haunting familiarity to Gerry's father, the fatherless, melancholic small-town druggist, long dead and buried; and yet, like all fathers, past and present, the name and memory of Will FitzGerald will fail to infiltrate the genteel boundaries of the Hartford Retreat.

As C.B. prepares to leave his friend, he consoles himself with the knowledge that today he has done his duty to the limit of his abilities. By his very presence, he has tossed a charitable coin into the tin cup of a spiritually homeless man and, for the moment at least, the gesture seems salutary.

═

August 14 psychiatric report:

> Dr. FitzGerald has shown a great deal of encouragement and some optimism since his visit with Dr. Farrar. However, as the days go by, and that visit becomes more distant in the

past, it is noted that the guest again shows his tendency to fluctuate into depressed moods. The guest has repeatedly stated that he realizes that any suicidal attempt would be foolish and in no way would benefit him or his problems. He is now facing his difficulties more on a diseased basis than upon an ethical or moral viewpoint.

═══

August 14, 1939

Dear C.B.,

There is no substitute for physical courage. I have tried to help others. I have just had letters from Mrs. FitzGerald and Jack. I am proud to think my son is doing so well. If his father could do even a little to overcome pathological fear. I have shown my condition to everyone here, so I cannot possibly regain courage, confidence, and self-respect.

Mrs. FitzGerald and Jack wrote such cheerful notes. It will be dreadful for them if anything more happens to me. If I were retired and no longer in an institution or in any position of responsibility, surely I would have peace of mind. I must get it. I will hope to hear within a few days, C.B. I have no delusions or false ideas . . .

August 25, 1939

Dear C.B.,

I am absolutely ashamed of myself, C.B. You have been and are so fine. So is Mrs. F. Then my brother Bill has written wonderful letters. Here I am so craven and chicken-hearted and so ineffective that quite properly I am looked upon with scorn and contempt by many fellow "guests."

It is just incredible, C.B., that anyone could be so spineless and weak. I can't help this outburst because it is true. When everyone—my wife, daughter, son, brothers, sister, you, Bob, my 74 year old aunt—all do everything within their power to keep and strengthen me, I make no effort practically in return. I feel that death should be the portion of one so little deserving of all the good things life and loving friends have given. Of course when I feel as I do part of every day, no place on earth will be comfortable.

Then I want to resign because of my failure and general incapacity. Some day of course I will have to return to Toronto and, C.B., I don't believe I will ever again direct the work of the laboratories. I have no confidence in my own judgment any more. I would then be in danger of leading a second rate existence which would be intolerable. I am just as fearful as ever and cannot picture the future at all. Where could I possibly go?

I do hope Mrs. FitzGerald isn't banking too strongly on my getting back, C.B. It is really tragic. She wrote yesterday, "It will be almost unbelievable to get back to the old norm again." So perhaps she does realize that we never will be back on that old basis. If however that thought was suppressed by what she wrote later in the letter—"C.B. came to tea and we were all so glad to see his last report from Dr. Burlingame . . . for they evidently consider that you are on the up"—then I'm afraid too much is being expected.

It is now Sunday at noon and I feel utterly hopeless and helpless. Nothing I do lightens the darkness or gives me any hope or confidence. There is no use writing anymore, C.B. Thanks as ever for your patience and kindness.

═

August 29, 1939

Dear C.B.,

With the world situation as it is, it seems absurd to write about the petty worries and problems of any single individual, however important the person may be. When it is a case of useless ones who have several others giving service, then it seems worse than ever. Never did I feel my incapacity more, nor as humbly. Then too, I feel it the more because in times of national peril, individual lack of courage seems to stand out even more gravely.

I can't even contemplate returning to work. Everything related to the lab and my life formerly seems to be severed from the present me. Not even pressure or threats would change me, I think. I have done my best but I have failed . . .

With the Nazi invasion of Poland on September 1, the long-feared horror of a second world war erupts in Europe. As in the Great War, the Connaught Labs faces a sudden and crushing demand to mass produce preventive medicines to immunize legions of mobilizing Canadian troops—millions of doses of gas gangrene antitoxin, diphtheria and tetanus toxoid, smallpox and typhus vaccine. Quitting his seaside vacation in Maine, Charles Best races back to Toronto to help prepare dried blood serum for the treatment of victims of wounds and surgical shock, recruiting donors from the staff of the School of Hygiene and students at the university. Repeating the scenes of a generation before, militia units practise bayonet thrusts on the lawns of the university. As the nightmare of human history turns full circle, rank upon rank of Connaught technicians rush to fill thousands of glass bottles with blood to shore up the ravaged hearts, veins, and arteries cut open by the mass madness of war. Back in Hartford, the distress that is consuming my grandfather doubles and triples, for in his own heart he knows he can never reclaim his former position of leadership.

═

> September 2, 1939
>
> Dear C.B.,
>
> War has come and here I am so self-absorbed that I am of no use whatever. It is a year ago this month since things got so bad. I can't stay here much longer. I should really be doing war work and thinking of other things than myself but I'm not. I just seem numb. Please don't ask me to remain here longer, C.B. Should I just pack up and go home? I feel defeated all the time practically and cannot get it out of my head that I have done something wrong. Where I am at the moment is a hopeless place . . ."

═

On September 9, C.B. consults Edna in Toronto about whether Gerry should be allowed to attend Molly's wedding the following Saturday. They decide against it. The news that he will be denied the traditional paternal role of giving away the bride—together with his loss of nerve, the declaration of war, the gearing up of the lab, his worries over Jack who may have to amend his plans to enter medical school at the University of Toronto within days—all rain down on Gerry like a calculated succession of blows.

On September 12, Gerry writes C.B.: "The final solution to the problem as I see it is the one attempted last February. This time it must not be unsuccessful. A suitable narcotic would give much needed sleep. There is no other way. To make this possible, assistance is necessary. Every other alternative has been thought of and there is only that way. You can make it possible. If you come with the necessary material, all will be well.

Please bring your hypodermic outfit with usual contents. There is no other treatment for cowardice . . .

═══

September 10 psychiatric report:

> Dr. FitzGerald has recently become worried that his daughter is to be married in the near future. He was at first rather upset about this fact and could not understand why it could be done when he was not there. However, all in all, it appears that he is taking this situation in a much more common sense and controlled basis than he has ever before manifested . . .

═══

On September 16, 1939, as Gerry begs C.B., the man with the life-ending syringe, to come down to Hartford, Molly gives her hand in marriage to Tom Whitley, a rising star within the Royal Bank of Canada. She chooses to hold the wedding reception on the storied lawns of the Connaught Farm, her father's seminal creation that is revered by the press as a national treasure. To an ungracious few, the time and place of the nuptials might seem like a veiled act of daughterly defiance against the near-simultaneous outbreak of global hostilities and the shameful collapse of her father. Why marry now, when her father is not free to attend?

Molly weds Royal Bank executive Tom Whitley at the Connaught Farm in September 1939 as Gerry languishes in the Hartford Retreat

The society wedding, attended by Canada's medical elite, is covered in the newspapers, as Molly's parents' had been a generation before. A 16mm colour film camera captures the bridal party, cut into jumpy, dreamlike images. There is my great-aunt Hazel, proud of the gown she fashioned for her only niece—white and gold brocade, empire-waisted with a gold and pearl headdress. There is my father, Jack, cracking wise in his cups, standing in for Gerry as the giver of the bride; even on this happy occasion, he calls his sister by his favourite nickname for her—"Tristesse." There is the bride herself, cradling an armful of cream-coloured gladioli as she cuts through the late summer heat; there is Canon Cody, the president of the University of Toronto, manfully suppressing the agonizing pain of his bleeding ulcer, the clergyman called upon to bless the union even as Dr. FitzGerald's increasingly desperate letters form an uncomfortable stack on his desk.

Exactly one week after the wedding, the father of psychoanalysis dies in London. Suffering the unspeakable ravages of cancer of the mouth, Sigmund Freud had asked his doctor to inject a series of lethal doses of morphine into a vein even as, an ocean away, Gerry pleaded for the same deliverance; but unlike C.B., Freud's doctor had complied. Among the last to see his mentor alive, the ever-loyal Ernest Jones delivers a eulogy that deeply moves the surviving members of the Freud family. To the end, Freud insists on the vital importance of lay analysis, especially in North America, where his revolutionary discovery is "nothing but a maidservant of psychiatry." Medical doctors have no historical claim to the sole possession of psychotherapy, he believes; those deeply versed in the humanities—literature and law, philosophy and art, anthropology and pedagogy—are best suited for the work. "The therapist," he insists, "should not be a doctor, must not be a priest." And given that eight of Freud's followers committed suicide, they had better be tough as well; Freud loved to quote the myth of the eagle that forced its young to look into the sun without blinking and disowned those who failed. The sun, he said, symbolized the father.

═

In a quaint Connecticut cottage, the struts of my grandfather's psychic life continue to buckle in concert with the wider world. The news is bad on all fronts: the Rockefeller Foundation has cut off all funding to the international movement of mental hygiene that Gerry had spent twenty years supporting. Unlike the Connaught Labs, the movement lacks an endowment to sustain itself. The systematic prevention of insanity is seen as too diffuse and grandiose, an idea a generation ahead of its time. A Rockefeller official puts it bluntly: "It should have done fewer things better." Even as the lifeblood drains from the movement, its founder, Clifford Beers, is spiralling down into his final madness.

Simultaneously, the tentacles of war are strangling the global dream of public health reform. As a bulwark against leftist extremism, the Rockefeller Foundation has recently opened a new School of Hygiene in Prague and installed the Johns Hopkins–educated Czech, Hynek Pelc, as its first director. But with the Nazi coup in March 1939, Pelc is quickly replaced by two Germans; some time later, my grandfather's Slavic counterpart is taken out on the front steps of his new school and shot.

Even as my father enters medical school, even as my grandfather suffers in an exclusive American psychiatric hospital, Adolf Hitler orders the launch of Aktion T4, a systematic, medically conducted cleansing of the German national gene pool. Nearly half of all German doctors have joined the Nazi Party, and physicians and psychiatrists compose the single largest professional group within the party. The high principles of preventive medicine—*vorsorge*—are cited as justification for eugenic sterilization programs of "racial hygiene"; under the smokescreen of world war, German psychiatry does not hesitate to shunt the locomotive of public health idealism engineered by my grandfather's generation—holistic, government-run programs to improve the health and higher good of the nation—down the dark track to the death camp.

The first test killings of German mental patients by carbon monoxide poisoning take place in the Brandenburg Sanitarium, managed as a branch of the Rockefeller-funded Kaiser Wilhelm Institute of Brain Research. "Operation Mercy Killing" is organized exactly as Alfred Hoche—the man who had worked in the same Freiburg autopsy room as my grandfather and deeply influenced C.B. Farrar—had laid out in 1920. The technical expertise gained from the extermination will be later transferred to the vast warren of concentration camps.

A lifelong Germanophile, my grandfather, of course, is unaware of the Nazi perversion of his own fervently held ideals; unaware, too, that had fate confined him to a German mental hospital, he would not have lived out the year, murdered by gas or lethal injection along with the entire population of German psychiatric patients—seventy thousand in all.

═══

On September 19, Dr. Farrar visits Gerry in Hartford for a second time, his presence once again lifting his friend out of his gloom. But the relief is short-lived.

Desperate to escape Hartford, Gerry asks to be transferred to Homewood sanitarium in Guelph, where he can be closer to his family; in his wildest imaginings, he would not guess that eventually both his children will be committed to Homewood. His legendary talent for persuasion is failing him, for C.B. refuses. With each refusal, Gerry feels the seethings of his disturbance deepen, a torturing, blood secret that both taunts him and stays hidden from his conscious mind. He mentions that he has written a "bad" letter to his wife who, following C.B.'s advice, has resisted his pleas to visit him in Hartford. In his shame, he will not mail the litany of despair, instead passing it on to Farrar. In the letter to Edna, he speaks for the first time of committing some unspecified sin.

> Mum darling,
>
> People here in the cottage, where I am and where I am obliged to sit at table, simply ignore me. They may or may

not nod when they come for lunch and dinner. Then they carry on conversations and simply leave me out. Others look directly at me and ignore me or look contemptuously and then turn away. Obviously I have been sent to Coventry.

Most of the doctors and aides behave similarly. Now all this can mean only one thing—a general determination (among those with whom I have to associate) to ignore me and show scorn and contempt.

You can imagine what it is to have to live in this even though I really deserve it. I must have been so cowardly that this is the result. Now both Dr. Farrar and the doctor here to whom I report daily insist that I imagine this. Of course I don't. It is true I suppose I should destroy myself but I just can't do that to you and Molly and Jack and all my family and the relations generally. It is dreadful.

If I go home, won't it be the same there because at least one member of the Board of Governors is here as a patient and he now avoids me when he sees me coming. You see how serious and dreadful it is. You see, I can't do well because of the remorse from which I suffer for wrongs done you and the family. And you know, Angel, remorse is a disease which God can't cure, much less man. I don't know whether I have committed some sin which makes me like this. Perhaps so.

I hate this place or rather hate myself for having to face such conditions. While I may exist for years yet, I have really ceased to live. And to think that not long ago, I occupied such honourable, responsible, and important posts. It is cruel, darling, because you say C.B. told you I was improving and it gave "your heart a lift." But I should die at once. I am an ungrateful, hopeless wretch and I am not improving. I have wronged you beyond redress. I know I have committed the unpardonable sin.

═

September 22, 1939

Dear C.B.,

You know, C.B., no one recovered from loss of courage—no one, no one. That's why in the Army they shoot them. They contaminate others as well as being a burden themselves, in the bondage of fear. I have been so restless and depressed all day. The letters from my friends make me feel so terribly ashamed of my dreadful weakness and incapacity and failure and I have committed some dreadful sin. Please, C.B., I can't remain here. Naturally people won't speak to one who has committed the unpardonable sin. It explains all the difficulties that I have . . .

═══

September 24 psychiatric report:

Dr. FitzGerald was transferred from a room on Terry II to the group nursing room on Butler I. The transfer was made because the guest was not able to meet the social requirements and demands which were made in a very pleasant environment of Terry II. He took his transfer with a considerable amount of resistance at first, but since then has accepted the situation in a much more reasonable manner. It is thought that perhaps another course of tincture of deodorized opium might be tried again because of the beneficial effects secured the first time.

═══

October 2, 1939

Dear C.B.,

I have ruined my family. I have committed an unpardonable

sin so I probably deserve all I receive. But to awake in the morning, get up, and dress at the behest of a callow youth and with three other men is not in my opinion likely to aid anyone. It certainly won't help me. I'm resentful and so are the men aides. I will get out just as soon as I can. . . . I can't get the idea out of my head that I have committed an unpardonable sin but I don't want to remain here. It isn't just regret and remorse, though they are bad enough. It's the other conviction that frightens me.

═

On October 8, Tom and Molly, the honeymooning newlyweds, try to visit Gerry as they pass through Hartford. But overcome with shame, he refuses to see them. As he confesses to C.B.: "I couldn't risk making them unhappy on their honeymoon. It was something I shall never forget nor forgive myself for. How can I?"

═

October 8 psychiatric report:

Dr. FitzGerald has continued to show a considerable amount of anxiety and depression. . . . Today he refused to attend his classes and refused to come to the office of the doctor in charge of his case for an interview. He did this because he thought that if he stayed in the hall he would not be subjected to the criticisms and laughter of the various people about him. Every effort is being made to overcome his stubbornness and the fixed ideas which surround his melancholy.

═

October 15, 1939

Dear C.B.,

I know you are my dear friend as well as doctor and I know

you are not deliberately having me suffer more than is absolutely necessary. I know I am guilty of the unpardonable sin but it is nothing less than sadistic to have me live in a room with two others who deliberately torture me. If I can't kill myself please let me be kept somewhere where I won't be deliberately subjected to continuous and studied and constant and aggressive insult, not only against me.

When I had a room alone, I at least had a few minutes peace by closing my door. Now that is impossible. Even when you have committed an unpardonable sin, cowardice, and desertion, you might have a room to yourself in an institution.

So for all these reasons, I again plead with you to please let me go to Butler if I can't go to Guelph or Whitby. Surely when I need to be protected from myself I can be in a place where if I keep my mouth shut (which I try to do here and promise I will do at Butler), it will be better and perhaps I will have a single room and so not contaminate anyone else. The wee bit of food I eat could be given to me on a tray.

I have disgraced my family. I have done everything that is dishonourable and I never expect again to be well and happy. Winter is coming and so it will be hell. Remember, I deserve it all and more. You and Bob are standing between the family and stark ruin. There is no doubt of that, C.B.

You once said my letters were exaggerations, C.B. This is not true at this time here. Jack's pleading letter of last July is my constant hell. Then a letter from Lorne Hutcheson, [the comptroller of the Connaught Laboratories] on Saturday, taken with Jack's letter—there is surely nothing in my hell now or in the future to equal that.

Now for Christ's sake don't, please don't, insist on my remaining here. You know I would far rather be dead. I only wish I were dead but please don't insist on my remaining here. Even one who has committed the unpardonable sins can only be asked to bear so much. Perhaps my cross is not heavy enough . . .

October 22 psychiatric report:

> Dr. FitzGerald has adjusted himself fairly well to the group nursing room on Fuller II. He continues to have numerous ideas about an unpardonable sin that he has committed which explains all of his feelings of the attitude of other people toward him. He still attends his classes although he requires much more persuasion and urging to remain active. In all of his activities, the hopelessness which he expresses so readily colors his behavior.
>
> The guest is receiving 10 units of insulin twice daily and extra feedings. He now weighs 131 pounds. The male sex hormone has been stopped.
>
> He has shown a considerable amount of agitation recently. He has also shown a tendency to have repetitive speech. When interviewed recently, the guest said:
>
> "I don't want to get well. I don't want to. I don't want to. I don't want to. I don't want to get well. I don't want to, I tell you. I don't want to . . ."

In the words of Freud, repetition is the price we pay for repression, and the repetition of the ancient FitzGerald family drama continues to unfurl in the microcosm of my grandfather's letters. He has reached a crisis point, a fever-pitched moment of truth, identical to the one his son, my father, will meet thirty years down a common path. As I read and reread the letters, I feel as if I am standing at the crossroads of my family history: what happens here, now, will infuse the fates of future generations. A crisis can become an opportunity, a path to salvation, but not here, not now, not this man.

As Gerry's agitation builds into a frenzy, I feel my own rage rising in my chest, twisting into flaming darts aimed at the cold reptilian brain of psychiatric science. I feel a fierce desire to smash through the prison wall of the past tense and burst boldly into the present. I know Gerry can't hear me, but he is dead-right not to want to "get well." Who defines what is "well"? The doctors continue to dismiss his "unpardonable sin," his cowardly "letting down of the side" as repetitive and delusional, lacking symbolic weight; yet surely the phrase means something, surely its deeper roots can be exposed and loosened through patient talking and listening. I sense what he is thinking and yes, it is a terrible thought: *If this is the best hospital in North America, if these are the best trained doctors, then the conclusion is inescapable. If they can't help me, who can?* I know there is nothing I can do to help my grandfather, the self-appointed, Christ-like saviour of public health who cannot be saved; nothing I can do to stave off the fate of my father. Even so, I feel I must keep trying, keep listening, even if only in my dreams; I must press against the bottomless depths of my own powerlessness, and save myself.

The doctor reaches for a syringe, jabs the point in Gerry's arm, and squeezes a tincture of deodorized opium into his bloodstream. It is a familiar, familial scene, one played out before him countless times as a boy back in Harriston—his druggist father, Will, slipping a needle under the soft skin of his invalid mother. As Gerry's head sinks back on his pillow, my protests from the future fall on deaf ears. I am a mere layman, lacking a medical degree; who am I to question the trained experts, so dedicated to their difficult work, their faultless willingness to help, doing whatever it takes? In this hospital room, where a grown man has been reduced to a powerless child, strong feeling is taboo, and slowly the burning in my throat dies into ashes, fusing with the blackness of my grandfather's silence. Fixing Fitz is futile.

═

October 25, 1939

Dear C.B.,

You say my reasons for wishing to move are sick thoughts

and that I won't like to hear it. That you are my old friend and wish to help me is certainly true. Now I know you can't take my sick thoughts away or give me courage. You will agree with that, I'm sure. Unfortunately, neither can I. But there we reach a blank wall. I will not call myself any names, beg for pity or show self-pity, nor indulge in self-blame. But your closing line which says, "the only one against you is you!" makes it clear that I'm to try and make myself ignore what it is impossible to escape or ignore for a moment except when I'm unconscious in sleep . . . God is love but I am cut off from God's love because of my sins, unforgivable and unpardonable . . .

═

November 14, 1939

Doctor Farrar,

I am very sorry I can't say, "Dear C.B." Today an aide deliberately tortured me in the toilet and soon they will make it impossible for me to urinate. But you had it in your power to take me out of this place which may be wonderful if you are not the victim of a horrible conspiracy as I am. You have made me desperate. Every night I am locked in at 5:30 and then when I asked to be allowed to try one other place, you were adamant though I told you things that were true and you said they were delusions.

Dr. Angus and Dr. Kennedy saw me yesterday but it did no good, just another lot of injections—useless. They want to drive me into suicide if they can. Even if I have committed the unpardonable sin, they have no right to do what they are doing. There is a conspiracy here among the aides to make it hell for me.

Here I am leading a life that drives one to wish for death every night and you know too about the horrible doghouse.

You see you not only ruin my faint hopes but what is more dreadful make it impossible for me to acknowledge that I did commit a horrible, unpardonable sin. That is worst of all and as far as can be seen goes on indefinitely. If Dr. Cody can't or won't help, surely the Ontario Government can. It seems dreadful that you will not visualize my life here and see how I am paying retribution and ruining the lives and happiness of others. So all I beg and pray is that I be allowed to do so elsewhere, far away from everyone.

If there had been any hope, I would not have written to Dr. Cody. Now that is done, surely you will let me leave here without further pleading. Remember, this is the substitute for suicide, the death penalty. I would rather be a penniless wanderer than attempt to go on with this. This is the end.

═

November 19 psychiatric report:

Dr. FitzGerald has been begging to have Mrs. FitzGerald come and visit him because he feels that he might be able to convince her if she were free from the influence of some of his friends in Toronto. . . . He had one poor day in which he wrote a letter to President Cody of the University of Toronto and mailed it without first acquainting the physician of his plan. Fortunately, it was possible to catch this letter in the mail, and it has been sent to Dr. Farrar.

═

December 15, 1939

Dear C.B.,

What in Christ's name is the use of your insisting on my remaining here under these conditions? Surely you want to help me. At least you say you do. Now prove it. Just to be told that I must remain in the doghouse is almost as bad as having to spend Christmas under such hellish conditions. Surely I have suffered enough . . .

═══

December 17 psychiatric report:

Dr. FitzGerald has been very tense, agitated, and is beginning to feel that an organized plot has been arranged against him. He has been working in horticulture and has really been accomplishing something there although he is very critical of his own results. His attitude toward his own physician and family varies from one of intense bitterness to a rather pathetic desire for sympathy. At one time, he says that they are the arch enemies of the plot against him and are preventing him from doing the only thing that might possibly help him. At another time, he says that they have been far too kind to him, and he wonders if they will ever condescend to see him and talk to him again after the way in which he has behaved.

═══

Over Christmas, my grandfather suddenly finds religion. He pleads with C.B. to find him a sanctuary in the Catholic Church, the faith his Protestant ancestors had abandoned 250 years earlier at the Battle of the Boyne. For Catholics practise something he needs: the liberating ritual of the confessional.

Gerry writes his colleague, the devout bachelor Dr. Defries, holding the fort at the Connaught. "Please do pray for me every day, Bob. Nothing else matters. I will spend the rest of my life doing penance and in prayer and I will not be in the old life again. . . . I am prepared for poverty and penance. I have brought my family to ruin and disgrace and soon to poverty. Please, prayers Bob. You have no idea how I really need prayers and supplications."

But Gerry is denied his newest wish, for his treatment is now taking a radical, and ironic, turn: insulin shock. In preparation, the psychiatrists prescribe a course of Percorten, but due to the drug's expense, Dr. Burlingame asks Dr. Farrar to arrange with Ciba, the private manufacturer, to supply it free or to invite the Connaught Labs to pay for it themselves. And so the latter is done—Gerry FitzGerald's dream of free medicines for the masses realized in an unexpectedly personal way.

On the early morning of January 6, 1940, two burly attendants strap my grandfather to a gurney and wheel him into the insulin coma ward. Insulin, the same magical substance that first flowed from the dream life of his friend Fred Banting nearly twenty years earlier; insulin, the same miracle drug that had made the reputation of his lab; insulin, a scientific achievement of near-religious import. If only Fred could see him now.

The needle slips into the vein; then come the violent convulsions and thrashings, the falling into coma, and the pulling out by glucose solution. The clockwork cycle will be repeated for weeks to come. Knowing that his team of psychiatrists must grapple daily with the harrowing experience of administering insulin shock, the protective Dr. Burlingame has consistently praised their selfless heroism. "There are few scenes that match the dramatic intensity of the insulin treatment room in a modern psychiatric center," he writes. "This skillful sparring with death, where a few moments of neglect, inattention, or inadvertence may cost a life." He readily admits that the human system "takes a terrific battering during the process of insulin shock. If it were not for the amnesia concerning these episodes, it would be difficult to get anybody to submit willingly to the treatment . . . one cannot help wondering what damage is possibly been done to parts of that body. Sufficient information is not yet available to answer that question."

Yet to his dying day, Burlingame will never cease to laud the brave new regimens of lobotomy, ECT, and insulin shock, if only for one, fundamental reason: "No longer does a psychiatrist *feel helpless* when a depressed or suicidal patient comes to him." For what could be worse than a patient triggering feelings of impotence in a doctor, the unassailable expert who is expected to know all the answers? Not for a moment does Burlingame stop to wonder why he insists such ugly, unbearable feelings must be eradicated in those under his sway; for perhaps he does so, not so much to relieve the helpless patient, but to protect the helping doctor.

═

January 6, 1940

Dear C.B.,

I started insulin today. It will do no good. I can never come back. The doghouse is so bad I can't read anymore. I am pressing to see the church representative who will come, I hope next week. I will tell him everything and press to be taken at once to a Catholic refuge. I will then send in my resignation for everything. I can't go on any longer and despite being in mental hell all day, I will go on living in a Catholic institution rather than resort to the only other alternative. Please, C.B., do all you can to get my wife to agree to this . . .

═

January 14, 1940

Dear C.B.,

The insulin treatment has been on now for eight days. Two days ago I refused and was dragged upstairs. You see Dr. Stolzheise told me I couldn't see a representative of the Catholic Church while under his care.

C.B., I have never confessed to you as I should have done when I first came under your care last year or been completely frank. That is why you think I have a condition from which I can benefit by further medical care. You ask why I can't recover as my brother and my cousin, Dr. Armstrong, did. You know that they were never in the doghouse and never had the past history I have.

I am absolutely in hell all the time now. I expect to continue to be also as long as I live but this is the position. You are my doctor. Dr. Burlingame will be guided by you. He will probably not allow Dr. Brennan to hear my story as a personal confidential matter. He will tell me tomorrow, Monday evening, about that. It is about certain that it will be that anything I wish to say must be to the doctor in charge and put in the hospital records. I simply can't have this material put in any hospital records.

Now I implore you, C.B., to come just as soon as you possibly can. Please don't delay any longer than you must. I realize you have been here very recently but only for a couple of hours. If it is humanly possible, I would wish you to give me one whole day or even two if you could to hear the sordid, shameful, disgraceful reason and then to advise me as to what to do. I can't read anymore or do anything useful or constructive. It is because of the fearful hell in which I am—not only the doghouse but everything else—that I do beg you to come and let me tell you the story and then hear what you think I should or might do in order to minimize the shame and disgrace and dishonour to the family as much as possible.

You see, as soon as you learn the truth from me, you will agree at once that neither insulin or any other human treatment will benefit me. I have not spoken to Angus or any of the others about this confession. I can't. I don't go to see Dr. Stolzheise because he thinks I have some endocrine disturbance whereas it is something entirely different. I have had eight insulin treatments, up to 70 units daily. I sweat and go through the rest of it but it cannot get me out of the doghouse or benefit fear, shame, and remorse. Never did I need you as I do now. Oh, C.B., please come soon.

═══

January 28 psychiatric report:

Dr. FitzGerald had his 19th insulin treatment yesterday with 180 units in divided dosage. He was quite cloudy through the later part of the treatment and perspired profusely. He did not go into shock and was able to drink his sugar solution at 11:30. In the afternoon, he was shaky and somewhat tired. He came over to the medical officer's office for an interview and upon arising from his chair he showed cyanosis of the fingers and lips and stated that he was cold and shaky. He was returned to the hall immediately and put to bed. Later in the evening, he ran a temperature of 101 degrees which subsided to 97.8 in the morning. This morning, the guest felt quite comfortable and was allowed to be up and around the hall. Emotionally, Dr. FitzGerald shows marked depression and self-condemnation, though his positiveness in this regard is somewhat less.

═══

On the morning of February 6, as Gerry surrenders his body to yet another battering of insulin shock, John Buchan—the man whose farewell, fatherly handshake precipitated my grandfather's dramatic collapse at the Connaught Farm only months earlier—lowers himself into a hot bath at Rideau Hall. When he arises, the Governor General slips and bashes his skull on the side of the tub; bleeding and unconscious, he lies undiscovered for an hour. An emergency trepanning—the cutting of a channel in his skull, a stone-age technique once used on the mad—is performed, but even the skill of the eminent neurosurgeon Wilder Penfield cannot save him. "The Last Victorian" dies six days later; when C.B. visits Gerry in Hartford shortly afterwards, he chooses to withhold the distressing news from his friend, who is showing signs of rallying.

Indeed, by March and the melting of the winter snows, my grandfather's attitude starts to shift. He no longer speaks of unpardonable sins; nor does he pine to withdraw into a Catholic monastery, an appalling prospect for the Catholic-loathing Farrar. As I struggle to read Gerry's mind between the lines of his letters, I wonder if the ordeal of insulin shock—the tremors, the sweats, the sobbings, the ravenous hunger—has, temporarily at least, made him more compliant, or perhaps satisfied his feelings of guilt and the craving to be punished. He no longer speaks of being an object of a horrible conspiracy, even though he was, in a fundamental way, dead right: everyone—doctors, nurses, aides, friends—seamlessly united in their dismissal of his own strongest thoughts and feelings. For my grandfather is now reversing his field, propelled by a newfound, crystal clear realization—that if he "gets well," if he behaves in a prescribed, agreeable way, if he conforms to an appearance of health, the doctors will grant him his fervent wish and let him fly free of Hartford, like a wasp escaping a bottle. It is a solution of such stunning simplicity, he is amazed that he didn't think of it sooner.

═

On March 19, 1940, Burlingame writes Farrar: "Of course you and I know that it is extremely unwise to throw our hat in the air because a man who is still under insulin appears to have cleared up. He may take a tumble. On the other hand, we have got to take a chance. I think the important thing is not to have any gap between his leaving here and going back to work. It sounds like a bit of a gamble but I'd feel uneasy about his rocking around on any vacation in-between. I believe that sojourns in the woods and all that would be bad news. What he needs is to go back to work . . ."

On the first day of spring, a euphoric Gerry writes C.B.: "I completed the insulin course today—57 treatments in all. Weight this morning 145 lbs. I was wrong when I said the treatment could do me no good. It has done me good!"

Over the next two weeks, he makes daily visits to downtown Hartford to shop, read in the public library, and go to church. On April 7, 1940, Dr. C.B. Farrar, Canada's pre-eminent psychiatrist, boards a train down to Hartford to pick up my grandfather and together they return to Toronto, buoyed by the spring breezes and the Easter myth of Christ's resurrection. Gerry arrives home on Prince Arthur Avenue in time to mark his thirtieth wedding anniversary with Edna two days later. In a cheerful farewell letter, Dr. Burlingame tells Gerry, "I'm glad your troubles are now permanently behind you." The American psychiatrist is accustomed to ending his correspondence with a signature phrase—"Yours without a struggle." But this time, he leaves it off.

TWENTY-SIX

Damn Clever, These Spooks!

> A man had as soon go to bed with a razor as to be intimate with a foolish friend.
>
> MARQUESS OF HALIFAX

As he had done in the fallow months before he beat his retreat to Hartford, Gerry pours out his heart to those nearest at hand. At work, it is Neil and Hollie McKinnon, tucked in the rural serenity of the Connaught Farm; at home on Prince Arthur, it is Edna, hiding her mixed feelings at her husband's return. Determined to return to work after his eleven-month exile, he promises Neil, "I'm going to be a man and conquer myself." But the willpower that once kept him going at all costs is being seduced by a longing to give up; he knows he can never be the man he once was. He is no longer the conqueror but the conquered, locked in the grip of what he thought he could grasp.

As the acting director of the Connaught, Bob Defries has given Hollie, grappling with a depression of her own, a part-time job; Gerry often asks to sit with her or take a walk in the woods. If he still believes people—not just any people, but the eminent mental experts of the Western world—can help him, the hope is fast draining away. Hollie confesses to her husband, "I don't have three children, I have four . . . he's a sick man." As Neil tartly observes, "She knows infinitely better than the doctors know."

The Connaught Labs is scrambling to immunize hundreds of thousands of Canadian fighting men. Over sixty members of the university teaching staff, mostly from the medical faculty, have signed up for full-time military duty. Lamenting that his expertise will not be needed overseas, Gerry forces himself back into the endless rounds of meetings and greetings, telephone calls, and letter writing of years past. In early May, he spends a weekend in Baltimore and Washington, D.C., seeing a chain of men in the government departments of health and the School of Public Health at Johns Hopkins. On his return home, he writes the executive of the Rockefeller Foundation, proudly citing the collective international influence of the 198 graduates of the School of Hygiene. A Rockefeller director responds to Gerry: "Your many friends in this country as well as in Canada have been made very happy indeed to know that you are back on the job, hale and hearty, and working like a Trojan."

With the war on, Fred Banting is working harder than ever, driven to prove himself, yet again, the man his father wanted him to be. That same week in early May, after a test on the eyes of rabbits, he spreads mustard gas on his thigh to gauge the effect of an ice-pack treatment, a secret formula that a refugee German chemist had recently turned over to him. Banting leaves for his home at 205 Rosedale Heights, taking the antidote with him, as mustard gas does not take immediate effect. The next day, as he tends his rose garden, a grass fire breaks out and the ice pack melts; he continues to work with a badly blistered, swollen leg. The deep wound takes six weeks to heal; the ugly six-inch scar on his thigh he dubs "the holy terror." In this new war, Sir Frederick Banting says, a man must be prepared to suffer all manner of pain and indignity "if he desires to have the privilege of the name of a research man."

On May 9, the Nazi blitzkrieg shatters the months-long calm of the "Phoney War," unleashing unholy terror across western Europe; the following day, Winston Churchill is made prime minister, offering the beleaguered British people nothing but blood, toil, tears, and sweat. Four days later, Gerry travels to Ottawa for three days of meetings of the Dominion Council of Health, as he has done so faithfully, twice each year, since its birth in 1919. The sixteen federal officials plough through a lengthy agenda, discussing the crippling rate of tuberculosis among the native population, the curbing of prostitutes pouring into Canada from Asia through BC, a twenty-year study on World War I veterans suffering from venereal disease, purification of oysters, and the control of excessive noise and vibration in Canadian cities. Gerry volunteers to chair a committee on wartime industrial hygiene. The chairman reads out an appreciation of seventy-three-year-old Dr. John Amyot, the country's first deputy minister of health who has recently died in Ottawa after a long illness. Gerry feels the loss keenly, for Amyot was his old ally who stood behind him in the pioneering years of the Barton Avenue horse barn—"the miracle in a stable."

Three thousand miles away in Vancouver, Claude Dolman, head of the lab's western arm, sympathizes with the travails of his rallying chief; when he invites Gerry to stay with him, C.B. encourages the move. But Gerry responds hesitantly: "I am endeavouring to get back into harness again. But with so much uncertainty, I find it impossible to make any plans at the moment . . . the absence of those in active service has thrown a very heavy load indeed on the senior members of the staff who are more than fully occupied. I am really in doubt as what is best to do . . . the war news is so bad, it makes it extremely difficult to decide." He does decide, however, to visit his brother Sid, who is struggling to pull himself out of a three-year depression, much of it spent in a rest home, listening closely, like his brother Gerry, to the reassuring bromides of Dr. C.B. Farrar. Like his brother, Sid is now struggling hard to re-establish himself in his work.

On June 14, the Nazi war machine goose-steps into Paris and hoists the swastika on the Arc de Triomphe and Eiffel Tower; the Rockefeller Foundation evacuates its staff from their offices at rue de la Baume where workers were struggling to create a workable penicillin. Like a plague of murderous germs, platoons of steel-helmeted Nazis converge on the gilded gates of the Pasteur Institute. Here, the sixty-four-year-old Joseph Meister is serving as caretaker—the same Joseph Meister made world famous at age nine when Pasteur saved his life with the first rabies vaccine. The loyal Alsatian tried to bar the descendants of the hated Prussians from disturbing Pasteur's sleep. Powerless to stop the Nazis from opening the door of the crypt, he flees home in despair. Two days later, Meister takes out his World War I service revolver and shoots a bullet into his brain.

That morning, Sunday, June 16, 1940, the sound of church bells toll through the dead streets of the city of Toronto. At his home on 18 Prince Arthur Avenue, my grandfather paces up and down the worn carpet of the living room, wringing his hands in a frenzy of agitation. Even in his clouded mind, he knows it is Father's Day, for he is habitually sensitive to meaningful dates and anniversaries, the synchronistic, symphonic collisions of human fates, the mysterious and sublime patterns that elude and defy the cold clockwork gears of rational science. He knows, too, that the Canadian Medical Association, an organization in which he has played a prominent role, starts its annual convention at the Royal York Hotel the next morning, but he cannot bear the prospect of putting forward a false face to colleagues he imagines as uniformly judgmental of his true inner state. And perhaps, even then, on this day of all days, Gerry is living out a line in a letter that his only son, Jack, struggling through his last term at Cambridge, had written Edna a year earlier, an enigmatic line encoded with shades of unconscious forboding:

"Dad is going to realize an old ambition next spring. Damn clever, these spooks!"

Standing by the living room window, my grandfather parts the curtains and gazes into the empty street, muttering that the University of Toronto is out to get him. Suddenly, he is engulfed by a rising wave of anxiety, the dreaded black panic he can no longer suppress, spiked with shafts of paranoia. The blitzkrieg that rages in Europe rages inside his own breast, here behind the cowardly yellow bricks of a silent Toronto house, as mournful church bells knell; it's a war he has declared on himself, an internal war as civil as it is savage.

As if tenuously held up by a hangman's wooden scaffolding, his body buckles and splinters under an intolerable weight. All that was long concealed is now revealed. Ascending the stairs to the bathroom, he opens the door of the medicine cabinet, oblivious of his reflection in the mirror, and removes the familiar blue bottle of Nembutol. He shakes out a handful of tablets, stuffing the empty bottle in the pocket of his dressing gown. Swallowing the pills, he lies down on the bed—not his own bed, but Jack's. There is always room at the top, he has never tired of saying, but here, now, in this room on the third floor, only doom awaits.

Once again, someone is meant to rescue him from the lip of oblivion; and this time it will be that very son, back home from his first year of medical school. For it is Jack who, like his daughter thirty years in the future, must climb the stairs to the room at the top of the house; it is Jack who, like his daughter thirty years in the future, must find the father here, sprawled on his bed, unconscious as death, his chest heaving, starved of air, his face as blue as a painted Celtic savage. Once again an ambulance, its banshee-siren wailing, must speed the father down University Avenue to the Toronto General Hospital, the place his son will meet him thirty years on.

Once again the compassionate visage of Ray Farquharson hovers over his bedside; once again, the cowardly pills have failed to do their duty. And once again, the coils of horror and grief, screaming for release, slip into the vast, frozen pool of silence where generations of abandoned children sleep. The father wants to die, he wants to be saved, he feels compelled to drag a living child down with him; yet even if his body survives his darkest desire, nothing in life can redeem his starving Irish spirit. In the chaos of the unconscious, time does not exist; the year is 1940, the year is 1970. The wheel turns; the child is the father, the father is the child.

Over the following days and nights, my grandfather slowly recovers. Long a believer in the healing power of sympathetic listening, Dr. Farquharson feels that no patient should ever be told that nothing can be done, no matter how desperate the situation; and his friend Gerry FitzGerald is no exception.

Lying in a brass bed in a private room in the Private Patients' Pavilion, Gerry turns his head on his pillow and, buried in a dark reverie, stares at the lace curtains billowing like sails in the humid breeze. He gazes out the French windows, wreathed in ivy, that he knows overlook the epicentre of his professional field, a world lost to him now. Buttoned down in a blue and white fine pinstriped dress, white cotton apron and bib, celluloid collar and winged cap, the night nurse glides into my grandfather's room like an angel of mercy and with practised efficiency pulls out her pocket watch and feels his pulse. If Gerry had been infected with diphtheria or smallpox, rabies or syphilis, he would have profited from the miraculous treatments he had done so much to champion and refine; but no lab has yet made a vaccine to stem the spread of fear and dread of emotional affliction; and that failure seems the unkindest cut of all.

Gerry's friends, all consumed by the dire war news and many rushing off to enlist, remain in the dark as to his true condition. Many, like Banting and Best, are furiously busy, and besides, they nurse troubles of their own. A man gripped by a suicidal melancholia is a drowning man; even the strongest of allies fear being dragged down into the swirling vortex. To expose oneself to the miseries of a friend means to risk feeling the sting of his anguish; even if they found the time to visit, this patient, like themselves, is not an easy man to comfort, an habitual caretaker who, even as he asks for it, reflexively pushes away the proffered hand.

One day passes, then two. Sitting up in bed, Gerry summons the strength to listen to the radio and scan the newspapers. On June 18, as the free world cowers under the dark wing of the gathering apocalypse, he hears the stirring broadcast of the freshly anointed British prime minister, Winston Churchill, a man unusual for admitting to a vulnerability to the "black dog" of depression, rallying his nation to fight the Battle of Britain. If we fail, Churchill thunders, then the whole world will "sink into the abyss of a new Dark Age made more sinister, and perhaps more protracted, by the lights of perverted science. Let us therefore brace ourselves to our duties, and so bear ourselves that, if the British Empire and its Commonwealth last for a thousand years, men will still say, 'This was their finest hour.'"

The next day is sticky and humid, the mercury rising into the eighties. An electric fan hums at my grandfather's feet as a sheet of rain drums on the French doors of his balcony; lightning forks its serpent tongue against the surly sky, and as each thunderclap bursts and rumbles, low in the belly of a vengeful God, he trembles like a child.

Unable to read or concentrate, the chronic insomniac yearns for the arms of Morpheus, the Greek god of sleep, to quell his fears stoked by the hateful June heat. The reaping of dreams, the royal road to the unconscious, the contrary voice of Freud boldly declares, may lead to the possibility of healing and salvation. But the grave weight of the culture urges my grandfather in the opposite direction—dreams are useless, even crazy, no more meaningful than the crackling static that spills from his bedside radio. So says secular science and so says the Christian religion. And if he is told that dreaming is a sin, so is the terrible act of self-murder. Because of the impossibility of repentance, Thomas Aquinas denounced it as the one unforgivable sin; in the Middle Ages, the corpse of the suicide was dragged to the edge of the village and abandoned at the crossroads with a stake driven through its heart. In Dante's *Inferno*, suicides were cast into a dark forest deep in the seventh ring of hell, below the greedy and the murderous.

But here in his living purgatory, my grandfather's dreams come, all the same, the heaviness of his caged feelings pressing like an incubus on his chest. Freud once said that a man's disturbance rides the strongest horse in the stable; and out it charges, leading the armies of days and nights, memories sweet and savage, three lifetimes packed into one. We all carry voices inside our heads; who can say which are our own? And so, pulsing through the anarchic blood of his brain, the spectral figures surge, one by one: the staunch men of bearded authority, born under the starched skirts of Queen Victoria, filing past his brass bed, a line of ghostly visitations, nodding grimly, face after face, name after name, trailing the scorching rays of adulation bestowed by the wider world.

In they cram and press, the atheists Huxley and Darwin, Pasteur the chemist and Catholic mystic, my grandfather's Olympian gods, standing like stout sentries by the door, influencing and being influenced, hard-headed men driving hard-headed ideas, difficult men all, men making good, men making better, men making the best of it. (Had not Huxley himself declared that it is a young man's game, that all scientists over sixty should be strangled?) Then comes Campbell Meyers, down from his gabled sanitarium on Heath Street where he is practising the rest cure on the pliable, affluent women under his care; next comes C.K. Clarke, over from the clammy, raucous wards of 999 Queen Street West, trailed by Hincks and Farrar, the redoubtable trio of mental hygienists who never doubt that all mental defectives are possessed of criminal intent. In from the twin towers of the Buffalo madhouse and the bewitched castle of Danvers tramp Meyer and Southard and Gay. Then the relentless flow of visitors quickens, the Americans, Japanese, Germans, Russians, English, and French; Osler, Flexner, Aschoff, Noguchi, Calmette, Park, Welch, and Metchnikoff; Roux, Russell, Rose, Ramon, Rudin and Rajchman; Bordet, Beers, Burlingame, Bruce, Bates, Blatz, Buchan, Banting and Best; and Balfour, unstrung from the noose that hung from the tree in his English asylum. The irrepressible moneymen, Rockefeller and Gooderham, arrive like absolute monarchs, erect in black limousines, then the loyal lieutenants of the cutting edge labs of Connaught, Defries, Fraser, McKinnon, and Dolman; then Amyot and McCullough, and Graham. Next appear the once-cherubic faces of stricken children, upturned like flowers to the sun, crowding into every last square inch of his hospital room, arms extended, tonguelike, to receive the sacrament of the diphtheria needle, given by the high priest of the cult of preventive medicine—*mens sana in corpore sano.*

Now the air is pierced by the plaintive whistles and horns of ocean liners and A trains, the chattering of Benares monkeys and the whimpering of insulin dogs. On the horizon, Gerry makes out the figure of his ancestor, James FitzGerald, fleeing on horseback from the humiliating defeat at the Battle of the Boyne, turning his back on his Catholic faith. Under a wrathful sky, my grandfather sees the immigrant coffin ships docking in Port Hope, disgorging the pox-faced Irish rabble, lowest of the immigrant low, stinking of cholera and typhus, his own thirteen-year-old grandfather he never knew lingering among the survivors. He glimpses the furrowed faces of his terrified children, Jack and Molly, shadowed by their aunt and uncles, Hazel, Sidney, and Bill, standing together as one, casting imploring looks of wordless prophesy. He sees his talented, tragic cousins, maternal and paternal: young William Woollatt, claimed by suicide, and Dr. Herbert Armstrong, tortured by sexual fears and longings in a string of sanitariums. He sees his father, Will, born fatherless, the same year as Sigmund Freud, bending over the counter of his Drayton drugstore, tediously rolling his pills by hand. He sees the face of his bedridden mother, Alice, twisting down the rabbit hole of an ancient melancholy, falling away from him forever, soft pale hands outstretched, her pain unhealed, a sin for which he can never atone, his life's work nothing but a thumb in the dike of the cruel certainty of death. Her eyes blaze with tacit accusation: *Who do you think you are? What have you done? For all your manifold networks and buildings and teachings and travels, the good works of Connaught won at best a pyrrhic victory, its Icarus-like strainings all for naught.*

Now the sly figure of Ernest Jones, the subversive Celtic carrier of the seed of Freud, the power behind the throne of psychoanalysis, elbows past the wall of implacable medical authority and, repeating the gesture of thirty years earlier, lays his unusual wedding present, *The Interpretation of Dreams*, at the foot of my grandfather's bed. Jones's unhappy sojourn in Toronto gave birth to a seminal paper on the agonizing Oedipal equivocations of Prince Hamlet, yet for Gerry FitzGerald, such words remain unread, such thoughts undreamt of. Who is this blind, mythical figure the Welshman speaks of, tortured by unconscious guilt for stealing the love of his doting mother away from his remote father? What does he have to do with me?

In my grandfather's sleep of unreason, monsters gather in final judgment, clad in stark crimson robes, rank upon rank, the somebodies and nobodies, the sensitives and the brutes, the profane and the mute, the patients he healed and failed to heal, a dense tribunal of accusation. A stentorian voice reads out a litany of indictments, real and imagined; no plea bargaining will be tolerated. The fingers of his multiple failings fuse into a massive mailed fist, *the* unpardonable sin, born in a Connecticut Yankee retreat, pounding and pounding on his thoughts like a gavel, failure upon failure, grief upon grief, shame upon shame. *Guilty* of cowardice and desertion, of letting down the side; *guilty* of falling into the kind arms of a woman not his wife; *guilty* of the loss of social status that so mortified his long-suffering Edna; *guilty* of the failure to attain some impossible standard of moral and scientific perfection; *guilty* of slitting open the brains of cadaverous Irish lunatics, probing the tissues with his gleaming lancet for the mythic microbe of madness; *guilty* of eradicating the immigrant ghetto of St. John's Ward, supplanted by the sterile hospital where he now lies; *guilty* for failing to bring his influence to bear against the policy of anti-immigrant eugenics, a man not so long detached from his own immigrant past; *guilty* of defying the forces of nature with his unnatural science; *guilty* of his ties with the robber baron Rockefeller and his blood money; *guilty* of feeling the germ of shame that blossomed into two attempts on his life, the shame merely redoubled when the attempts failed; *guilty* of hyperrationality and hubristic overreaching; *guilty* of torturing thousands of lab animals; *guilty* of making tainted vaccines that damaged the lives of innocent children; *guilty* of abandoning his own children to his God of Science, tainting them with an insatiable craving for the touch of his flesh, the seedbed of their addictions; *guilty* of trampling his true nature in the mad rush to please others; *guilty* of the unmanly stigma of emotional collapse and the loss of self-sufficiency, an ideal personified in the institutions he had created; *guilty* of succumbing to the unpardonable sin of despair.

And yet, even as the gavel pounds, a consoling thought slips through the din. As if anticipating the loss of all hope and the inevitability of his plight, he has shrewdly slipped an ace up his sleeve, the one last chance to redeem himself and prove his courage. In one final, clean, incisive act of will, he can, like a Roman noble falling on his sword, control the time and manner of his own end, a line of attack as logical as it is inevitable. A man who has spent a lifetime obliterating an ancient disease needs only a moment to obliterate himself.

Thursday, June 20, 1940, dawns fair and cool, the lashings of rain having washed clean the streets of Toronto the Good. As the hands of the wall clock click towards the summer solstice, the boldface headline of the *Toronto Star* delivers the dire news: "Pétain tells why France quit." That evening, the night nurse knocks on the door of my grandfather's private room—a nurse trained in a school deemed by the Rockefellers as the best in the world—and places a tray of food on his lap. With the dying of her footsteps, he pauses for a moment and hears the sharp, insistent intrusion of a familiar voice, the same tyrannical trumpet of conscience that makes cowards of us all, the selfsame force that catapulted the man-on-a-mission to the airless, grandiose summit of his profession. The root of the ancient word "sin", he knows, comes from archery: missing the mark. *Next time, I'll do it right.*

The words I pulled from John Hamilton, the frail octogenarian doctor in the Vancouver retirement home, flow back to me once more, melting into the archival footage I have spent years playing over and over in my daydreams, the primal scene I am "murdering to dissect," the son remembering what the father forgot, the tomb of pain where generations past fuse with generations future, repeat offenders entangled in the killing jar of the family unconscious. "We looked for healing," thundered the Old Testament prophet Jeremiah, "but we found terror."

My grandfather picks up the dull, sterling silver dinner knife off the tray and runs his fingertip over the short, serrated tip of the blade. The shaft gleams ice-cold, but no colder than the silverware his wife laid out over the tablecloths of a thousand genteel dinner parties, the cherished heirlooms engraved with the crest of the armoured, medieval Knight of Kerry, mounted on horseback, wielding his doubled-edged broadsword high above his head, a head that loathes cowardice as the summit of all sins. Pulling up the pristine white hospital gown, Gerry feels for the femoral artery in his thigh. Marshalling the stoic willpower of his father, Will, and the willful Irish fathers before him, he raises the knife and stabs the flesh of his groin again and again and again, cutting deep to the thigh bone. His life's blood spurts and gushes over his hands, over his legs, over his genitals, the pale, limp organs of regeneration; as the scarlet sap seeps into the snow white linen, he successfully stifles a cry; as he knew it would, the excruciating pain gives way to the melting consolations of shock, the rush of the natural morphines, ringing the citadel of the body against the merciless blows of the invader, blows that mirror the thrustings and stabbings and gushings of the purest male lust.

The visionary closes his eyes. In the five minutes it takes my grandfather to achieve the perfection of death, his fifty-seven years are borne on a quicksilver train of glass-lantern slides, an eternity of chaotic images and voices fluttering and flickering under his eyelids, thrashing like flightless birds, straining to quit the prison cells of memory. It feels strangely sweet to fall into the ruby mouth of death, the sex-fed pleasures of the flesh he has denied himself for so long.

The images and voices drift and melt, falling and fading into a chasm of nothingness. Flayed by a demonic rage, the fleeing shadow of my grandfather, the horseman of the apocalypse, straddles the jet black diphtheria mare, Crestfallen, and streaks across the night sky, the animal's legs flailing against the vacant air, far above the prison stall of the Barton Avenue barn. Clutching its mane, the doctor smells the rank, hot breath of the beast; and from its wide neck he feels a stream of lukewarm blood gushing from a vein severed by his knife. He allows himself one last thin smile as the healing blood drips from the celestial vault, falling on the open lips of a quivering, ashen-faced boy, alone in a desert as vast and arid as the Balmoral nursery. Perhaps my grandfather knows a single drop of his own blood carries the genetic secrets of his antique race; but racing onward, never for a moment does he stop to ponder how high and low a man will go to feel worthy of a father as unloving as a stone.

If some jagged fragment of truth has set him free from the shell of a false self, no human soul will ever know. Men say no good deed goes unpunished; alone and bereft as the sacrificial lamb, my grandfather pronounces the sentence fit and just, the bloodletting a kind of cure. Slipping from the rim of the pagan summer solstice, sins unpardoned, John Gerald FitzGerald takes final leave of the body he had abandoned long ago.

TWENTY-SEVEN

Occam's Razor

> God is dead. God remains dead. And we have killed him. How shall we, the murderers of all murderers, comfort ourselves? What was holiest and most powerful of all that the world has yet owned has bled to death under our knives. Who will wipe this blood off us?
>
> FRIEDRICH NIETZSCHE
> *The Gay Science*

To prevent a public scandal, the official cause of my grandfather's death is recorded as a hemorrhaging duodenal ulcer. Suicide, after all, is a criminal offence. No record of the autopsy—the final cut—survives.

At the peak of his influence, the hospital's physician-in-chief, Duncan Graham, is only one among many who colludes in the pall of silence that engulfs the medical profession, the city, and the country. And, of course, the family. No one issues a formal edict or memo to keep mum about the tragedy; if people speak at all, they keep the dialogue internal or whisper rumours that leak through the cracks of half-open doors. The contradictions abound: How could a man of such high attainment and stainless reputation stoop to the base act of self-murder? The explanations, if they come at all, seem superficial and inconclusive, trailed by an elephantine silence. The doctors, especially, remain mute; had he appeared as a mind-reading ghost at his own funeral, Gerry would have found confirmation of his worst fears in the thoughts of his closest friends, for few doubted he—an Olympian figure in whom so many had invested so much hope—had shamefully let down the side. People do not like their gods to be human. Nor is it easy to admit that it might have been the fifty-seven injections of insulin, the powerfully restorative preventive medicine that made the man's name, if not a fortune, that drove him deeper into a toxic hopelessness that in turn compelled his last incisive act of will. What else but a tainted heredity, an Irish madness, could explain it? What else but bad blood? What else?

At his request, Gerry's ashes were placed inside his bust in the library at the School of Hygiene; on Edna's death eighteen years later, Jack interred his parents' remains in Mount Pleasant Cemetery

On Saturday, June 22, 1940, a casket draped in a Union Jack solemnly bears the body of my grandfather onto the stage of Convocation Hall, the site of many of his past public lectures. Hundreds of mourners, many drawn from the Canadian and international medical elite, turn to catch a glimpse of Sir Frederick Banting and Charles Best among the distinguished pallbearers; only one hundred yards eastward stands the lab where the stubborn secrets of insulin were first revealed to the world. Banting nurses a slight limp, for the ugly six-inch burn scarring his thigh is only now healing. The sturdy exemplars of the Connaught—Defries, Fraser, McKinnon, Hutchison, Moloney—stand erect as rows of medicine bottles; behind them are arrayed another thirty honorary pallbearers, including Ray Farquharson, Duncan Graham, and Gordon Bates; and of course, prominent among all the ardent seekers of scientific truth looms the precise, slender figure of C.B. Farrar, all the more conspicuous for the black sliver of a Hitlerian moustache.

My father, Jack, a twenty-three-year-old medical student, stands silently beside his sister, Molly, and his mother, Edna; Hazel, a habitual wearer of black, is flanked by her brothers, Sidney and Bill. Before the university president, Canon Cody, delivers the eulogy, the lid of the casket remains open, against the express wishes of the widow. This small betrayal is puzzling; the family can only guess the university's decision was a tacit way of showing the public that the body was not disfigured. When people are invited to view the remains, only a single soul—Albert Double, the elderly farm superintendent who had loyally worked for Gerry since the beginning, risking his life injecting diphtheria toxin into the bloodstreams of retired police horses—shuffles up to the stage and gazes at his face. Raising her white-gloved hand to her mouth, Edna suppresses a cry of anguish. No one else comes forward.

On the day of the funeral, Fred Banting writes in his diary: "This has been the worst week imaginable." But he does not mention his old friend and confidant Gerry, rather laments the war in Europe and his sexual travails with his new wife, Henrietta. In her own diary, Margaret Best, though fond of Gerry, also makes no mention of the suicide.

As he had wished, my grandfather—ever the hygienist, ever thinking outside the box—is cremated and his ashes placed inside his bust in the library of the School of Hygiene. In his will, he left a total estate of $2,354 in cash and stocks.

Then come the letters, the obituaries, the memorials. "A man of rare gifts and rarer vision," the Academy of Medicine writes, "has been cut off at an age at which rich fruits of a ripened judgment might have been the expected harvest for many years to come." Don Fraser writes: "To few are given the qualities of executive ability, singleness of purpose, imagination, and vision, combined with gentleness, modesty, and charm of character in such large measure as he possessed." John McCullough, the loyal ally in the early days of the Connaught, recalls: "I saw him last exactly seven weeks before the day of his death. He called one Sunday afternoon and I remarked how well he looked. He talked and laughed in the same old manner that I knew so well. There did appear to be a trace of illness about him."

Sir Wilson Jameson, a future architect of the British national health service, eulogizes in *The Lancet*: "Preventive medicine had no more ardent or skilled advocate. The fact that Canada has probably the best government-controlled scheme for the preparation and distribution of biological products is largely due to FitzGerald. He was a delightful companion—I once travelled with him for three months—always cheerful, the possessor of a keen sense of humour and a highly developed critical faculty. On his many journeys, he was always on the lookout for talented youngsters and many a young man has to thank him for his first real start in life. He had friends all over the world; they will feel his death as a great personal loss—the younger perhaps even more than the older, for FitzGerald was himself a boy at heart."

From her home at 22 Bernard Avenue, Molly writes C.B. Farrar: "It would be impossible to express to you in words our gratitude to you for all you have done for my father. I would just like to try to thank you on behalf of Jack and myself for your great patience and understanding during these last hard months. We will never forget how you were always at hand to help and encourage us all for the duration of Dad's illness. Also, for your unfailing service to him—he had such a tremendous admiration for you. It was only the malice of the disease, I know, which prevented him from expressing his profound gratitude for your kindness and care."

Jack chooses not to write a letter of his own.

Eight months later, in February 1941, Sir Frederick Banting, having performed clandestine experiments on bacteriological warfare with members of the Connaught Labs, travels to a snow-covered airfield in Gander, Newfoundland. As he prepares to fly to England on a secret wartime mission, his eleven-year-old son, Billy, worries the plane might crash, but Fred reassures the boy that he can survive in freezing water as long as anyone. He has never flown the treacherous Atlantic in mid-winter; only one plane has made it across that year. But he famously symbolizes the fearless spirit of the pioneer and there is no choice but to fly.

The twin-engine Lockheed Hudson bomber rises into a pitch black sky lashed with driving snow; only twelve minutes out, both engines fail. The pilot hopes to skid across a frozen lake, for half of Newfoundland is composed of water; but by a quirk of fate, as the plane crash-lands, a wing strikes the only tree within miles. The navigator and radio operator die instantly; the forty-nine-year-old Banting suffers grievous injuries to his head, arm, and ribs. Wrapped in a silk parachute, numbed by the cold, the hero of insulin commands the pilot to take dictation. Flitting in and out of consciousness, he deliriously spouts a stream of technical war secrets, surreal gibberish that mimes the encoded ravings of a schizophrenic. Finally, the pilot heads off for help; when he returns in the blue light of dusk, he finds Banting fifteen feet from the wreck, lifeless in his "teddy bear" flying suit, a final burst of horselike energy spent. He survived for nearly twenty hours; if not for the outline of the corpse, the stark scene could pass for one of his Arctic paintings.

Like his friend Gerry FitzGerald only months earlier, Sir Frederick Banting's body lies in state in Convocation Hall, the very spot where he received his medical degree; Canon Cody, who stood by Banting during his highly publicized divorce scandal, eulogizes him as "a man loyal to his friends, especially his old friends." Yet the father of insulin never found the time to visit Gerry in his last months.

With Banting's death, the University of Toronto discontinues its five-year program of insulin shock experiments on psychiatric patients.

═

In the late spring of 1942, my twenty-five-year-old father, Jack, mounts the stage of Convocation Hall, the site of his father's funeral nearly two years earlier, to receive his medical degree. With his father gone, Jack could have walked away from medical school and the killing weight of expectation, perhaps to pursue a bohemian career in jazz journalism—the Duke of Ellington trumping the Duke of Connaught—but the paternal "Memorandum of Wishes" will be followed to the letter. That same spring, Farrar uses his influence to secure my father an internship at his alma mater, Johns Hopkins, launching his friend's son on eight years of rigorous postgraduate medical training.

It is in Baltimore, of course, that my father beds Dr. X even as he is wooing his first wife, Caroline. When Jack marries the slim, red-headed American that September, the guest list, naturally, includes his bachelor uncles, Bill and Sidney, but both decline to attend the ceremony. In particular, Sidney is becoming increasingly removed from his social circle, inching outwards to the shadiest branch of the family tree. He misses his dead brother and the Sunday afternoons when the jet black Connaught limousine pulled up to his door and delivered him to family picnics at the windy edge of the Scarborough Bluffs. Within the past decade, Sidney has moved his residence half a dozen times within a half mile radius, from Vaughan to St. Clair West, Alcina, and Hilton. Not long after his brother's death, he settles in as a tenant in a distinctive three-storey Gothic revival house at 6 Wychwood Park, perched on the Davenport escarpment overlooking the lake.

Sidney is naturally drawn to the place, for the pattern of yellow brick, originating in Ireland and unique to Ontario, is redolent of childhood memories of the family home back in Harriston. Besides, Wychwood Park is a twenty-two-acre island of bucolic calm, a serviced private road lined with three-hundred-year-old trees. His landlord, Marmaduke Matthews II, is the headmaster of St. Alban's, an all-boys Anglican cathedral school, and 6 Wychwood serves as a boarding house for several schoolboys and teachers who seasonally swim in and skate on the pond opposite the house. Happily, Gerry's loyal associate, Don Fraser, lives next door at number 7 Wychwood, where he hosts lively parties for university academics and the extended family of the Connaught Labs; sometimes, Sid slips over for a sip of sherry and sharp, refreshing jags of conversation, an axe to the invisibly tightening tentacles of emotional isolation.

On the morning of Thursday, September 17, 1942, Sidney Maurice FitzGerald sets off to his law office on Vaughan Road, a few blocks north. The war is going badly; the morning headline of *The Globe and Mail* reports the Nazi siege of Stalingrad and lists over six full pages of names of casualties inflicted during the ill-fated Canadian raid on Dieppe only weeks earlier on the shores of France. Few dare say it, but the five thousand doomed Canadian shock troops had been used by the British as guinea pigs for the planned D-Day invasion of Europe two years hence. Stepping outside the front door of 6 Wychwood, Sidney passes the tree-shaded pond fed by Taddle Creek, the genesis of a once-pristine watercourse paddled by Aboriginal peoples. Urban filth forced sanitary engineers to bury the creek, and so now it snakes underground like a femoral artery, cutting under Prince Arthur Avenue, running parallel to his dead brother's School of Hygiene, then drains into the lake.

Normally Sid would head straight to the office on Vaughan, but on this morning he decides to keep going, for much weighs on his mind. The previous day, his niece, Molly, celebrated her third wedding anniversary; only two weeks earlier, his nephew, Jack, committed himself to the institution of marriage with his glamorous, wealthy American bride. Sidney walks on, lingering for a moment outside his Anglican church, St. Michael's and All Angels, where he dutifully serves as a warden each Sunday. He presses a few blocks up Wychwood Avenue, entering the neighbourhood of Cedarvale.

As Sidney comes upon Connaught Circle, a grassy roundabout named after the Duke of Connaught, my great-uncle reflects for the briefest of moments that no matter where he walks, his older brother's ghost encircles him. Then he spies his destination: the hundred-yard-long wooden footbridge that spans the Cedarvale ravine. Over the path of his life, Sid looked up to Gerry and his "room at the top" with a mixture of admiration and envy, but on this day his eyes are cast downward. Edging to the middle of the bridge, Sidney stops and looks down on the muddy footpath one hundred feet below; here, on the grassy glades and twisting creek where Aboriginal tribes once travelled, he fixes his last gaze. If he had chosen to live a generation longer, if he had stood on this spot on a summer Saturday in the early 1960s, he would likely have seen me, a skinny, crewcut kid, playing war games with my brother, killing imaginary Nazis with juvenile ferocity; with any luck, we might have looked up and seen the shadow-man whose genealogical map will one day point me to my Irish ancestral village, a harmless, doddering old man, leaning on a cane, waving down at us with the hint of a wordless smile.

Sidney Maurice FitzGerald (1884–1942)

But no: He swings one leg over the railing, then the other. Then he joins his brother in the silence.

He dies shortly after being admitted to the Western Hospital, the place of my birth eight years later, almost exactly to the day; he is fifty-eight. A short, euphemized item in *The Globe and Mail* reports that Sidney Maurice FitzGerald, possessed of his "usual health," was "stricken" while walking to work. For the rest of his life, my father never utters the name of his quiet, melancholic bachelor uncle so obsessed with a lineage traceable to a ruthless Irish medieval knight that he changed his middle name to Maurice, a name that no one else in the family would let fall from their lips, the name of a cipher of a man who preferred the oblivion of eternity to the prospect of witnessing his only, fallen-in-love nephew standing bravely at the altar, returning the steady gaze of the radiant young woman whose perfect lips and hair flamed red as fire.

==

With my grandfather's death, Bob Defries assumes the reins of both the Connaught Laboratories and the School of Hygiene. In 1944, sixty-four-year-old Albert Double, the loyal employee who had shuffled up alone to Gerry's open casket, celebrates his thirtieth anniversary working at the lab. One November day at the farm, Double's nephew cleans a septic tank and is overcome by sewer gas. Double races to the rescue, but he too is overcome and collapses. Both men die.

By 1950, the year of my birth, the Connaught Labs is producing a state-of-the-art rabies vaccine that has become the most widely used in the world. That same year, Hollie McKinnon dies of a Nembutol overdose. Lorne Hutchison, the fifty-three-year-old Connaught comptroller who served as an RCAF squadron leader during the war, had once prevented Hollie from plummeting off the Bloor Street viaduct, but this time he can say or do nothing to dissuade her from her darkest wish. The role of failed saviour is a familiar and painful one for Hutchison, as his first wife had killed herself twenty years earlier. In the spring of 1952, his turn comes when he leaps from the St. Clair viaduct, not far from the site of my great-uncle Sidney's plunge. Once again, the blade of silence falls; it were as if members of the tight-knit Connaught family, intimately linked by the most taboo of thoughts, had been hand-picked by the Fates to pay the ultimate price; as if, from the grave, my grandfather had bequeathed to certain susceptible admirers tacit permission to follow their leader.

In September 1957, my seventy-one-year-old great-uncle, Bill FitzGerald, is admitted to Wellesley Hospital—the place where his father and namesake, William, had died forty years earlier—and given seven rounds of electric shock that trigger grand mal seizures in his brain. After repeating suicidal ideas and the fear he will go mad, he is moved to 999 Queen Street West, the crumbling mental hospital where his idealistic older brother had laboured under C.K. Clarke exactly half a century earlier. Bill will stay for over a year.

A thin, stooped, scarecrow-ish six-footer, my great-uncle has retired from the bank and lives a subsistence existence off a forty-six-dollar monthly pension in Thornhill, together with his wife, Doris McLean, a Bishop Strachan old girl who had written a history of the town, and her son, David, from a previous marriage. Tortured by ulcers, Bill can neither eat nor sleep and lacks all interest in the world; overcome with anxiety and self-pity, he is terrified of choking to death in the night. When asked by the admitting psychiatrist why he feels depressed, he replies that his two older brothers, Gerry and Sidney, had killed themselves, and he fears he might do the same.

At Queen Street, Bill receives a further regimen of electroshock, but reacts badly and the treatment is discontinued. He is evasive about the content of his depression, blaming it on lack of sleep. "It is clear that this man has a lot of repressed hostility," writes the attending psychiatrist, "but he is extremely defensive about this."

Before and after his confinement at 999 Queen Street West, Bill remains financially dependent on Edna and Hazel and later, his adult stepson, David. Passing his last years in Thornhill in a century-old house on John Street—which bears the same name as the street he grew up on—Bill gardens, plays Scrabble, and makes rugs, a skill he learned in occupational therapy; hooking all day long, he piles them up by the hundreds. Nibbling bag after bag of humbug candies, he obsessively records each day's temperature, rainfall, and humidity levels in a weather journal for years on end. His stepson regards him as dull, inept, and withdrawn, a sad sack who tightly reins in his emotions, simply smiling, nodding, and

The FitzGerald brothers Gerry, Sid, and Bill, Drayton, 1890

agreeing with everyone to avoid a fuss. His stepson's wife agrees: "I didn't think he was deep enough to feel anything. He was just kind of locked in neutral."

After Gerry's death, Edna returns to 186 Balmoral Avenue, the gloomy house she had built with her inheritance in 1914. Perhaps it is no coincidence that with her husband gone, her "vapours" immediately disappear. Edna never speaks of her husband, at least not until the fall of 1955 when, in the midst of international publicity about the polio vaccine, June Callwood, a young *Maclean's* journalist, interviews her about the Connaught Labs. The article is entitled "The Miracle Factory That Began in a Stable." For the record, the normally tight-lipped Edna is unusually candid. She recalls telling her dynamic husband: "Slow down, Gerry . . . 100 years from now, who'll care? It's not that important." Then, Callwood reported, the widow reconsidered her words: "I guess Gerry knew all along. A hundred years from now, people *will* care. It really *was* that important."

As for her husband's premature death, Edna, the proper Edwardian, could be forgiven for choosing not to reveal to the *Maclean's* readership that Gerry had killed himself. "He suffered a physical breakdown two years before he died and then a mental breakdown," she allows. "He simply wore out." Perhaps my grandmother reveals more than she intended when she adds: "I was married to an idea, not a man." When, against her wishes, Jack sells the inherited Balmoral house he had hated as a child, she dies shortly thereafter, in January 1958, her heart broken. Whether she overdosed on Nembutol was never made clear to me.

Less than two weeks later, seventy-nine-year-old Ernest Jones, suffering from terminal cancer, swallows a pill that ends his life; his ashes are placed next to Freud's in Golders Green. Serving for fifty years as Freud's "indomitable fighter" and idea-disseminator across the world stage, Jones had laboured for over a decade writing the classic three-volume biography of the father of the talking cure, the capping of a lifetime of achievement that had narrowly averted ruin in his early years in Toronto and London.

"Genius," he once declared, "consists in an infinite capacity for enduring pain—and inflicting it."

In July 1963, C.B. Farrar publishes a four-page memorial, "I Remember J.G. FitzGerald" in the *American Journal of Psychiatry*, which, astonishingly, he has been editing since 1931, the longest tenure in its history. In the piece, Farrar wonders what might have happened if my grandfather had chosen the path of "psychological medicine" instead of public health.

"Dr. FitzGerald knew how to handle problems and men," writes Farrar. "He was calm, soft-spoken, kindly, judicial, and tolerant, and withal he was a lovable person. There was something charmingly sweet in his personality, an almost feminine tenderness in his concern for others, especially for any who might need his help. And he loved beautiful things. His standard was *arête*. [A Greek word for the achievement of excellence without undue strain or stress, a kind of Mozartian dictation from God.] One of his most felicitous characteristics was his confident facility in delegating responsibility, and he was not likely to be disappointed in such decisions."

Farrar chooses not to discuss Gerry's early connection to Ernest Jones and the untaken road of psychoanalysis; nor does he mention the suicide. "Maybe in his extraordinary career," he ends the article, "FitzGerald's initial years in psychological medicine helped a little—maybe."

From her cottage on the Burlington escarpment, Hazel responds: "It is a wonderful tribute, so finely done, so concisely billing all the hopes and aspirations and the final accomplishments. It helps one to bear the sadnesses. You have remembered his lovableness, his sense of fun, his desire always to help, and finally his achievements. There were heartbreaks in the early days in the basement of the medical building but we were young and difficulties were only things to be overcome. And fun we had too—Dr. McCullough, who was with us so often, sharing our box lunch and taking lessons with me in bacteriology!"

Semi-retiring in 1947, Farrar maintains a meticulous office in a red-brick medical building at 200 St. Clair Avenue West at

Avenue Road, opposite the rival enclaves of Jungian and Freudian analysts that line "angst alley"; his window overlooks the statue of Peter Pan on which I climbed and played as a boy, as had my father before me. An avid collector, C.B. stuffs his Oriole Road house, three blocks from my childhood home, with Oriental rugs and works of art. His young, live-in companion, Joan, gratefully dips into his extraordinary library—five thousand books on art, religion, astronomy, medicine, pornography, and the literary classics lining glass-fronted bookcases. "Oh, the culture!" Clarence Hincks, the pioneering mental hygienist and eugenicist receiving intermittent treatments of ECT for his depression, exclaims one day while standing in his friend's living room, windmilling his arms with characteristic abandon. Over the years, C.B. kept all of his old passports, notebooks, meerschaum pipes, baby boots, opera programs, glass-lantern slides, a black, gold-crested case for preparing brain slides, a photo of the death mask of Cardinal Richelieu, autographs of famous doctors, stones taken from Sir William Osler's English garden, and letters to his estranged first wife and two daughters. The enormous collection, of course, also includes the sixty letters from Gerry FitzGerald in Hartford, destined to pass under my eyes; so, too, is a quirky sample of C.B.'s black humour, written in the 1940s, not long after Gerry's death. Working one day at the Toronto Psychiatric Hospital, C.B. jotted down a list of twenty-three "Rules and Regulations" for his patients, that presumably were never made official policy, but included:

> "If you wish to commit suicide, please protect the rugs."
>
> "Implements for suicide or homicide will be found in the entrance hall."
>
> "If you wish to cut your throat with a razor blade, a written request form must be handed to the barber two weeks in advance. Only three blades are available per week."

"When you intend to kill another patient, do not tell him. Surprise him."

"Sword drill will be held once a week and methods of committing suicide will be taught the third Friday of every month."

"Always talk to the Superintendent if you have a delusion that he is sane."

D. Mentia, Superintendent.

Eugenic-minded to the end, Farrar remains convinced that tainted heredity is the chief cause of madness, despite the intrusive fact that the mentally disturbed reproduce at low rates; if insanity were an inherited disease, it would have been wiped out centuries ago. Indeed, his critics ask if any human can claim to be entirely free of "taint"? Farrar never ceases to mock the Freudian enterprise, arguing for the training, control, and direction of emotions, not giving them loose rein; there is, he insists, no such thing as free association.

"The pathology of memory must not be overlooked," Farrar tells an American audience, "but we may as well remember too that there is such a thing as healthy forgetting. Moreover, it is useful to keep in mind the real danger of inducing too much introspection in the therapeutic process . . . we should not feel the need to resort to complicated or speculative interpretations when simple or more obvious ones will serve. We have an ancient term for this rule—Occam's Razor. This cutting instrument is quite useful in psychiatry and especially in psychotherapy, an excellent pruning knife for verbal excess baggage."

In 1969, Roland Michener, the Governor General, presents Farrar with an Order of Canada at his home on Oriole Road as C.B. is too frail to travel to Ottawa. At ninety-four, he is the oldest person to ever receive the award, even as people never cease remarking on his youthful, Peter Pan–like appearance. Farrar dies a year later in June 1970, nearly thirty years to the day after the death of his friend Gerry FitzGerald; within weeks, three short blocks westward, his old friend's son, Jack, will climb the stairs to the third floor of our Dunvegan home, carrying a syringe.

═

Through the postwar years, Charles Best drinks deeply from the cup of fame, hobnobbing with movie stars and heads of state, travelling on international lecture tours, enjoying the trappings of celebrity. But as he hits his sixties in the 1960s, the bloom is fading from the rose. He thinks he is out of date, unable to keep up with younger scientists, falling short of his high standards; gradually, he is overcome by paralyzing waves of indecision and self-blame. Margaret Best does not understand her husband's crushing depression and never mentions it in her diary; trying to spare his wife, Charles manfully bottles up his anxieties, and for his trouble, he steadily worsens. In 1955, when 18 Prince Arthur, the former home of his old mentor Gerry FitzGerald comes up for sale, Best considers buying it, but Margaret, born of stout Ulster Irish blood, vetoes the idea. "Suicide was not to be mentioned in our house as it disturbed her," her son Henry told me. In the absence of an open, frank explanation, a belief persists that Dr. FitzGerald had died in the house; others whisper rumours he had jumped off the roof of the School of Hygiene.

Charles stoically bears another source of deep distress: his eldest son's chronic marital and financial problems. Elected a member of parliament at age twenty-five, Sandy Best admires wealth and power and lives a fast-paced life, making and losing money in repetitive cycles. Charles has done all he can to help, including paying off his debts, but the troubles persist. For both good and ill, Sandy has inherited his father's precocity and extraordinary powers of concentration, the ability to mentally shut out other people at will; theirs is a father-son relationship of unbroken intensity.

On an early spring day in 1964, Charles Best's black state of mind plunges to suicidal depths and he is hospitalized. Days earlier, he had opened sample boxes of potentially lethal pharmaceuticals and lined them up in a neat row on his bedside table in his home on Old Forest Hill Road. Charles is given a series of twenty-three electroshock treatments, against the wishes of his son Henry. "I fought it," he wrote in his memoir of his father. "I was not convinced that this made any sense." Henry does not receive a sympathetic hearing from an attending doctor, but a wordless glare that speaks volumes: "Who the hell are you? You don't know what you're talking about."

Unaware of the drama raging a ten-minute walk north of our house on Dunvegan, my father, Jack, as he walks the dog each day through Forest Hill, routinely passes his godfather's house; such is the shame and stigma of depression that sufferers dare not reach out to those in a similar plight. Isolation is the symptom of the age; both Best and my father wear their aloneness like a rusty suit of armour, an emotional vaccination against unwanted human touch or word. Not that some people—particularly sons—do not try. One day, twenty-nine-year-old Henry, who played in the sandboxes of the Connaught Farm as a child, finds his father sitting in a chair holding a book upside down; he looks like a shell-shocked cavalryman with his horse shot out from under him. To pierce his father's gloom, Henry saddles up Lady, a police horse, and walks her back and forth in front of his father. Suddenly, Charles looks up and snaps: "What the hell are you doing with my horse?" Grabbing the reins, he jumps in the saddle and gallops off.

By the 1970s, Best's moods level off somewhat, but his oldest son's troubles persist. Now forty-seven, Sandy has left his pregnant second wife; shortly afterwards, on the verge of a third marriage, he is felled by a massive heart attack, although there are whispers of a suicidal aspect to his lifestyle, as if the son, bested by the father, was waving the white flag. The following evening, March 26, 1978, as Charles is writing a tribute to his son, he is cut down by a brain aneurysm. "Don't let them take me to the hospital!" he pleads with Henry, but he does.

Six days later, early on the morning of Easter Monday, the day of his son's memorial service, Charles Best dies, months short of his eightieth birthday. His wife personally answers over four hundred letters of condolence. One day a few years later, when Henry asks if his children finished first in their class, his wife rolls her eyes and exclaims: "God save us from brilliance." Henry will spend a decade writing the only existing biography of his father, defending Charles against criticisms of over-ambition that "smack of the common Canadian compulsion to cut down the tall poppy." Henry will die of complications arising from diabetes, refusing treatment with both of his father's great achievements—insulin and Heparin.

═

In 1980, my aunt Molly is sent to Homewood, the private sanitarium in Guelph once run by C.B. Farrar and recently occupied by her brother, Jack, his addled brain poised precariously close to the knife-happy lobotomists. Having split with Tom Whitley in 1976 (my mother separated from my father a month later), Molly had isolated herself in an apartment on Lonsdale Road. Now a hopeless, brain-damaged alcoholic engulfed by anxiety attacks, she is prone to wild tirades and impulsive sprees, buying up thousands of dollars worth of bathroom fixtures and bidding cabbies to deliver bottles of booze. More than once, the police are called to investigate her disturbances. One of her several psychiatrists is bewitched by her Irish charm, which she can turn on and off like a tap, and he nearly convinces her to run off with him. Molly is the fifth FitzGerald in two generations to be institutionalized and over the last decade of her life, her Homewood file thickens to five hundred pages of Meyer-like detail, ceasing with her death in 1990 at age seventy-six. I never gave a thought to visiting my aunt Molly in her private sanitarium, for to me she was as ethereal as a fairy story; if I had counted the number of words we exchanged in our lives, they would not have filled a handful of formulaic greeting cards. The family consoles itself with wry anecdotes of Molly at her best, her spontaneity, style, and sophistication, her dry, witty asides. To the end, a sphinxlike figure suffering in furs, Molly denies she ever had a drinking problem; like her brother, Jack, she never learns the true nature of their father's death—that he had taken his life with a dinner knife.

═

My grandfather liked to say that institutions are more important than men; I would tend to say *as important* as men. On a blustery spring day in 2005, repelled by the mocking blankness of the computer screen, I quit my downtown apartment and take a long walk through the receding snows of the city, a strategem to shake loose memories of my dead father, to soothe myself with the alchemy of words bubbling in my head. I no longer jog or sprint, as I did as a kid, pitted against my father's stopwatch; I am now a confirmed walker. As if pulled by a magnet, I find myself, within the hour, on the doorstep of 25 Leonard Avenue, the plain, four-storey, concrete medical office block my father built in 1960. It was not my conscious intention to venture that far downtown, yet here I am. Designed to unite the medical specialists of the Western Hospital under one roof, the place embodied my father's ambition that he hoped would rival, even surpass, the integrating power of his father's achievements. Immediately, I notice something is different: The name has been stripped from the entrance and shabby blankets, passing for curtains, hang from several windows. It turns out that the hospital sold the building in 1998, its forty doctors' offices converted into one-bedroom apartments. "The House That Jack Built" is now a multi-faith shelter for homeless women.

I think of the old, profit-driven allergy lab, tucked down in the cinder-blocked basement in Suite 001, that I visited on Sunday afternoons as a child—the place where my father toiled in his white lab coat, the womblike place that increasingly delivered little joy. I imagine the solitary, faceless women who occupy it now. Canada's social safety net—together with his Aunt Hazel's dress shop bequest—partly saved my suicidal father, exiled from family, friends, and himself, from dereliction on the streets. He had also, early in his career, heavily stocked up on private health insurance as if he anticipated an ever-darkening future; his choice of London Life, the hometown of his mother, seemed symbolic to me. For reasons I can still only guess at, he chose to grind out his final years alone in a spare one-bedroom apartment, his existence reduced to a few simple possessions—not so different from the shelter from the urban storm I now behold, mere yards from my own birthplace.

As I stand alone on the curb, the familiar, fragmented feelings for my father simmer in my chest. I cannot avoid the feeling that the levelling hand of Nemesis has served up on the crested dinner plates of the overreaching FitzGeralds one more slice of poetic justice.

TWENTY-EIGHT

The Archaeology of Silence

You have to do your own growing, no matter how tall your grandfather was.

IRISH PROVERB

What's past is prologue. Year by year, I have scoured the archives and laboratories of the Western world and plumbed the memories of the silver-headed doctors who knew my father and grandfather as well as they were allowed to know. Out came facts and up flowed dreams, for dreams are a kind of fact, clues to the secrets I stalked, the self-murder mystery I knew I could never entirely solve. Year by year, the contemporary witnesses die off, one by one, till next to none remain, the returning silence mocking the sleuthing of the insolent grandson. Yet in the gathering up of my coincidences, in the dreaming of my dreams, in the making of my meanings, I have won something wonderful: I have lived out Nietzsche's wisdom, "What was silent in the father speaks in the son; and often I found the son the unveiled secret of the father." I have told myself the story never told me.

A seminal fact survives: how a single human being can have such a powerful impact on his family—and his country—far beyond his own lifespan. Even so, my grandfather remains unknowable, as we all do, slipping free of the cage of a final diagnosis; God is Dad. But the jagged pieces of the jigsaw puzzle leave me with more than enough to contemplate to the last of my days. They give me an unfinished and unfinishable portrait of a charming, tortured man at war with himself, possessed of gifts equal to his flaws: rigid and obsessive, open and gentle, ruthless and generous, witty and dour, passionate and repressed, sane and insane, visionary and blind. The entrepreneurial Canadian socialist; the radical, fiery Irishman tempered by the ice-cool, imperial Brit. Some might see a Nietzschean madman driven by a "creative malady," believing himself to be God, carrying a lantern into the marketplace, the disturbance of blood firing the engine of intellect. Perhaps others see a Christ-like avatar eaten alive by a hero worship he could never reconcile with the human frailty hidden under his skin; still others, a cautionary tale of a true believer in the perfectibility of man cut down by original sin. All carry partial truths and all are open to the fallible light and shade of interpretation.

The word "science" means "to know." Perhaps a more fitting meaning would be "to doubt"—being able to tolerate the state of not knowing, even accepting that there are deep ways of knowing without the tools of science. My grandfather's medical ideal—help "within reach of everyone"—remained unreachable, as he was. Nature makes us imperfect; we suffer and die, our reach exceeding our grasp. All along, I knew I would fail to crack the ancient FitzGerald protection racket or wipe clean the dynastic wound, but that did not stop me from trying—and learning how to heal. The Irish don't know what they want in life, but they'll fight to the death to get it. Suicide, of course, is the perfect crime; and my perfectionist grandfather got away with it. The murderer and the victim turned out to be one and the same.

═

My daily walks through the city often take me over to the traffic-clogged intersection of College Street and University Avenue, the pulsing arteries of my grandfather's fevered dreams. The university has swelled to seventy thousand students, many the children and grandchildren of hard-driving immigrants, not so different from my keen young grandfather who first set foot on the campus in September 1899. The University of Toronto now forms the hub of the third-largest medical research centre in North America, after California and Massachusetts. Back in 1919, Toronto academics agonized over whether a public institution should accept tainted corporate money from the likes of the Rockefeller family, but we voice few such qualms of conscience today. We live in a material world transformed, and who am I to judge if we are not the better for it.

I pass the newly renovated Toronto General Hospital at 101 College, the place of my father's birth and my grandfather's death, with its original 1913, brainlike dome respectfully preserved; directly across the street stands the Banting Institute, soon to be disestablished, and I reflect on the fate of its founder, the reluctant hero of insulin, his battered body sprawled on the autopsy table. Behind the Banting looms the former psychiatric hospital of C.B. Farrar, its thick, brick walls standing like mute witnesses to the long reign of the pill, the electrode, the lobotomy knife, the needle of insulin shock. Moving on, I find myself on the steps of the ivied, brown-brick building at 150 College Street, the former School of Hygiene now over eighty years old, and I contemplate with pride the fiercely dedicated, long-lived doctors inside its labs, clad in pure white, labouring through the decades for the greater good.

I know that since my grandfather's death, much has changed; and much has not.

In the postwar years, the Connaught Laboratories and School of Hygiene thrived as a unique, innovative, and unconventional organization, emanating the mystique of a national icon, emblematic of the very best of Canada. Continuing to achieve a series of scientific firsts, the lab maintained a tradition of helping other nations reap the benefits of its achievements—insulin in the 1920s, diphtheria toxoid in the 1930s, Heparin in the 1940s, penicillin and polio vaccine in the 1950s, combined vaccines in the 1960s. Billions of doses of Connaught products saved countless lives, domestically and overseas; insulin alone has saved more people than were lost in both world wars. Working with the World Health Organization, the lab acted as the driving force behind the decades-long global campaign to wipe out smallpox, a dream that came to triumphant fruition in 1979. It was the first disease—and to date, *only* disease—to be globally eradicated. In fact, it was the first problem in human history to be solved by organized effort on a global scale. Yet paradoxically, like the pathogens they stalked, the heroes of public health have remained largely invisible. As *The Varsity* editorialized: "Throughout Canada and the world, thousands of people who owe their lives to the laboratories may not know that they even exist."

As I wander the halls of the old school, I know that if my grandfather had survived to the age of ninety, as did many of his robust confreres, he would have seen his dream cruelly die. In 1972, in a controversial move, the University of Toronto sold the Connaught Laboratories to a newly created federal Crown agency, the Canada Development Corporation, for $24 million. Noted one observer: "It was like hearing that the Vatican was becoming lax in its liturgy." The university worried about potential liabilities in the wake of the thalidomide scandal, increasing government regulatory powers, and pressure from commercial drug makers who resented Connaught's tax-free status. Besides, the university knew a sale would deliver a windfall.

Three years later, the multiple scientific disciplines within the School of Hygiene were broken off and swallowed by the various departments in the Faculty of Medicine, dissolving what critics had viewed as a "shotgun marriage" instituted by Gerry. With the steep decline in the incidence of infectious diseases, others saw the lab and school as a victim of their own excellence. Without a forceful personality to balance the perpetual tension between the School of Hygiene and the medical faculty, my grandfather's unique institutional vision of public health had run its natural (or unnatural) course—fifty-eight years—almost exactly the length of his own life. Of all people, it was Dr. John Hamilton—the pathologist who had trained my father and reluctantly revealed to me the true details of my grandfather's suicide—who endorsed the breakup of the school in his role of dean of medicine.

With the stroke of a knife blade, a lifetime of rigorous integrative work—"a monument more lasting than bronze"—disintegrated overnight. Critics called the dissolution of the monastery of public health short-sighted and unimaginative; the school should have been "strengthened, not killed." Miscommunications between the university administration and the faculty of the School of Hygiene lay at the root of the problem; the bureaucrats just did not grasp what the school—or Gerry's overarching preventive-minded vision—was all about. When the original 1920s Rockefeller-FitzGerald correspondence stipulating the permanent union of the school and lab could not be found, Neil McKinnon thundered: "That was a hell of a poor excuse! I'd have taken FitzGerald's word for it."

In the years following 1972, the Connaught Laboratories wore the suit of a radically new corporate personality, an international biotech company steered by a Harvard MBA. A hostile foreign takeover in 1989 was followed by another a decade later, giving birth to the new multi-billion-dollar leviathan, Aventis Pasteur. In an end-of-the-millennium moment in December 1999, the legendary name of Connaught was erased, reduced to an historical footnote by the riptides of globalism. Gerry, of course, had originally wanted the lab to be named in honour of Pasteur. In 2004, the French pharmaceutical giant Sanofi made a hostile takeover of Aventis Pasteur, eventually forming Sanofi Pasteur, now the largest vaccine maker in the world.

Within months, a landmark ruling by the Supreme Court of Canada paved the way to a two-tier health care system, opening the door to private health insurance, an Americanized system out of the reach of millions of the poor and marginalized. With the wielding of machetes over the sacred cow of universal health care and the severing of the word "public" from the term "public health," the message bled through loud and clear: Some lives matter more than others. Yet health is not a consumer product to be bought and sold; it is a public good, a ubiquitous golden thread of social interdependency.

The medical revolutions of the twentieth century bred a sense of complacency that is now returning to haunt us. Our current high life expectancy is due to the labour of the early public health pioneers, not medical interventions; yet preventive medicine continues to march to the rear of the "curative" camp that keeps sick, even terminally ill, people alive at staggering financial and emotional cost. Magical thinking sustains the myth of the magic bullet, a pill-for-every-ill. Our belief in the final conquest of infectious disease has proven a myth, too; the emergence of new plagues like AIDS, SARS, Ebola, and new strains of influenza—over thirty new pathogens since 1970—remain as certain as death and taxes. Contemporary public health workers push the proverbial Sisyphean boulder to the top of the hill only to have it roll back. Today, that rueful Darwinian maxim—"Mother Nature bats last"—stands heroically confronted by Louis Pasteur's famous dictum, "My strength lies solely in my tenacity."

As I leave the old School of Hygiene, these and allied thoughts trouble my mind. Descending the semicircular stone steps that my grandfather so often trod, I am suddenly engulfed by the angry rattle of jackhammers, an infernal racket that I know Gerry's sensitive ears could not have tolerated. Gazing up, I realize I am flanked by a pair of sleek, thirteen-storey glass and steel towers, one devoted to pharmacology and the other to cellular and molecular biology, which now rise over my grandfather's former domain. Here, a brave new generation of Huxleyan scientists will artificially encode virus genomes to attack disease; perhaps their idealism will drive some to seek the unholy grail of immortality, delivering to us the perfect, disease-free superman, immune to sickness and decay. Yet the paradox will never die: If we deny the very flaws, imperfections, and vulnerabilities that make us human, we dam up the source of our greatest human strengths—our intuition, creativity, playfulness, humour, spontaneity, tenderness, spirituality. Our capacity to grieve the inevitable shocks of life. If we eradicate all sadness and suffering and imperfection, we eradicate our depths and meanings; we wipe out our human selves.

Slipping into the medical school next door, I walk with the students down a long corridor, passing the busts of Banting and Best. Both walls bear rows of glassed portraits of graduating classes dating back to the nineteenth century. Under 1903, I quickly locate the face of my twenty-year-old grandfather; turning to look for my father in the class of 1942, I find him, in another in a long line of coincidences, directly across from his father, the two symbolically aligned as they were not in life. Their eyes stare at each other as if across an uncrossable abyss, and for a moment, I realize I am slung between them, like a rope bridge swaying in the wind.

As I return to the clamour of College Street, I ponder, for the thousandth time, my grandfather's achievement, in all its complexity and magnanimity. And my father's as well. For most of us, the impossible idealism of preventive medicine remains elusive and abstract, demanding a willful leap of imagination to fully grasp, let alone practise. Sometimes I accuse myself of being morbid or self-absorbed if I dwell too long on the catalogue of tragedies, the agonizing, hidden private cost of the public health triumphs—a kind of pyrrhic victory, strangely invisible to the world they serve—that my grandfather exacted from himself and his own family. Even as I think of the countless, faceless strangers whose lives were saved by the utilitarian efficiency of the Connaught Labs—"the greatest happiness of the greatest number"—my mind shifts to the legacy willed to us by our great-grandfather Will: the image of my grandfather (less of a stranger, but a stranger all the same), bleeding to death in his hospital bed; the body of his brother Sidney, the desolate, self-effacing bachelor, smashing onto the bottom of the Cedarvale ravine; his brother Bill, huddled in the shock wards of 999 Queen Street West; his daughter, Molly, a brain-addled dipsomaniac, confined to a sanitarium in her last years; Jack, the perpetual son, cast into the death-in-life of self-exile, his body having tolerated enough drugs to kill a horse. I could point to the loveless, body-snatching Victorian nursery as the source of it all, but of course I cannot produce the hard proof demanded by a world ruled by science. From the beginning, I was invited to look the other way and let sleeping dogs lie, for that was the Toronto way; and yet there was enough of a questing scientist alive inside of me that I refused.

Pressing westward for another block, the old Clarke Institute of Psychiatry at 250 College is now upon me. I cannot fail to meet the image of my sister, over thirty years earlier, sprawled on the front lawn of Homewood, wailing in anguish, unseen and unheard by the doctors within. In 1998, the four Toronto institutions through which my father, the hot Irish potato, was passed in the 1970s—the Clarke, the Donwood, the Addiction Research Foundation, and 999 Queen Street West—were consolidated under a single organization with the Orwellian name of the Centre for Addiction and Mental Health, where doctors struggle to get half the occupants off drugs and the other half on drugs. I think of my gowned father pacing the crazy floors above my head, the voiceless specimen squirming inside his Kraepelinian box, pressed flat on the pages of his Meyer-esque clinical file; I think of the failure of the specialized scientific mind to grasp the humanly obvious: that my father laboured "under the influence" of his influential father as much, or more, than any drug.

The University of Toronto's Department of Psychiatry is now seven hundred strong, the largest in the world, the fulfillment of C.K. Clarke's century-old dream of scientific psychiatry, secure in its belief that the human condition is a medical condition, mechanically reducible to metaphor-free diseases of the brain. Historically, our medical treatment of the insane has been, for the most part, just another form of insanity. Despite the ongoing good intentions of legions of decent people, I suspect the capacity of psychiatric thinking to heal and restore the human soul has evolved little over the past generation. Rarely do we find the word "soul" as part of a psychiatrist's vocabulary; they study emotions, yet do not have them. By 2020, the World Health Organization projects that depression will be the leading cause of all disability; meanwhile, doctors, its putative "fixers," paradoxically rack up suicide rates twice that of the general population they serve. In the labs, the quest for a magic bullet continues apace, just as my grandfather aimed to do a century ago; meanwhile prescription drugs continue to kill more people than street drugs. Turning on the amnesiac wheel of history, earnest men and women parse genomes, fashion new pills, slice open human brains, scan darkened tissues, or dream of creating a vaccine for the virus of madness or the cravings of addiction. In 2003, out of the Toronto Western Hospital, the place of my birth, came the latest in a long line of aggressive interventions: pacemaker-generated electrodes surgically implanted in the brains of severely depressed "treatment resistant" patients of my father's ilk. As long as people suffer, they will seek and receive desperate attempts at cures; simultaneously, I indulge in dreams of a utopian alternative—that an enlightened twenty-first-century philanthropist will one day found a truly therapeutic "holding environment" for the nervous and the mad, staffed solely by tough, compassionate, psycho-dynamically trained laypeople, a place purged of dissection-minded medical doctors. There are some places the hand of science has no business being.

Berkeley, California, 1922

As I continue my walk westward, a kinder, more hopeful train of images arises. I come upon 1 Spadina Crescent, the former nineteenth-century theological college, a gothic island of land encircled by streetcar tracks, where the Connaught Labs once pumped out heroic quantities of blood plasma, penicillin, and polio vaccine. It is heartening to think that the idea that my grandfather hatched in 1916—free distribution of vaccines across Canada—is carried out by Sanofi Pasteur and others to this day and covered by the medicare system. I fantasize an image of my grandfather, a benign old man enjoying a peaceful retirement, picnicking here on the lawn with his grandchildren, or in the natural beauty of the Connaught Farm, mugging for the camera as he cultivates his garden, not "playing to win," just playing. I wonder how he would have viewed the cultural revolution of my own privileged '60s generation, which flagrantly rejected his values of duty and self-denial in favour of sex, drugs, and rock and roll. I still dream of rich conversations unfolding between us, the ones I never had with my father. I assume he would have been puzzled, as was my father, as to why my siblings and I never chose to marry or have children; that our line of the FitzGerald name will die with us. My choice to resist the pressures of social convention—and the fear of emotionally poisoning my offspring—cost me the experience of being a father or grandfather. Under the natural laws of Darwin, I am marked a eugenic fiasco; yet, of course, I feel no such failure. If the child is father to the man, then I have, with help, fathered—and grandfathered—myself as best as I can.

For if we care to look, consolations abound. When I interviewed Dr. John Hamilton who as dean of medicine had presided over the breaking-up of my grandfather's legacy twenty years earlier, he said he regretted the decision in hindsight; a prominent doctor even suggested, half-seriously, that instead of merging the School of Hygiene into the university's Faculty of Medicine, the Faculty of Medicine should have been merged into the School of Hygiene. A tangible form of vindication arrived in 2008 when, in the wake of the 2003 SARS crisis—a new infectious disease that killed dozens of Torontonians and revealed the depths of neglect and disorganization to which Canada's once incomparable public health system had sunk—the University of

Toronto founded—or refounded—with twenty million dollars of philanthropic assistance, a powerhouse School of Public Health that promises to reclaim a lost world-class status. The new school embodied a welcome restoration of my grandfather's vision and values, and I was thrilled when its forward-and-backward thinking director responded enthusiastically when in 2008 I proposed the creation of an exhibit inside the school devoted to my grandfather's memory. That same day, as I walked through the university's medical school and passed the plaque commemorating the site of the discovery of insulin, my happiness was, for a moment, cut short: mere feet away, a Tim Horton's doughnut shop now stands as a sugar-fuelled testament to the overriding influence of the corporation—and the irony-deficient.

The wheel of eternal recurrence spins on. I am reminded of the relay races I used to run on the oval track at school, the passing-off of the baton from runner to runner, my heels and Achilles tendons aching in protest; how, along the lines of my own paternal legacy, the baton secretes a vial of some unnamable poison, capable of giving both life and death. I see my great-great-grandfather John, a freighted, frightened thirteen-year-old boy struggling inland from the Irish ship docked in Port Hope harbour, branded by the harsh immigrant winters, destined to die in the wilds only days after his eighth and last child, William, is born. On comes Will, wresting a prosperous life from the making and selling of handmade painkillers. Yet in the black hole of his fatherless rages, he must pour poison on the heads of his three sons; impotent to save his invalid wife, he must capitulate to toxic, irredeemable despair. And so his eldest son, Gerry, the gifted one, takes up the baton, chasing an impossible ideal founded on the noblest of motives, as if in the bleeding of horses he can fly above his father's tainted descent. But this was earned at a price, for while his amazing antitoxins wiped out disease, they sometimes invited death-dealing allergic reactions. And so his only son, Jack, unconscious as night, cut through the thickets of allergy and immunology, embraced the profit model, and cleaned up after his father's mistakes. And yet what did it profit him? There is, as the poet says, more to life than having everything.

And so comes James, the last of the FitzGerald line, sentenced to writing the last of his lines. Each in his own way, the men of my family were pathologists, makers of our own truth serums, set on our cyclical paths of conquest and reparation, going against the grain. But it fell to me to uncover the poisoned well, the unmarked burial plots, the empty holes inside us where the secrets lay; someone had to dig down, then fill them with words. Looming ahead on the cindered track, the tape stretches across the finish line, the tape I am told I must cut in "first place"; and as my father extends the baton, silent to the last, I know what I must do.

═

In my dream, I am standing in the second-floor den of 75 Dunvegan Road, the room where my father, drugged and inert, passed much of the 1960s slumped in his armchair, staring at the smooth glass of the television screen, showering him with endless talk shows. This was the room where I sometimes docked beside him in weird, talk-less kinship; this was the room where my father gave me "the silent treatment."

But in this dream, in this room, it is my grandfather, not my father, whom I meet. As always, he is wearing a tie and jacket; in all his pictures, he dresses formally, even while communing with the cathedral of old-growth trees and breezy beaches of Vancouver Island. I recognize his profile from his pictures, the nose and chin mirroring the jutting peninsulas of western Ireland; he emits an unworldly aura, hyperreal, like a Renaissance oil painting, like the mystique all longing children imbue in a remote parent. He moves from the bay window to the bookshelf and back again, restless and edgy. I am anxious to interview him, to prime a flowing conversation, but I'm careful not to press too hard.

I want to ask 1,001 questions, verify my own intuitions, as if fact-checking the book I am writing. But I rein in my excitement at meeting him in the flesh and blood, resisting the temptation to feel for the Christ-like gash in his side. I want to find a way to tell him to cut Jack some slack, to stop pressuring him to become a doctor, a copy of himself; that if he did, he might alter the course of my own destiny, his unborn grandson.

Delicately, I pose the first question:

"What was it like working in your father's drugstore?"

"Boring." His voice sounds like my own.

I expected the opposite response; I had convinced myself that he loved every minute of it.

I ask if the family "oral history" was true—that his sister, Hazel, had to take care of their invalid mother, Alice, preventing Hazel from enrolling in Bishop Strachan, the private girls' boarding school.

"Hazel went to Havergal," he corrects me, dodging my implicit invitation to talk about his sick mother.

I know his wife, Edna, went to Havergal, not his sister; I'm beginning to realize his "facts," even the insignificant ones, are not reliable. It's as if he's deliberately trying to put the hound off the scent. But I am determined to keep up the questions, warming him up, hoping, stealthily, to lead up to the mother of them all:

"What was the unpardonable sin?"

But that moment never comes; abruptly, the emotional force of the dream snaps me awake, flapping like the torn sprockets of a strip of film.

I lie still, the shafts of dawn light cutting across the foot of my bed. I turn the dream over in my head, then groggily reach for my pad and pen, as I have done countless times before, as if emerging from an insulin coma. I think of the pilot feverishly transcribing the last stream of words, the war secrets, the dying Banting poured out in the Newfoundland snow.

In the dream, I'm most aware of the fact that my grandfather cannot hear me or take me in; my credentials as his grandson gain no purchase on his time. It seems I'm fated to play the role of the archaeologist of silence, slipping around the surfaces of the past, deferentially questioning the man of vision, the man of action, the man of despair, who can't, for the life of him, see me.

===

I suspect the ghost of my grandfather hovered beside me as I made those sporadic intrusions, year after year, into that forlorn apartment of my father on Orchard View Boulevard. Only after his death did I realize that in my father's room, there was no room for me; yet strangely, apartment 1107 was the place where I felt most like a father myself. Only now do I understand why he could not rise to the status of fatherhood: He always saw himself as a son. (Gerry's pet name for him was "Sonny Boy.") His father was simply too big.

Jack and Janet in sunnier days, before I was born

"His parents were so remote," my mother once told me, "he had no one to hold on to." It was a truth, of course, that could have fit us all. Nearly twenty years after Jack's death, random images still course through my mind: How he silently skimmed leaves from the blue surface of our backyard swimming pool and fiddled obsessively with its "chemical imbalance"; how he habitually sorted the sections of the daily newspaper, keeping sports and business and throwing out the others; how he walked the dog through the streets of Forest Hill, alone with his tumbling thoughts; how he never dropped the cool, knowing posture of the hipster who feels no need to explain himself. I think of how, on my irregular visits to his apartment, the routine never varied: We'd shuffle out to a fine French restaurant on Yonge Street for dinner; often he'd wolf down his food and fall into a choking fit, then bid the waiter to bring the bill before we finished the main course. I'd see him back up the elevator to his apartment, where he'd sink into his worn leather chair, his pallid, stubbled face bathed in the sickly blue light of the TV screen, an unsteady hand grasping the remote control like a child. As I slipped out the door, he'd spoon up vanilla ice cream straight out of the carton, as if regressing in search of the breast he never had. My visits seemed like futile puffs of mouth-to-mouth resuscitation; yet words, my solitary strength, failed to move him. Sometimes I feared I had the power to stab him to death, as if my tongue was the sharpest of needles. Perhaps I was expecting far too much, flogging a dead horse; after all, my father's brain was scrambled by grand mal seizures ignited by the ECT, the tributaries of his bloodstream polluted with decades of toxic drugs, the thousand unnatural shocks his flesh was heir to. His most attractive quality—his spontaneity—had long since been crushed out of him. Perhaps it was vain of me to think I could teach an old dog new tricks.

Every Christmas, my father hated the ritual exchange of presents, manically crumpling up the cast-off piles of wrapping paper and thrusting them into the raging fireplace, violently disturbed by the human business of giving and receiving. For all that, he remained a creature of memory. One year, I made him a tape of Django Reinhardt music, including a cut of "Sweet Chorus," the cherished rare recording that I had accidentally broken as a toddler at Balmoral. He perked up for a moment, craning his neck to see where the melody was coming from, then receded behind his mask of mute defiance, once again declining, as it were, to face the music. In hindsight, I'm not sure what I hoped to see change. Who was I to suppose I possessed such influence? I was made to respect his refusal to be known like the pages of an open book; he was free to spurn what other people defined as happiness. My brother, sister, and I made separate visits to our father, for we knew he could not tolerate the presence of all three of us for long; nothing had changed from our—or his—childhoods on Balmoral. Like King Lear, my father was, in the end, a man "who but slenderly knew himself . . . more sinned against than sinning."

I suppose he wanted what most of us want: to be loved for who we are, not what we do; our truest selves ultimately seek not pity, but understanding and acceptance. And yet, as I circled the dead planet of my father like a monkey in a sputnik, why could I never shake the feeling that, congealed in the narcissism of his own victimhood, he knew exactly what I wanted from him? That he was willfully withholding the truth as he knew it? The cagey glimmer in his eyes made me think I wasn't the only one doing all the thinking.

My father was such a difficult man that I suspect that even if he had found the right kind of help, he would have found ways to resist it. And yet, and yet. I will, of course, never forget that overcast autumn Sunday in 1969 when he drove down to my undergrad digs at Queen's, weeping scalding tears of self-loathing, and wrapped his unfamiliar arms around me for the first and last time in our lives; how I stood as cold and still as a statue; how he turned round and headed back down the highway into the arms of narcotic oblivion, his darkness extinguishing his brilliance. Only years later did I open myself to the intuitive conviction that he had felt compelled to repeat the cyclical, unconscious drama of the third storey; that his father had done exactly the same thing when *he* was a university undergrad. At the moment my father opened up to me, at the moment he was most human and most reachable, it was too late. And that, of course, is a tragedy.

An historical tragedy, too, that the Freudian movement suffered an aborted childhood in Canada, all the more ironic given that my grandfather, an earnest colleague of Ernest Jones, was uniquely placed to plant the seed of psychodynamic psychotherapy in Toronto as early as 1908, a seed that might have borne and harvested mature fruit by my father's time. But my grandfather loved hard science too much to be swayed by a method that smacked of art. Freudian thought would not penetrate the Canadian Shield until the 1950s, and only then fatally cast in medical garb, in defiance of Freud's own desire for the supremacy of lay therapy.

In the end, I can forgive my father's silence, for now I understand his sadness. Less easy to forgive are the men who failed to even try to understand him, the men who failed to grasp—many of them willfully and heartlessly—the elementary power of the unconscious mind and the "undreamt of importance" of the transference, the men who declined to stand in awe of the eternally recurring drama of father and son; the men who—had they risked opening themselves up to my father in his vulnerability, risked feeling what he felt—risked facing their own tangled feelings for their own fathers. "We act out," Freud said, "until we remember."

For all its inherent limitations, talk therapy retains an indispensable virtue: A real person, made of flesh and blood, is there to listen, to bear witness, to open up and experience the suffering of the other; that we all possess majestic, poetic depths, that we all relive the past in the present, that we carry meanings inside us, infused by the transformative power of language. My grandfather—the blinkered, high-strung, winning thoroughbred racing past all thoughts of loss—could not find the time to recline on a couch and shine a headlight on the dim, unexplored rooms of his inner life. And yet, perhaps, if he had, the path of fate might not have compelled his only son, a generation on, starving for a dead father's love, to climb those stairs like a sleepwalker, to cross the threshold of that room at the top, to lie flat on that son's bed, to reach out, not for a kind word, but for a syringe.

If my father and grandfather could not forgive themselves for their unmanly collapses, their unpardonable loss of control, I can and I do. Only recently did I come to appreciate that both were radical men for their time and profession in that they openly sought help and did not deny their problems, even in their deepest shame. Even as I was appalled by the bunglings by two generations of unconscious psychiatrists, Gerry and Jack implicitly gave James paternal permission to explore himself on his own terms—a healthy part of my inheritance. My father and grandfather's terror of their own pasts ultimately destroyed them; their fears gave me the courage to confront my own. Year by year, it was the regenerative cycles of dreaming and listening and talking, without a single drug or shock, that broke the unconscious compulsion to repeat and restored me to myself; only now do I know for certain that without the ever-unfolding legacy of Sigmund Freud, I would likely have not lived past my fifth decade. Nor would I have succeeded in radically reducing the painful inflammation in my guts, my forty-year-long bout of Crohn's Disease, had I not, in 2007, consulted a naturopathically inclined MD. He prescribed me a kinder, gentler diet that is gradually removing my dependence on a prescription drug. Aptly, the doctor works at the foot of Balmoral Avenue, where it meets Yonge Street, the symbolic locus of my beginnings on earth.

Perhaps that's why I have so closely embraced the multiple ironies of my life as friends. If my grandfather had been a different man, driven by personal greed over public need, I might have inherited part of a vast fortune earned by a family-owned private lab; such cushioning wealth would have likely undercut any impulse for self-reflection. But fate—and something inherently self-effacing in the Canadian character—dictated otherwise. Insulin revenues sustained the Connaught Labs during the rigours of the Great Depression, but its founder failed to profit from insulin, nor insulin shock, in his own great depression. (Nor has the country: For years, no Canadian lab has manufactured insulin; and my grandfather's old American partner, Eli Lilly, has long since grown fat on massive profits from Prozac).

But in my own struggle, I did inherit—and learn—something priceless. Skimming across the inscrutable face of my father's silence, I learned that words heal.

═══

One blazing hot day in the summer of 2005, I headed off for my regular swim in the Hart House pool, my chance to be alone with my thoughts, cleansed by the cooling blue water. Resembling a Roman bath, arched and tiled and columned, with a curving barrel skylight, the pool is the same one my grandfather frequented in the 1920s and 1930s. The lanes are divided into three speeds—fast, medium, and slow—and as I slip into the water, I imagine the speed-demon doctors, Best and FitzGerald, cutting through the space ahead of me, as if life were a perpetual race, an escape from intolerable feelings of mortality. Even though I am a strong swimmer, even though I am tempted to catch up, I choose to hew to the middle way.

On this day, I left my gold signet ring, inscribed with the crest of the sword-wielding Knight of Kerry, in my locker; when I returned after my lengths, I discovered a thief had broken the lock and stolen the ring and cash from my wallet. Naturally, I felt angry, and I wondered if I could afford the price of a new ring. But then I paused and thought: the loss seemed a kind of sign, a fit end to a strange story, this fable of familial estrangement. Over the years, my grandfather's grip on me had gradually weakened, inch by imperceptible inch, like the dying of a passionate yet unrequited love affair; the ring had symbolized an inherited neurotic tie, a kind of shotgun marriage to the Irish apparitions. After years of work, I was nearing the end of the first draft of my book, and with it a dim feeling of divorce and rebirth; it felt like the right time to cut Gerry, and myself, loose. Even as I strolled out of Hart House and across the tree-shaded lawns of Queen's Park, I knew it was the right thing to do, as surely as I knew my own name. As if for the first time, I sensed I was a changed man, and changing still. Then came the epiphany: I was daring myself to be happy.

Some months after our mother died in early 2006, I showed my working manuscript to my sister; I knew my rendering of the horror of September 1970, as she climbed the third-floor stairs of Dunvegan, was my creation, not hers. Our relationship had been tense for some time and I hoped words on the page might heal something between us. And they did. Shelagh told me that my version of events was not entirely accurate—somehow the details of the searing double betrayal, first the mother, then the father, had not fully penetrated my consciousness, partly due, I suspect, to my mother's dominating influence. Now I felt I understood something fundamental for the first time: Incredibly, over thirty-six years had to pass before Shelagh felt free enough to divulge the *full*, painful story to my brother and myself, aided, no doubt, by the recent death of our mother. Such was the stunning power of the family drama to hold our lives in check; such was the stunning power of the antidote of truth-telling to strip off the layers of myth we build over pain like spikes on a moated castle.

As I write these words, I am sitting alone in my office, tapping at my computer as if it were the baby grand piano that once stood in the living room of Balmoral. It's no coincidence that jazz is my least favourite genre of music, but today a live cut of Duke Ellington's anthem, "Take the 'A' Train," is once more flowing from the speakers; as I harvest middle age, I remain forever amazed by the uncanny power of music to fling us back into the smoky alcoves of the past. My mind dwells on the thousands of train trips taken by my hard-charging, Type A, globe-trotting grandfather; then I think of his son, spinning his wheels to keep up, with nothing but his jazz to carry him.

Another train of thought returns me to the day in 1998 when, as part of my ongoing exorcisms, I visited 186 Balmoral Avenue. The house remains, of course, the ultimate archive, the tomblike casebook, the womb-matrix of contagious nightmares common to fathers and sons and grandsons. Everything I ever needed to know I knew as a boy there, although half my life would pass before the son would at last remember what the father had forgotten.

The new owners, a smiling, childless couple in their thirties—the same age as my parents in the 1950s—kindly allow me to wander alone from floor to floor. Faced with their cheerful amiability, I cannot bring myself to reveal the dark history

Shelagh, Michael, and James at Balmoral, 1953

of the house they seem to love. The rooms appear virtually unchanged—the same hardwood floors, the cast-iron radiators, the squinting dormer windows, the staircases echoing with the mingled footsteps of three generations—and as I pass through the museum of my conception and the nursery of my night terrors, I might as well be climbing over crumbling Norman ruins in the Irish mist. The melancholy mood is broken by the sight of a newly installed skylight, flooding a shaft of sun into the first-floor foyer. Its rectangular, booklike shape makes me think of my own struggle to shed enlightening words on the darkness and a smile rises to my lips. Suddenly I feel like Dickens's Pip, ripping down Miss Havisham's rotting curtains to let in the light. I say the words aloud: *I am here; I am here; I am here.*

For years, I had a recurring dream that I took to my therapy sessions: Revisiting Balmoral for the first time since childhood, I encounter friendly new owners who are renovating the place. I had lost count of the number of times I had dreamt this scene, now unfolding before me in my waking life. I am literally experiencing a dream coming true.

Preparing to leave, I thank my hosts for their graciousness; then I realize I am standing on the threshold of the living room door where the boy who would be good must teeter for a lifetime, unseen and unheard, the child as perpetual guest; no coincidence, I muse, that my grandfather suffered the role of "guest" in his Hartford hell.

"Was this a happy house?" comes the voice of the beaming Balmoral wife, and I can only telegraph back a wordless smile. I take a last look at the child-hating Victorian crypt that filled hundreds of my dreams, the room now so small and innocuous and banal and unmasked, and the flitting apparitions competing for my waning attention. By the orange glow of the fireplace of "Club 186" lies the prostrate figure of Count Basie, stoned and down for the count; by the baby grand, Banting and Best play charades with my sad-eyed, pyjamaed father; on the windowsill sits his talismanic plaster statue and on the wall hangs the winter landscape painting, dead as dust. The images flicker, then ember-like fade away for good. In this living room of ghosts, there was no room for the living child; and so my adult body turns and steps across a once uncrossable threshold, outside into the life-giving air of the city, into the wide open world, into myself being reborn.

═══

I know my work is never done, but it is time to rest for a spell from the ceaseless business of saving and being saved. My wall calendar reads September 2007, the month of my fifty-seventh birthday, the same age as my grandfather when he plunged the sword in his side, the same age that my father, emerging from the soul-crushing grind of the psychiatric mill, gave up the ghost, sitting alone, like Saturn, the god of melancholy, devouring his own children. My grandfather was a man who lived for the future, obsessively hurtling ahead, down the twisted, crowded street of his fifty-seven years, trying too hard for his own good. I was similar, then I was different: Turning in the opposite direction, I found my life in the past; and it is here that we meet again, at the intersection of our time travels, in the flesh of the words I write, here at the eclipse of the father by the son.

I am still standing; but I am not standing still. I am walking alone through a moonlit dusk in late winter, perhaps a dream, perhaps not. I am heading north towards the Avenue Road hill where, in centuries past, the waves of a vast glacial lake lapped at its base. Here, the former homes and apartments of four generations of the FitzGerald family cluster and resonate—the hill where my widowed great-grandfather, Will, once lived alone with his killing melancholy; the hill down which my three-year-old brother once fled the chill vacuum of the Balmoral nursery; the hill my sister, in this year of 2007, has decided to come home to, "like Dorothy," occupying an apartment directly across the street from our grandmother's bedroom where, fifty years earlier, she died of a broken heart.

But on this day, the old neighbourhood feels like a changed place, an excavated world that, had I not dug it up, would have buried me. As I walk, I feel lucky to be alive, secure in the knowledge that if my grandfather was a victim of his success, I am a beneficiary of his failure. In the distance, I see the tall, thin object of my former obsession coming down fast from Balmoral Avenue, like a steaming A train, over the snow-topped brow of the Avenue Road hill, even as I am striding up towards him, up and up and up, my heart beating like mad. I am coming closer and closer and he grows taller and taller, sharper and sharper. I see his keen, green Irish eyes shooting deep into mine, yet I know my grandfather, my intimate red-headed stranger, the mythical, wounded Fisher King with the medieval lance piercing his groin, does not know me; nor that I now know him far better than he ever knew himself.

In the thinness of a moment, we are exactly the same age; and as we pass, I feel the cold breath of mortality brush my cheek. I feel a fleeting impulse to stop, perhaps to speak, perhaps to listen, perhaps to rave in praise, to rage in anger, perhaps to throw my arms around him in a moment of anguished catharsis. But I feel I have done all these things, over the slow wheel of the years, week by week and page by page, in my own way, shedding my identification with him and finding my own self. And so I choose to simply nod, wordless, and head up towards the crest of the hill.

Over my shoulder, my grandfather grows smaller and smaller, diminishing in stature, dissolving, like sutures of grief, into a blurred, childlike shape, small and alone in the distance. I no longer feel like a fragile, italicized presence hovering on the margins of the past, riding the bold coattails of my grandfather's jet stream, dreaming his dreams; I have earned my own. Little by little, I have cut up the windblown deadfall, strewn across the Western world, and laid to rest the ghost that has haunted my body since birth. He is dead and I am alive; my conversations are now with the living.

I do not look back, but press north, walking at my own even pace, step by step, looking forwards and upwards, past the skeletal remains of my own survival guilt, open to the scything winter winds, edging towards the once-impossible prospect of loving and being loved. I now know with a quiet clarity that when I reach the unknowable moment of my own end, it need not be an unnatural one; even as I age, I grow. As I carry on, I feel in the flesh of my body the true lightness of my soul, as strong and free and vital as I have ever known, bearing me up—up and over the frozen, bone white pavements of the city of Toronto.

James on the Avenue Road hill, 2009

Acknowledgements

I am indebted to the many historians and scholars whose published and unpublished works, listed in a separate bibliography, guided me through the labyrinthine medical careers of my father and grandfather. In particular, Michael Bliss's books on the life of Frederick Banting and the discovery of insulin proved invaluable.

I cannot recount in full the dozens of generous people whose friendly conversation and support leavened the intensity of an arduous odyssey, but I will single out a prominent few: Bob Banting, Henry Best, Bill Brown, Christine Buijs, Sarah Byck, Kathleen Byrne, Jane Bryans, Liz Calvin, Carmela Circelli, John Court, Frank and Tanis Cherry, Mark Czarnecki and Liz Kalman, Peter and Judy Dales, Lee Daugharty, Craig Defries, Doug and Jane Densmore, Craig Doyle, Ian Easterbrook, Eric Evans, Joan Farrar, Leonard Roy Frank, Haigan Gobalian, Cyril Greenland, Naya Kee, James Laidlaw, Jennifer Lem, Leah Lucas and Damian Da Souza, Ken and Susan Ludlow, Margaret McCaffery, Don McKibbin, Peter Murdoch, Terry and Janice Priddy, Geoff Reaume, Christopher Rutty, Robert Sandler, Jack and Anne Schaffter, Ned Shorter, Nancy Simpson, David White, Don Weitz, and Margaux Williamson.

Of the many medical professionals I interviewed, I'd like to acknowledge in particular Alice Briggs, Roman Bladek, Irv Broder, Stan Epstein, Ken Ferguson, Don Fraser, Jr., Steven Fried, Jack Griffin, John Hamilton, John Hastings, Barbara Hazlett, Robert MacBeth, Gordon Romans, and John Toogood.

The following institutions provided research material: the Buffalo State Hospital Archives; Centre For Addiction and Mental Health Archives, Toronto; Chesney Medical Archives, Johns Hopkins Hospital; City of Toronto Archives; Hartford Institute of Living Archives; Harvard University Archives; National Archives of Canada; Homewood Health Centre; Ontario Archives; The Rockefeller Foundation Archives; Sanofi Pasteur Archives; University of California, Berkeley, Archives; University of Toronto Archives; and Wellington County Archives.

I am grateful to the Canada Council, the Toronto Arts Council, the Lil Hewton Memorial Bursary of the Centre For Addiction and Mental Health, and Dr. Barbara Hazlett for their financial assistance.

In 2004, Mark Lievonen and Luis Barreto of Sanofi Pasteur were instrumental in the conversion of my grandfather's Barton Avenue horse barn, the seedbed of Canada's modern public health system, into a museum; that same year, they played a key role in the induction of Dr. FitzGerald into the Canadian Medical Hall of Fame. I am also grateful to Dr. Jack Mandel, director of the University of Toronto's Dalla Lana School of Public Health, whose generous historical-mindedness spawned the establishment of The FitzGerald Reading Room at the University of Toronto.

The compassionate ruthlessness of my editor Anne Collins I found indispensable; without her extraordinary acumen, enthusiasm, and good humour, the book would not have assumed its ultimate form. Additional thanks to Craig Pyette for futher attention to detail. Thanks to my friend Helen McLean for suggesting the title, *What Disturbs Our Blood*, drawn from the poetry of Yeats. The ongoing guidance of my agent, Suzanne DePoe, I continue to appreciate.

A book about family members, living and dead, risks the creation, or deepening, of rifts; more so a book that aspires to honesty. I am fortunate that my own family has accepted my questings in a spirit of healing. For their inseparable gifts and curses, I thank my late parents, Jack and Janet. For their tolerance and understanding, I thank my sister, Shelagh, and brother, Michael; my paternal cousins, Anna and Patrick Whitley; and my half-brother, John Leuthold. Finally, I thank my partner, Katy Petre, whose constancy, generosity, and love saw me through my best and worst moments.

═

Audio-visual material related to *What Disturbs Our Blood* accessible at: www.jamesfitzgerald.info

SELECT BIBLIOGRAPHY AND SOURCES

BOOKS:

Akenson, Donald Harman. *The Irish in Ontario: A Study in Rural History*. 1st ed. Montreal: McGill-Queen's University Press, 1984.

Andre, Linda. *Doctors of Deception: What They Don't Want You to Know About Shock Treatment*. New Jersey: Rutgers University Press, 2009.

Ardagh, John. *Ireland and the Irish: Portrait of a Changing Society*. London: Penguin Group, 1995.

Barry, John M. *The Great Influenza: The Epic Story of the Deadliest Plague in History*. New York: Penguin Group USA, 2004.

Bator, Paul and Andrew Rhodes. *Within Reach of Everyone: A History of the University of Toronto School of Hygiene and the Connaught Laboratories*. 2 vols. Ottawa: Canadian Public Health Association, 1990, 1995.

Beers, Clifford W. *A Mind That Found Itself*. Pittsburgh: University of Pittsburgh Press, 1981.

Best, Henry B. M. *Margaret and Charley: The Personal Story of Dr. Charles Best, The Co-Discoverer of Insulin*. Toronto: The Dundurn Group, 2003.

Blair, John S. G. *In arduis fidelis: Centenary History of the Royal Army Medical Corps 1898–1998*. Edinburgh: Scottish Academic Press Ltd., 1998.

Bliss, Michael. *The Discovery of Insulin*. Chicago: University of Chicago Press, 1982.

———. *Banting: A Biography*. Toronto: University of Toronto Press, 1984.

———. *William Osler: A Life in Medicine*. Toronto: Oxford University Press, 1999.

Braceland, Francis J. *The Institute of Living: The Hartford Retreat, 1822–1972.* Connecticut: Hartford, 1972.

Breggin, Peter. *Toxic Psychiatry: Why Therapy, Empathy, and Love Must Replace the Drugs, Electroshock, and Biochemical Theories of the "New Psychiatry."* New York: St. Martin's Press, 1991.

Brome, Vincent. *Ernest Jones: A Biography.* New York: W. W. Norton & Company, 1983.

Bryer, Jackson and Cathy W. Barks, eds. *Dear Scott, Dearest Zelda: The Love Letters of F. Scott and Zelda Fitzgerald.* London: Bloomsbury Publishing PLC, 2002.

Buchan, William. *John Buchan: A Memoir.* London: Buchan & Enright Publishers Ltd, 1982.

Bulfinch, Thomas. *Bulfinch's Mythology.* New York: Avenel Books, 1979.

Burlingame, Clarence Charles. *A Psychiatrist Speaks: The Writings and Lectures of Dr. C. Charles Burlingame, 1885–1950.* 1959.

Cassel, Jay. *The Secret Plague: Venereal Disease in Canada, 1838–1939.* Toronto: University of Toronto Press, 1987.

Chernow, Ron. *Titan: The Life of John D. Rockefeller, Sr.* New York: Vintage, 1997.

Connor, J. T. H. *Doing Good: The Life of Toronto's General Hospital.* Toronto: University of Toronto Press, 2000.

Cooter, Roger, Mark Harrison, and Steve Sturdy, eds. *Medicine and Modern Warfare.* Amsterdam: Editions Rodopi B. V., 1999.

Cosbie, Waring Gerald. *The Toronto General Hospital, 1819–1965: A Chronicle.* Toronto: MacMillan of Canada, 1975.

Dain, Norman. *Clifford W. Beers: Advocate for the Insane.* Pittsburgh: University of Pittsburgh Press, 1980.

Davenport-Hines, Richard. *The Pursuit of Oblivion: A Global History of Narcotics 1500–2000.* New York: W. W. Norton & Company, 2002.

Davies, T. G. *Ernest Jones, 1879–1958.* Cardiff: University of Wales Press, 1979.

Debré, Patrice. *Louis Pasteur.* Baltimore: The Johns Hopkins University Press, 1994.

Defries, Robert Davies. *The First Forty Years, 1914–1955: Connaught Medical Research Laboratories.* Toronto: University of Toronto Press, 1968.

de Kruif, Paul. *Microbe Hunters.* New York: Harcourt, Brace and Co., 1926.

Dowbiggin, Ian Robert. *Keeping America Sane: Psychiatry and Eugenics in the United States and Canada, 1880–1940.* Ithaca, NY: Cornell University Press, 1997.

Duffin, Jacelyn. *History of Medicine: A Scandalously Short Introduction.* Toronto: University of Toronto Press, 1999.

El-Hai, Jack. *The Lobotomist: A Maverick Medical Genius and His Tragic Quest to Rid the World of Mental Illness.* Hoboken: John Wiley & Sons, Inc., 2005.

Farley, John. *To Cast Out Disease: A History of the International Health Division of the Rockefeller Foundation (1913–1951).* New York: Oxford University Press, 2003.

Fedunkiw, Marianne. *Rockefeller Foundation Funding and Medical Education in Toronto, Montreal, and Halifax.* Montreal: McGill-Queen's University Press, 2005.

FitzGerald, John G. *An Introduction To The Practice of Preventive Medicine.* St. Louis: The C. V. Mosby Company, 1922.

Forbush, Byron and Bliss Forbush. *Gatehouse: The Evolution of the Sheppard and Enoch Pratt Hospital, 1853–1986: A History.* Philadelphia: J. B. Lippincott Company, 1986.

Frank, Leonard Roy. *The History of Shock Treatment.* 1978.

Frayn, Douglas H. *The Clarke and Its Founders: The Thirtieth Anniversary: A Retrospective Look at the Impossible Dream.* Toronto: The Clarke Monograph Series, No. 6, 1996.

———. *Psychoanalysis in Toronto: Historical Perspectives.* Toronto: Ash Productions, 2000.

Friedland, Martin. *The University of Toronto: A History.* Toronto: University of Toronto Press, 2002.

Gamwell, Lynn and Nancy Tomes. *Madness in America: Cultural and Medical Perceptions of Mental Illness Before 1914.* Ithaca: Cornell University Press, 1995.

Geison, Gerald. *The Private Science of Louis Pasteur*. Princeton: Princeton University Press, 1995.

Griffin, John D. *In Search of Sanity: A Chronicle of the Canadian Mental Health Association, 1918–1998*. London (ON): Third Eye Publications, 1989.

Grosskurth, Phyllis. *The Secret Ring: Freud's Inner Circle and the Politics of Psychoanalysis.* New York: Addison-Wesley, 1991.

Hare, Ronald. *The Birth of Penicillin and the Disarming of Microbes.* London: Allen & Unwin, 1970.

Harris, Seale. *Banting's Miracle: The Story of the Discoverer of Insulin*. Philadelphia: J. B. Lippincott Company, 1946.

Hincks, Clare. *Prospecting For Mental Health: The Autobiography of Clare Hincks.* Unpublished, 1962.

Hurd, Henry Mills. *The Institutional Care of the Insane in the United States and Canada*. Baltimore: The Johns Hopkins University Press, 1917.

Jackson, A.Y. *Banting as an Artist*. Toronto: Ryerson Press, 1943.

Jonas, Gerald. *The Circuit Riders: Rockefeller Money and the Rise of Modern Science.* New York: W. W. Norton & Company, 1989.

Jones, Ernest. *Free Associations: Memories of a Psychoanalyst*. London: Hogarth, 1959.

Karlen, Arno. *Man and Microbes*. New York: Tarcher (Penguin Group USA), 1995.

Kerr, Robert and Douglas Waugh. *Duncan Graham: Medical Reformer and Educator*. Toronto: The Dundurn Group, 1989.

Kolata, Gina. *Flu: The Story of the Great Influenza Pandemic of 1918 and the Search for the Virus That Caused It*. New York: Farrar, Straus and Giroux, 1999.

Leavitt, Judith Walzer. *Typhoid Mary: Captive to the Public's Health*. New York: Putnam Publishing Group, 1996.

Levine, Israel E. *The Discoverer of Insulin: Dr. Frederick G. Banting.* New York: Julian Messner, 1959.

Lifton, Robert Jay. *The Nazi Doctors: Medical Killing and the Psychology of Genocide.* New York: Basic Books, 1986.

Lownie, Andrew. *John Buchan: The Presbyterian Cavalier.* Edinburgh:

Canongate Books, 1995.

MacDougall, Heather. *Activists and Advocates: Toronto's Health Department, 1883–1963*. Toronto: The Dundurn Group, 1990.

Maddox, Brenda. *Freud's Wizard: The Enigma of Ernest Jones*. London: John Murray, 2006.

McLaren, Angus. *Our Own Master Race: Eugenics in Canada 1885–1940*. Toronto: University of Toronto Press, 1990.

Montagnes, Ian. *An Uncommon Fellowship: The Story of Hart House*. Toronto: University of Toronto Press, 1969.

Nikiforuk, Andrew. *The Fourth Horseman: A Short History of Plagues, Scourges, and Emerging Viruses*. Toronto: Penguin Group, 1991.

O'Brien, John. *The Vanishing Irish: The Enigma of the Modern World*. New York: McGraw-Hill, 1953

O'Driscoll, Robert and Lorna Reynolds, ed. *The Untold Story: The Irish In Canada*. 2 vols. Toronto: Celtic Arts of Canada, 1988.

Oliver, Wade W. *The Man Who Lived for Tomorrow: A Biography of William Hallock Park*. New York: E. P. Dutton, 1941.

Oppenheim, Janet. *Shattered Nerves: Doctors, Patients, and Depression in Victorian England*. New York: Oxford University Press, 1991.

Parkin, Alan. *A History of Psychoanalysis in Canada*. Toronto: Toronto Psychoanalytic Society, 1987.

Paskauskas, R. Andrew, ed. *The Complete Correspondence of Sigmund Freud and Ernest Jones, 1908–1939*. Cambridge: Harvard University Press, 1993.

Pettigrew, Eileen. *The Silent Enemy: Canada and the Deadly Flu of 1918*. Saskatoon: Western Producer Prairie Books, 1983.

Pickens, Donald K. *Eugenics and the Progressives*. Nashville: Vanderbilt University Press, 1968.

Porter, Roy. *A Social History of Madness: The World Through the Eyes of the Insane*. London: Weidenfeld & Nicholson, 1987.

———. *The Faber Book of Madness*. London: Faber & Faber, 1991.

Raymond, Jocelyn Moyter. *The Nursery World of Dr. Blatz*. Toronto: University of Toronto Press, 1991.

Real, Terrence. *I Don't Want to Talk About It: Overcoming the Secret Legacy of Male Depression*. New York: Scribner, 1997.

Reaume, Geoffrey. *Remembrance of Patients Past: Patient Life at the Toronto Hospital for the Insane, 1870–1940*. New York: Oxford University Press, 2000.

Roazen, Paul. *Freud and His Followers*. New York: Alfred A. Knopf, 1971.

Roland, Charles G. *Clarence Hincks: Mental Health Crusader*. Toronto: The Dundurn Group, 1989.

Roustang, François. *Dire Mastery: Discipleship From Freud to Lacan*. Paris: Les Éditions de Minuit, 1976.

Schwartz, Joseph. *Cassandra's Daughter: A History of Psychoanalysis*. London: Penguin Group UK, 1999.

Scull, Andrew. *Madhouse: A Tragic Tale of Megalomania and Modern Medicine*. New Haven: Yale University Press, 2005.

Shorter, Edward. *TPH: History and Memories of the Toronto Psychiatric Hospital, 1925–1966*. Toronto: Wall & Emerson, 1996.

———. *A History of Psychiatry: From the Era of the Asylum to the Age of Prozac*. 2nd ed. Hoboken: John Wiley & Sons, Inc., 1997.

Shortt, Samuel E.D., ed. *Medicine in Canadian Society: Historical Perspectives*. Montreal: McGill-Queen's University Press, 1981.

Showalter, Elaine. *The Female Malady: Women, Madness and English Culture, 1830–1980*. New York: Pantheon Books,1985.

Stevenson, Lloyd. *Sir Frederick Banting*. Toronto: Ryerson Press, 1946.

Szasz, Thomas. *Coercion As Cure: A Critical History of Psychiatry*. Edison (NJ): Transaction Publishers, 2007.

Taylor, Kendall. *Sometimes Madness Is Wisdom: Zelda and Scott Fitzgerald, A Marriage*. New York: Ballantine, 2003.

Tracey, Patrick. *Stalking Irish Madness: Searching for the Roots of My Family's Schizophrenia*. New York: Bantam Books, 2008.

Turnbull, Andrew. *Scott Fitzgerald*. London: The Bodly Head Ltd., 1962.

Warsh, Cheryl Krasnick. *Moments of Unreason: The Practice of Canadian Psychiatry and the Homewood Retreat, 1833–1923*. Montreal: McGill-Queen's University Press, 1987.

Wear, Andrew, Ed. *Medicine In Society: Historical Essays.* New York: Cambridge University Press, 1992.

Whitaker, Robert. *Mad In America: Bad Science, Bad Medicine, and the Enduring Mistreatment of the Mentally Ill.* New York: Basic Books, 2002.

Wilson, David. *The Irish In Canada.* 2 vols. Ottawa: Canadian Historical Association, 1989.

Wrenshall, G. A., G. Hetenyi, Jr., and W. R. Feasby. *The Story of Insulin: 40 Years of Success Against Diabetes.* Bloomington: Indiana University Press, 1962.

ARTICLES:

Barreto, Luis, and Christopher J. Rutty. "The Speckled Monster: Canada, Smallpox and Its Eradication." *Canadian Journal of Public Health*, 93, no. 4 (August 2002): Special insert.

Bates, Gordon. "Lowering the Cost of Life-Saving: How the University of Toronto Is Performing Active Public Service." *Maclean's*, August 1915.

Bator, Paul. "Saving Lives on the Wholesale Plan: Public Health Reform in the City of Toronto, 1900–1930." PhD thesis, University of Toronto, 1979.

Brown, Thomas E. "Living with God's Afflicted: A History of the Provincial Lunatic Asylum at Toronto, 1830–1911." PhD thesis, Queen's University, 1980.

Callwood, June. "The Miracle Factory That Began In a Stable." *Maclean's*, October 1955.

Cameron, Donald. "The First Donald Fraser Memorial Lecture." *Canadian Journal of Public Health*, 51, no. 9 (September 1960): 341–348.

Connor, J.T.H. and Felicity Pope. "A Shocking Business: The Technology and Practice of Electrotherapeutics in Canada, 1840s to 1940s." *Material History Review* 49 (Spring 1999): 60–70.

Connaught Laboratories. *Annual Reports: 1914–1972.*

Defries, Craig. "Now Is the Time: The Early Years of Dr.

Robert Defries." master's thesis, Carleton University. 1993.
Defries, Robert. "Thirty-Five Years: The Connaught Medical Research Laboratories." *Canadian Journal of Public Health* 39, no. 8 (August 1948): 330–344.
Dolman, Claude. "Landmarks and Pioneers in the Control of Diphtheria." *Canadian Journal of Public Health* 64, no. 4 (August 1973): 317–336.
Dowbiggin, Ian. "C K. Clarke" in *Dictionary of Canadian Biography*, edited by Ramsay Cook et al, 15 (2005): 212–215.
Dubin, Martin David. "The League of Nations and the Development of the Public Health Profession." presentation, annual conference of the British International Studies Association, University of Warwick (December, 1991).
Edwards, Frederick. "A Peacetime Munitions Plant: Where Science Fashions Its Weapons Against the Armies of Disease." *Maclean's*, January 1928.
Farrar, Clarence B. "Psychiatry in Canada" (unpublished manuscript, CAMH Archives, 1952).
Farrar, Clarence B. "I Remember J. G. FitzGerald." *The American Journal of Psychiatry*, 120, no. 1 (July 1963): 49–52.
Fedunkiw, Marianne. "German Methods, Unconditional Gifts, and the Full-Time System: The Case Study of the University of Toronto, 1919–1923." *Canadian Bulletin of Medical History*, 21, no. 1 (January–June 2004): 5–39
Glasser, Ronald J. "We Are Not Immune: Influenza, SARS, and the Collapse of Public Health." *Harper's*, July 2004.
Godfrey, Charles. "C. K. Clarke: Biographical Essay." (unpublished manuscript, CAMH Archives. 1960s).
Greenland, Cyril. "Charles K. Clarke: A Pioneer of Canadian Psychiatry." Toronto: Clarke Institute of Psychiatry, 1966; 32 pp.
Greenland, Cyril. "Ernest Jones in Toronto, 1908–1913, Part I." *Canadian Psychiatric Association Journal*, 6, no. 3 (June 1961): 132–138.
Greenland, Cyril. "Ernest Jones in Toronto, 1908–1913, Part II."

Canadian Psychiatric Association Journal, 11, no. 6 (December 1966): 512–519.

Greenland, Cyril. "Ernest Jones in Toronto, 1908–1913, Part III." *Canadian Psychiatric Association Journal*, 12, no. 1 (February 1967): 79–81.

Greenland, Cyril. "Three Pioneers in Canadian Psychiatry." *JAMA*, 200, no. 1, (June 1967): 833–842.

Griffin, John D. and Cyril Greenland. "Treating the Mentally Ill in Ontario: A Documentary History of the Development of Mental Health Services in Ontario from 1867–1914" (unpublished manuscript, CAMH Archives, 1979).

Griffin, John D. "The Amazing Careers of Hincks and Beers." *Canadian Journal of Psychiatry*, 27, no. 8 (December 1982): 668–671.

Hopwood, Catherine. "Canadians and Psychoanalysis: The Early Years." graduate paper, department of sociology, McMaster University, 1991.

Johnston, Penelope. "Dr. C.K. Clarke: Fifty Years of Service to the Mentally Ill." *Medical Post*, December 1996.

Kindquist, Cathy. "Migration and Madness on the Upper Canadian Frontier, 1841–1850." The Canadian Papers in Rural History, 8, University of Guelph, 1992.

Lamb, Susan. "Nothing Human He Dare Ignore: Approaches to Psychiatry at the University of Toronto, 1927–1967." paper, department of history, University of Toronto, 2003.

Lewis, Jane. "The Prevention of Diphtheria in Canada and Britain, 1914–1945," *Journal of Social History*, 20 (1986): 163–176.

McConnachie, Kathleen. "Science and Ideology: The Mental Hygiene and Eugenics Movement in the Inter-War Years, 1919–1939." PhD thesis, University of Toronto, 1987.

Museum of Mental Health Services (Toronto). "The City and the Asylum." exhibition notes (1993).

Nicolson, Murray. "The Other Toronto: Irish Catholics in a Victorian City, 1850–1900," *Multicultural History Society of*

Ontario, 1984.

Pos, Robert, et al. "D. Campbell Myers, 1863–1927: Pioneer of Canadian General Hospital Psychiatry." *Canadian Psychiatric Association Journal*, 20, no. 5 (1975): 393–403.

Price, Gifford C. "A History of the Ontario Hospital, 1850–1950." master's thesis, University of Toronto, 1950.

Rutty, Christopher J. "'Do Something! . . . Do Anything!' Poliomyelitis in Canada, 1927–1962." PhD thesis, University of Toronto, 1995.

Rutty, Christopher J. "Robert Davies Defries, 1889–1975." In *Doctors, Nurses and Practitioners*, edited by L. N. Magner, 62–69. Westport, Conn.: Greenwood Press, 1997.

Rutty, Christopher J. "Personality, Politics and Canadian Health: The Origins of Connaught Medical Research Laboratories, University of Toronto, 1888–1917." In *Figuring the Social: Essays In Honour of Michael Bliss*, edited by E.A. Heaman, Alison Li, and Shelley McKellar, 273–303. Toronto: University of Toronto, 2008.

Rutty, Christopher J. "'Couldn't Live Without It': Diabetes, the Costs of Innovation, and the Price of Insulin in Canada, 1922–1984." *Canadian Bulletin of Medical History*, 25, no. 2, (2008), 407–431.

Seeman, Mary. "Psychiatry in the Nazi Era." *Canadian Journal of Psychiatry*, 50, no. 4 (March 2005).

Smith, Clifford. "Memories of the Connaught Laboratories: 1926–1969," unpublished manuscript, Sanofi Pasteur Archives (1983).

Wherrett, John. "A History of Neurology in Toronto, 1892–1960, Part I" *The Canadian Journal of Neurological Sciences*, 22, no. 4, (November 1995): 322–332.

Wherrett, John. "A History of Neurology in Toronto, 1892–1960, Part II" *The Canadian Journal of Neurological Sciences*, 23, no. 1, (February 1996): 63–75.

BROADCASTS AND TAPED INTERVIEWS:

O'Connell, Mary, producer. "Keeping This Young Country Sane." Radio broadcast, *Ideas*, CBC, October 13–14, 1999.

Clarence Hincks interviewed by Cyril Greenland, December 1959, CAMH Archives.

Gordon Bates interviewed by Cyril Greenland and Jack Griffin, December 1965, CAMH Archives.

Neil McKinnon interviewed by Valerie Schatzker, July 1981, University of Toronto Archives.

PERMISSIONS

Every effort has been made to contact copyright holders. In the event of an inadvertent omission or error, please notify the publisher.

Dedicatory poem, "Almost Human" by Cecil Day Lewis, copyright © 1957, reprinted by permission of Random House UK.

Quotation from "On the Therapeutic Action of Psychoanalysis" by Hans Loewald, copyright © 2000, reprinted by permission of University Publishing Group, Hagerstown, Maryland.

Quotation from "The Celtic Consciousness" by William Irwin Thompson, copyright © 1981, reprinted by permission of George Braziller, Inc., New York, NY.

Quotation from "In Memory of Sigmund Freud" by W.H. Auden, copyright © 1940, reprinted by permission of Random House, Inc., New York, NY.

Quotation from "The Greek Myths" by Robert Graves, copyright © 1955, reprinted by permission of Carcanet Press, Manchester, UK.

"Here at the Western World", Words and Music by Walter Becker and Donald Fagen, Copyright © 1978 Jump Tunes BMI and Feckless Music (BMI). International Copyright Secured. All Rights Reserved. Reprinted by permission of Hal Leonard Corporation, Milwaukee, Wisconsin.

Quotation from "Little Gidding" by T.S. Eliot, copyright © 1942, reprinted by permission of Faber & Faber Ltd., London, UK.

Quotation from "Worstward Ho" by Samuel Beckett, copyright © 1983, reprinted by permission of Grove Press, New York, NY.

Quotation from Jan Morris, copyright © 1990, reprinted by permission of Judith Erhlich Literary Management, New York, NY.

Quotation from "Brave New World" by Aldous Huxley, copyright © 1932, by permission of Estate of Aldous Huxley, Georges Borchardt, Inc., New York, NY.

PHOTO CREDITS

All photographs and documents are the property of the author, except where used courtesy of the sources indicated below:

Centre For Addiction and Mental Health Archives (CAMH)
City of Toronto Archives (CTA)
Family of Henry Best (HB)
Sanofi Pasteur Archives (SPA)
Sheppard Pratt Hospital (SPH)
University of Toronto Archives (UTA)

Page 93 CAMH
Page 98 CAMH
Page 145 UTA
Page 167 SPA
Page 171 CAMH
Page 184 SPH
Page 195 CAMH
Page 198 CAMH
Page 200 CAMH (both)
Page 210 CAMH
Page 236 SPA
Page 247 SPA
Page 250 SPA
Page 268 CTA
Page 276 UTA
Page 285 SPA
Page 290 SPA (both)
Page 292 SPA
Page 305 HB
Page 306 SPA
Page 340 SPA
Page 341 HB
Page 345 SPA
Page 363 SPA
Page 387 CAMH
Page 426 SPA
Page 467 Christine Buijs

INDEX

Page numbers in italics refer to material featured in photographs and their corresponding captions.

Adler, Alfred, 216
American Psychiatric Association, 356
Amyot, Dr. John, 164, 224, 232, 233–34, 267, 414, 419
Analysis Terminable and Interminable (Sigmund Freud), 357
ancestry, FitzGerald family, 119–126, 420
anthrax, 163
Aquinas, Thomas, on suicide, 418
Armstrong, Dr. Herbert, 150–51, 408, 420
Armstrong, Louis, 334, 347
Aschoff, Dr. Ludwig, 220, 419

Balfour, Sir Andrew, 320, 419
Banting, Dr. Fred: 146, 270–72, *276*, 279, 282, 283–84, 288, 295–97, 302, 304, *305*, 320–21, 324, 325–26, 364, 413, 417, 419, 426, 427, 450
  accidental death of, 428–29, 457;
  Banting Institute, 356, 368, 446
  development and production of insulin, 144, 273–76, 277, 278–79, 280, 281–82, 307–8, 309
  dissolution of marriage to Marion, 309, 325
  insulin shock treatments, 280–81, 282, 330, 356–57, 406, 429
  knighthood of, 333–34
  Nobel Price in Medicine recipient, 286–87, 295
  painter, 16, 270, 295–96, 303, 309, 312
  soldier in World War One, 262
  See also Best, Charles; diabetes
Banting, Marion (nee Robertson), 295, 296, 297, 309, 320, 325
Basie, Count, 30–31, 359, 465
Beautiful Losers (Leonard Cohen), 98
Beers, Clifford, 254–57, 312–13, 314, 315, 375, 376, 378, 382, 395, 419
Beers, George, 314–15
Beers, William, 314
Bernard, Dr. Leon, 285
Besselar, Hank and Haans, 36
Best, Charles, 144, 272, *276*, 279, 302, 304, *305*, 320, *341*, 342–43, 345, 363, 370–71, 386, 391, 417, 419, 426, 439–40, 450, 462, 465
  Banting Institute, 356, 368, 446
  Best, Margaret (wife), 320, 427, 439, 441
  Best, Sandy (son), 320, 440–41, 479
  death of, 440–41
  development and production of insulin, 144, 273–76, 277, 278–79, 280, 281–82, 307–8, 309
  Nobel Prize in Medicine, 287

See also Banting, Dr. Fred; diabetes; insulin
Betty Ford Center, 97
Bishop Strachan (private girls' school), 6, 59, 434, 456
Bladek, Dr. Roman, 53–54, 90, 97
Blatz, Dr. Bill, 10–11, 293–95, 313, 419
Briggs, Alice, 54, 97
Brill, Dr. A.A. (Abraham), 209, 210, 211, 313
Brown Public School, 31–32, 39, 242
Brown, Dr. Hurst, 51, 70, 92
Buchan, John, 365–66, 374, 409, 419
Buffalo State Hospital/Buffalo State Asylum for the Insane, 175, 178, 210, 374–75, 419
Bulletin of the Ontario Hospitals for the Insane (psychiatric journal), 198–99, 214–15
Burnside estate, 6
Butler Hospital (Providence), 375, 400

Cambridge University, 8, 28, 57, 58, 118, 177, 341, 357, 359, 361, 363, 367–372, 374, 381–383, 425
Canadian Academy of Allergy, 60
Centre for Addiction and Mental Health, 451
Churchill, Sir Winston, 13, 413, 418
Clark, Daniel, 165–66
Clarke, C.K., 101, 173, 191, 194, 195–96, 198–99, 203, 204, 211, 213, 232, 250, 253, 254, 255, 419
Clarke Institute of Psychiatry, 71, 99, 146
  Jack FitzGerald, patient, 91–94, 96–97, *98*, 104, 146
  McCulloch, Dr. Don, 71–72
Collip, James Bertram "Bert," 267–77, 275, 278, 279, 280, 287
Connaught Laboratories, 84, 99, 144, 241, 250–52, 267, 272, 282, 285, 291, 305, 312, 333, 338–39, 340, 344, 347, 349, 358, 364, 382, 385, 387, 391, 395, 405, 406, 412, 419, 426, 428, 433, 435, 446–47, 448, 450, 453, 461
  diphtheria toxoid, 291–93; flu vaccine, 264–65
  Insulin Division, 275–76, 277, 278–79, 280, 307–8, 309, 356, 386
  production of serums during the world wars, 263, 413
  University Farm, *250*, 273, 285, *294*, 311, 319, *326*, 341–43, 348, 365, 366, 393, 409, 412, 440, 453
Cooke, Robert "Pops," 27–28, 50, 52–53
  "Cooke units" of allergy antigens, 50
Cotton, Dr. Henry, 298–99
Craig, Reverend Canon, 44

Dante, 183, 323, 329, 418
Danvers State Insane Asylum/Danvers State Hospital, 207–8, 378, 419
Deer Park (Spokane), 22, 24
Deer Park (Toronto), 7, 169, 170, 212, 242
Defries, Robert, 232–33, 234, 245, 246–47, 264, 267, 282, *285*, 338–39, 340, 364, 384, 405, 412, 419, 426, 433
diabetes, 144, 270–72, 275, 308–9, 328, 441.
  See also Banting, Dr. Fred; Best, Charles; insulin

diphtheria, 18, 27, 202, 233, 293, 344, 417, 427
antitoxin/toxoid, 27, 219, 222, 224, 234, 235, 248, 291, 292, 340, 466
Edwin Klebs and Johannes Loffler, 163
Emile Roux and, 218
Gerry FitzGerald's role in fighting, 144, 219, 221, 234–39, 282, 290, 292–93, 366, 384
immunization programs for, 292–93, 391, 420
spread of through immigration, 221
Dolman, Claude, 344
Dominion Council of Health, 269, 327, 414
Douglas, Tommy, 52
Doyle, Dorothy, 282
Dr. Strangelove, 69
"Dr. X," 21, 22, 25, 430
Dunlop, Mabel (FitzGerald family housekeeper), 46, 61

Eastman, George (Eastman Kodak), 283, 324
Eaton, Timothy, and family, 6, 43, 137
crypt at Mount Pleasant Cemetery, 45
Timothy Eaton Memorial Church, 45, 74
Eli Lilly drug company, 280, 281, 309
Ellington, Duke, 30, 56–57, 106, 141, 318, 358, 367, 373, 430, 463
Elora, Ontario, 166, 206, 212, 305
Empress of Scotland, The, 305
English Dance of Death, The (Combe), 166
Epstein, Dr. Stan, 54–55, 90, 97
Ewart, John (Janet FitzGerald's great-great-grandfather), 16, 130

Falconer, Robert, 245, 279
Farquharson, Dr. Ray, 150, 369, 372, 416, 417, 426
Farrar, Dr. Clarence B. "C.B.", 101, 146–47, 148–49,151, 203, 299–301, 313, 323, 406, 410, 411, 414, 419, 426, 428, 430, 436–38, 441, 446
American Journal of Psychiatry, The, 320
Bill and Sidney FitzGerald, treatment of, 150, 315–17
correspondence between Gerry FitzGerald and, 146–48, 198, 382–83, 383–84, 385, 386–87, 389–391, 392, 398–400, 402–405, 407–409, 410–411
electroshock treatments and, 261
friendship with Gerry FitzGerald, 101, 148–49, 183, *184*, 185–86, 191–92, 297–98, 310, 311, 320, 370, 372, 374, 376, 386, 387–88, 396, 397
Homewood and, 101
insulin shock treatments and, 330, 354, 356
interview of a "mad" patient, 186–90
"I Remember J.G. FitzGerald," 436
lobotomies and, 354, 355–56
Order of Canada, recipient of, 438–39
Sexual Sterilization Act and, 301
Sheppard Pratt, at, 195, 256
views on eugenics and euthanasia, 220, 300–301, 314, 324, 396, 438
views on Sigmund Freud, 211,

294, 300, 388
working relationship with Gerry FitzGerald, 186–190
Farrar, Geraldine, 185–86
Fenton, William "Billy," 234, 236, 237
FitzGerald, Alice, nee Woollatt (Gerry FitzGerald's mother), 139, 154–55, 157, 159, 165, 174–75, 202–3, 330, 420, 456;
FitzGerald Memorial Research Fellowship, 53–54. *See also* Bladek, Dr. Roman
FitzGerald, Bill (brother of Gerry FitzGerald), 67, 109, 150, 154, 156, *157*, 202, 216, 225, 310, 315–17, 389, 420, 427, 430, 433–34, 450
FitzGerald, Edna (Jack FitzGerald's mother wife of Gerry FitzGerald), 7, 30, 327, 336, 375, 456, 136, 302
as matriarch of Balmoral, 34–35, *37*, 244, 334, 369–70, 435
courtship and marriage to Gerry FitzGerald, 203, 212, 215–16, *221*, 411
death of 44–46, *426*
family background, 34
financial support of Jack, Gerry, and Bill FitzGerald, 21, 26, 28, 223, 358, 434
Gerry's gradual breakdown and, 365, 369, 372, 373, 392, 412, 421, 427, 435
in Berkeley with Gerry, 220, 222, 282
involvement in Gerry's work, 236, 237, 253–54, 320, 349, 365
marital strife, 370, 396–97
motherhood, 236, 241, 242, 249, 322, 358, 371, 415
travels to Europe with Gerry, 295–97
Fitzgerald, F. Scott, 43, 318, 348, 369
FitzGerald, Gerry (full name: John Gerald father of Jack FitzGerald)
birth and childhood, 155–60
courtship and marriage to Edna FitzGerald, 202, 212, 215–16, *221*, 411
Hartford Retreat, time as patient there, 146, 377–393, 396–411, 412, 437, 464
Harvard, time working there, 206–7, 208, 212
insulin shock therapy, recipient of, 386, 406, 407, 408, 409
marital strife, 370, 396–97
999 Queen Street West, time working there, 146, 165–66, 195–96,198–99, 203, 323
role in fighting diphtheria, 144, 219, 221
role in fighting rabies, 144, *218*, 219, 317
FitzGerald, Hazel (sister of Gerry FitzGerald), 65–67, 88, 91, 109, 123, 139, 155, 156, *157*, 202–1, 216, *225*, 237, 245, 316, 327, 335–36, 337, *370*, 394, 420, 427, 434, 436, 442–43, 456
FitzGerald, Jack (full name: John Desmond Leonard; author's father)
Addiction Research Foundation, patient at, 99–100
admiration of J.F. Kennedy, 67–68
birth, 19–20
Clarke Institute of Psychiatry, patient at, 71– 72, 91–94, 96–97, *98*, 104, 146

courtship and marriage, to Janet, 13–14, *26*, 34, 61
Donwoods Institute, patient at, 97
electroshock therapy, recipient of, 93–94, 104, 146
Homewood, patient at, 100
insulin therapy, recipient of, 102, 146
marital strife, 8, 47, 60–61, 72, *72*, 80, 90, 99, 101, 133, 441
morphine, addiction to, 81, 82, 87, 208
999 Queen Street West, outpatient at, 106, 451
Queen's University, 28, 149
Toronto Western Hospital, time working there, 13, 33, 50–51, 70, 92
Upper Canada College, time at and classmates, 20, 25, 66, 70, 80, 302–3, 307, 311, 315, 327, 334
See also psychiatric treatments
FitzGerald, James (author; son of Jack and Janet), 38, 139, 303, 455, *467*
birth and childhood, 5–9, 10, 12, 15, 16, 17–19, 30, 35–36, 37–38, 39–40
Crohn's Disease, 78–79, 82, 101, 461
Old Boys: The Powerful Legacy of Upper Canada College, 136–37, 139, 142–43
Queen's University, 77, 78, 79–80, 82–83, 94, 98, 459
time spent in Ireland, researching family history, 119–126
Upper Canada College, time at, and classmates, 47–49, 62–64, 69–70, 75, 76, 107
FitzGerald, Janet (wife of Jack FitzGerald; author's mother), 14, 19, 20, 24, 25–26, 35, 36, 44–45, 46, 55, 66, 74, 78, 84, 85, 101–2, 109, 114, 117, 119, 135, 149, 457
"set up" of Shelagh, 85–87, 89, 90, 462–63
birth and childhood, 11–12
courtship and marriage, to Jack FitzGerald, 13–14, *26*, 34, 61
death of, 16, 462, 463
early years of motherhood, 6, 8–9, 10, 15–16, 17–18, 28, *32*, 32–33, 36–37, *37*, 39, 40, 41, 47, 49, *49*, 69
Jack's gradual breakdown and, 82, 97, 98, 99, 103, 260
marital strife, 8, 47, 60–61, 72, *72*, 80, 90, 99, 101, 133, 441
teen and young adult years, 12–14, 24, *457*
FitzGerald, Michael (author's brother), 12, 83, 94, 106, 126, 137, 138, 139, 141–42
birth, childhood, and teenage years, 8, 10, 17, *32*, 33, 35, 36–37, 43, 46, 47, *49*, 56, 57, 61, 66–67, 77, 79, 85, 89, 432, 463, *464*, 466
FitzGerald, Molly (sister of Jack FitzGerald), 24, 34–35, *37*, 44, *49*, 64–65, 66–67, 82, 92, 117, 139, 241, 242, 273, 276, *281*, 282, *294*, 295, 305, 309, 310, 315, 317–18, 319, 322, 326, 336, 348, 373, 374, 392, 397, 399, 420, 427, 428, 431, 441–42, 450
marriage to Major Tom

Whitley, 393–94
FitzGerald, Shelagh (author's sister), 6, 8, 9–10, 15, 28, *32*, 37, 44, *49*, 60, 74, 77, 83, 85–87, 89, 90, 94, 103, 117, 138, 463, *464*
FitzGerald, Sidney, 67, 119–20, 128, 139, 150, 155, 156, *157*, 202, 216, 225, 310–11
FitzGerald, William (Gerry FitzGerald's father), 131, 132–33, 134–35, 139, 151, 153–54, 156, 157–58, 159, 199, 202–3, 216, 225, 226, 252–53, 301, 366, 388, 402, 420, 423, 433, 450, 454, 465–66
Fitzgerald, Zelda, 318–19, 348
Fleming, Alexander, 344
Fleming, Grant, 334
Flexner, Dr. Simon, 182, 251, 285, 419
Forest Hill neighbourhood (Toronto), 7, 42, 47, 59, 63, 65, 66, 75, 81, 108, 208, 226, 335, 440, 457
"Four Horsemen, The," 167, 181–82, 285–86, 289–303.
  See also Johns Hopkins teaching hospital and medical school (Baltimore)
Fraser, Professor Donald, 339, 340, 419, 426, 427, 430
Freeman, Dr. Walter, 355
Freud, Anna, 360, 362
Freud, Sigmund, 27, 65, 86, 114, 115–16, 131–32, 138, 145, 152, 170, 172–73, 174, 180, 197, 198, 204, 205, 213, 214, 215, 224, 294, 299, 304, 328, 353, 363–64, 388, 401, 418–19, 420, 435, 459–60, 461
  Analysis Terminable and Interminable, 357
  anti-Freudians, 10, 79–80, 172–73, 298, 300, 379, 380, 438
  Auden, W.H., on, 141
  death of, 394
  fleeing Nazi invasion of Vienna, 360, 362
  Freud, Anna (daughter), 360, 362
  Freudians, 47, 109, 210–11, 298, 313
  Interpretation of Dreams, The, 27, 79, 162, 209, 300, 437
  Mourning and Melancholia, 253; on suicide, 216
  Question of Lay Analysis, The, 313, 377
  visit to North America, 210–11
  See also Clarke, C.K.; Farrar, C.B.; Jones, Ernest; Kraepelin, Emil; Meyers, Campbell

Gay Science, The (Friedrich Nietzsche), 153, 425
Gay, Dr. Frederick, 212, 419
Gerald's couturier shop, 66, 335–36
Ghosts (Henrik Ibsen), 357
Gilchrist, Dr. Joe, 280
Gooderham family (Gooderham & Worts distillery), 11
  Colonel Albert, 245–46, 250–51, 287, 305, 341, 343, 419
Goodman, Benny, 318, 358–59
Grace Church on-the-Hill, 6, 44
Graham, Dr. Duncan, 26, 27, 315, 419, 425, 426
Gregg, Dr. Alan, 330, 342
Grubbe, Janet. *See* FitzGerald, Janet
Grubbe, Talbot, 11–12
Gulliver's Travels (Jonathan Swift), 124

Halsted, Dr. William, 181

Hamilton, Dr. John (contemporary of Gerry FitzGerald), *33*, 149–50, 422–23, 447–48, 453
Hamilton, John (contemporary of Jack FitzGerald), 357, 358, 359
Hamlet and Oedipus (Jones), 214
Hartford Retreat, *148*, 185, 226, 255–56, 315, 374, 375, 376
  Gerry FitzGerald's time there as a patient, 146, 377–393, 396–411, 412, 437, 464
Harvard University, 166–67, 184, 285, 448
  Medical School, 184, 206, 209, 287
Hastings, Charles, 232, 264, 268, 285
Havergal College (private girls' school), 34, 64, 78, 95, 203, 317–18, 456
Health Organization of the League of Nations, 266, 308, 315, 316, 321, 327, 348, 362
Hearst, Phoebe, laboratory at Berkeley, 220, 282
Hemingway, Ernest, 98, 287
Hincks, Clarence, 254, 255, 257, 294, 301, 313, 314, 356, 419, 437
Hines, Earl "Fatha," 57
HMCS *Annan*, 23
Hoche, Dr. Alfred, 184, 220, 396
hookworm, 284–85, 331
Hunter, Dr. Robin, 93
Hutchison, Lorne, 342, 367, 400, 426, 433

Immunology, 37, 38, 63, 64, 169, 228, 316, 464
influenza, 144
  Spanish flu, 11, 263–65
insulin, 271–72, 284, 285, 297, 305, 307–9, 341, 343, 366, 386, 441, 446, 447, 454, 461
  shock, 276, 280
  shock therapy, as psychiatric treatment, 102, 328–30, 354, 356–57, 361, 362, 363, 379, 386, 406, 407, 408, 409, 410, 426, 429, 446, 461
International Health Board, 285–86
Interpretation of Dreams, The (Sigmund Freud), 27, 79, 162, 209, 300, 437

James, William, 184, 185
Jameson, Sir Wilson, 331, 428
Jenner, Edward, 163, 293
Johns Hopkins, teaching hospital and medical school (Baltimore), 21, 162, 167, 182, 184, 185, 333, 349, 395
  "Dr. X," 21, 22, 25, 430
  "The Four Horsemen" (Halsted, Kelly, Osler, and Welch), 167, 181–82, 285–86, 289–303
  Gerry FitzGerald's time training at, 181, 183
  Jack FitzGerald's internship at, 21, 28, 340
  School of Public Health, 413
  See also Farrar, Dr. C.B.; Halsted, Dr. William; Kelly, Dr. Howard; Osler, Dr. William; Rose, Dr. Wickliffe; Welch, Dr. William "Popsy"
Jones, Ernest, 146, 204–5, 209–11, 213–15, 224, 261, 299, 327, 360, 362–63, 388, 394, 420–21, 435–36, 459
Jung, Carl/Jungians, 40, 197, 210, 213, 253, 437

Kaiser William Institute for Psychiatry, 313, 396
Kellogg, Dr. John Harvey, 283
Kelly, Dr. Howard, 181–82, 285–86
Kennedy, John Fitzgerald, 20, 67–68, 90
Kennedy, Joseph, 362
Kennedy, Robert, 76, 90
Kennedy, Ted, 78
King, Martin Luther, 76
Klebs, Edwin, 163
Koch, Robert, 163, 201
Kraepelin, Emil, 171–72, 173, 179, 184, 186, 197, 204, 300
Krogh, August, 286. *See also* Nobel prize

Lang, Dan, 25
Leonard, Charles (Edna FitzGerald's father), 223
Leonard, Elijah (Edna FitzGerald's grandfather), 34
Leonard, Ibbotson (Edna FitzGerald's cousin), 45
Les Rives des Prangins, 319
Leuthold, Caroline (Jack FitzGerald's first wife), 20–24, 181, 438
Leuthold, Johnny (Jack FitzGerald's son with Caroline Leuthold), 23, 24, 25, 83
Lister, Joseph, 298
  Lister Institute, 220, 306
Loffler, Johannes, 163
Lord Tweedsmuir. *See* Buchan, John

Macdonald, Stuart, 23
Mackenzie, William Lyon, 166; "Mack the Knife," 57, 69
Macmillans, "Jick" and "Miggs" (neighbours of the FitzGeralds), 10
Maharaj, Dr. Neil, 97–98
Mallon, Mary ("Typhoid Mary"), 201, 202
Mangeolais, Anna (Molly's roommate), 326
March, Sydney and Vernon, 296, 302
Massiah, Reverend Bert, 106
McCullough, Dr. John, 232, 234, 235, 238, 419, 427, 436
McGill University, 28, 167, 331, 334–35, 347, 349
McKenzie, Dr. Kenneth, 355–56
McKinnon, Hollie, 341–42, 370, 412, 433
McKinnon, Neil, 340–42, 367, 368, 370–71, 374, 375, 377, 378, 380, 382, 384, 412, 426, 448
McLean Asylum (Boston), 185, 377
McLean, Doris (wife of Bill FitzGerald), 434
Meister, Joseph, 163, 234, 277, 415
Metcalfe, Dr. William, 196
Metchnikoff, Dr. Elie, 218–19, 419
Meyer, Dr. Adolf, 179–80, 199, 210, 211, 220, 256, 298, 299, 313, 319, 375, 419, 441, 451
Meyers, Dr. Campbell, 151, 170–71, 173–74, 175, 176, 212–13, 419
Mind That Found Itself, A (Clifford Beers), 254
Montgomery, Lucy Maud, 23
morphine, 23, 158, 181, 214, 258, 259, 262, 271, 328, 394, 423
  as psychiatric treatment, 323
  Jack FitzGerald's addiction to, 81, 82, 87, 208;
Mothercraft, 6
Mount Pleasant Cemetery (Toronto), 45, 138, 156, *426*
Mountain Sanatorium (Hamilton), 24

Mourning and Melancholia (Sigmund Freud), 253

National Committee for Mental Hygiene, 254, 257, 267
Nazis/Nazism, 13, 23, 141–15, 184, 220, 314, 339, 360, 364, 391, 395–96, 413, 431
Nembutol, 160, 352, 365, 369, 375, 382, 426, 443, 445
Nietzsche, Friedrich, 153, 195, 197, 425, 444, 445
999 Queen Street West Provincial Lunatic Asylum, 14, 16, 168, 213, 355, 356, 451
- Bill FitzGerald's time there as a patient, 433, 434, 450
- *Bulletin of the Ontario Hospitals for the Insane* (psychiatric journal), 198–99, 214–15
- Clark, Daniel, 165–66
- Clarke, Dr. C.K., 101, 173, 198–99, 215, 419
- Gerry FitzGerald's time working there, 146, 165–66, 198, 203, 323
- Jack FitzGerald's time there as an outpatient, 106, 451
- *See also* Jones, Ernest

Now We Are Six (A.A. Milne), 37, 38–39

Oaklands (Toronto), 44, 225
Oakley, Helen, 318, 326, 358
Old Boys: The Powerful Legacy of Upper Canada College (James FitzGerald), 136–37, 139, 142–43
On the Nightmare (Ernest Jones), 214
Osgoode Hall, 16
Osler, Sir William, 73, 146, 167, 182, 184, 185, 288, 419, 437
Oxford Group, 13
Oxford-Cambridge boat races, 8, 360, 373

Peerless, James, 199–200
Park, Dr. William, 201–2, 221–22, 419
Pasteur, Louis, 159, 160, 163, 218, 246–47, 270, 277, 296, 330, 419, 448, 449
- Gerry FitzGerald's time working at the Pasteur Institutes (Brussels and Paris), 217, *218*, 223, 242, 274, 297
- Pasteur Institutes, 216, 306, 415
- Sanofi Pasteur, 448, 453

Pelc, Dr. Hynek, 395
Penfield, Dr. Wilder, 409
Permission to Terminate Life Unworthy of Life, The (Alfred Hoche), 220
Plante, Jacques, 50
Prospect Cemetery (Toronto), 139, 202
psychiatric treatments:
- electroshock, 98, 261, 298, 328, 361, 363, 379, 440, 446
- insulin shock therapy, 102, 328–30, 354, 356–57, 361, 362, 363, 379, 386, 406, 407, 408, 409, 410, 426, 429, 446, 461
- lobotomy and leucotomy, 105, 328, 354–57, 361, 363, 379, 406, 441, 446
- Metrazol, 328, 354, 356, 361, 379

Queen's Park/Queen's Park Crescent (Toronto), 229, 232, 234, 287, 368, 462
Question of Lay Analysis, The (Sigmund Freud), 313, 377
Quintet of the Hot Club of

France, 57, 359

rabies, 217, 244, 277, 332, 417
Gerry FitzGerald's role in fighting, 144, *218*, 219, 317
immunization program for, 248
vaccine, 163, 224, 234, 415, 433
See also Pasteur, Louis
Reinhardt, Django, 31, 359, 458
Robertson, Eileen, 66
Rockefeller: Center, 13
fortune, 11, 239, 446
Foundation, 239–40, 275, 285, 293, *306*, 313–20, 321, 327, 330, 342, 368, 395, 413, 414, 448
Institute of Medical Research, 182, 251, 341
John D., Junior, 239–40, 284, 291, 368, 419
John D., Senior, 239, 240, 256, 286, 421
-sponsored circuit-riding missions, 284–85, 331, 348, 353, *363*
-sponsored Kaiser Wilhelm Institutes, 313, 396
Roosevelt Hospital (Manhattan), 27
Roosevelt, President Franklin, 328
Roosevelt, President Teddy, 256, 378
Rose, Wickliffe, 284–85, 286, 331, 419
Roux, Emile, 217–18, 221, 246, 419
Royal Canadian Navy, Jack FitzGerald's time in, 22, 25, 67
Royal College of Physicians and Surgeons, 53
Rudin, Ernst, 313–14, 419
Russell, F.F., 331, 419

Sanofi Pasteur, 448, 453
School of Hygiene (Toronto), 84, 91, 99, 144, 291, 296, 297, 304, *305*, 306, 307, 317, 320, 321, 327, 330, 331, 339, 345, 357, *363*, 368, 387, 391, 413, *426*, 427, 433, 439, 446, 447, 448, 449, 453
Scott, David, 278
Seconal (narcotic), 9, 44, 81, 93
Sexual Sterilization Act, 301
Sheppard and Enoch Pratt Hospital, "Sheppard Pratt" (Baltimore), 185, 195, 319, 377
syphilis, 144,163, 202, 214, 248, 263, 317, 357, 417
as cause of insanity, 180–81, 219–20
Nietzsche, Friedrich and, 197
treatment of, 219
smallpox, 144, 158, 163, 202, 219, 221, 268, 290, 332, 447
vaccination against, 231, 247, 248, 268–69, 293, 391, 417
Somerville, Duke, 63
Southard, Dr. E.E., 206, 207–8, 209, 211, 212, 419
Spanish flu, 11, 263–65
Spohn, Peter, 20, 26, 70
St. George School for Child Study, 293
St. John's Ward (Toronto), 229–230, 421
St. Paul's Anglican Church (Toronto), 48
Stancer, Dr. Harvey, 93, 96
Steele, Mable Ewart (Janet FitzGerald's mother), 11, 13
Stekel, Wilhelm, 216
Stephenson, Sir William, 13

Tender Is the Night (F. Scott Fitzgerald), 318

tetanus, tetanus antitoxin, 62, 144, 241, 244–45, 246, 248, 391
39 Steps, The (John Buchan), 365
Thompson, Leonard, 276, 277
Time Machine, The (Wells), 69
Timothy Eaton Memorial Church (Toronto), 45, 74
TItanic, 6, 336, 360
Toronto General Hospital, 26, 87, 135, 162, 173, 204, 229, 232, 241, 249, 254, 276, 295, 315, 367, 416, 446
Toronto Western Hospital, 6, 8, 82, 87, 103, 452
  Jack FitzGerald's work at, 13, 33, 50–51, 70, 92
Toronto Western Medical Building, 51, 92
Trenton State Hospital, 298, 299
Trinity Hall Boat Club, 350, 371
tuberculosis, 24, 130, 144, 163, 221, 231, 268, 271, 343, 414
typhoid, 144, 202, 232, 241, 248, 332
  Mallon, Mary ("Typhoid Mary"), 201, 202

University of Toronto, 25, 43, 45, 53–54, 71, 87,92, 99, 109, 149, 161, 166–67, 195, 202, 211, 212, 214, 224, 238–39, 250, 253, 267, 271, 279, 284, 291, 309, 339, *341*, 349, 357, 368, 382, 394, 404, 415, 446, 447, 451
  Medical School, 20, 160, 161–63, 237, 238–39, 240, 247, 254, 392, 429
  See also School of Hygiene (Toronto); Connaught Laboratories
Upper Canada College, 16, 17, 45, 47, 88, 100, 169, 212, 348
  Charles Leonard at, 34
  Jack FitzGerald at, and classmates, 20, 25, 66, 70, 80, 302–3, 307, 311, 315, 327, 334
  James FitzGerald at, and classmates, 47–49, 62–64, 69–70, 75, 76, 107
  Michael FitzGerald at, 85
  Old Boys: The Powerful Legacy of Upper Canada College (James FitzGerald), 136–37, 139, 142–43
  Preparatory School, 47

Welch, Dr. William "Popsy," 167, 182–83, 221, 285, 313, 419
Wells, H.G., 69, 314
When We Were Very Young (A.A. Milne), 37–38
Whitley, Anna (author's paternal cousin), 11, 35, *37*, *73*, 109, 139
Whitley, Major Tom (husband of Molly FitzGerald), 35, 64, 373, 399, 451
  marriage to Molly FitzGerald, 393–94
Whitley, Molly. *See* FitzGerald, Molly
Whitley, Patrick (author's paternal cousin), 11, 35, *37*, *73*, 139
Wigraine (narcotic), 29, 81
Wilson, President Woodrow, 266
Windy Ridge nursery school (Toronto), 10, 11, 295
Winters, Jonathan, 57, 226
Wodehouse, Sally, 59, 76
Woollatt, Bill (cousin of Gerry), 310, 420
Wychwood Park (Toronto), 164, 339, 430

JAMES FitzGERALD is a journalist and author whose first book, *Old Boys: The Powerful Legacy of Upper Canada College*, was a controversial inside look at the attitudes and mores of Canada's ruling class. Revelations of the sexual abuse of boys at the school, first published in the book, led to the charging and conviction of three former teachers and the launching of a class action lawsuit against the college in 2002. The article that sparked *What Disturbs Our Blood* won a National Magazine Award. James lives in Toronto.